IET RENEWABLE ENERGY SERIES 17

Electrical Design for Ocean Wave and Tidal Energy Systems

Other volumes in this series:

Volume 1	**Distributed generation** N. Jenkins, J.B. Ekanayake and G. Strbac
Volume 6	**Microgrids and active distribution networks** S. Chowdhury, S.P. Chowdhury and P. Crossley
Volume 7	**Propulsion systems for hybrid vehicles, 2nd edition** J.M. Miller
Volume 8	**Energy: resources, technologies and the environment** C. Ngo
Volume 9	**Solar photovoltaic energy** A. Labouret and M. Villoz
Volume 10	**Scenarios for a future electricity supply: cost-optimized variations on supplying Europe and its neighbours with electricity from renewable energies** G. Czisch
Volume 11	**Cogeneration: a user's guide** D. Flin
Volume 13	**Offshore wind turbines: reliability, availability and maintenance** P. Tavner
Volume 16	**Modelling distributed energy resources in energy service networks** S. Acha

Electrical Design for Ocean Wave and Tidal Energy Systems

Edited by
Raymond Alcorn and Dara O'Sullivan

Editorial Coordinator
Darren Mollaghan

The Institution of Engineering and Technology

Published by The Institution of Engineering and Technology, London, United Kingdom

The Institution of Engineering and Technology is registered as a Charity in England & Wales (no. 211014) and Scotland (no. SC038698).

First published 2013

The Institution of Engineering and Technology
Michael Faraday House
Six Hills Way, Stevenage
Herts, SG1 2AY, United Kingdom

www.theiet.org

British Library Cataloguing in Publication Data
A catalogue record for this product is available from the British Library

ISBN 978-1-84919-561-4 (hardback)
ISBN 978-1-84919-564-5 (PDF)

Typeset in India by MPS Limited
Printed in the UK by CPI Group (UK) Ltd, Croydon

Contents

List of contributors

Raymond Alcorn
Hydraulics & Maritime Research Centre,
University College Cork,
Ireland

Judicaël Aubry
ESTACA
France

Jochen Bard
IWES Fraunhofer
Germany

Hamid Ben Ahmed
SATIE Laboratory, ENS Cachan
France

Mohamed Benbouzid
University of Brest
France

Seifeddine Benelghali
University of Aix-Marseille
France

Anne Blavette
Hydraulics & Maritime Research Centre,
University College Cork,
Ireland

Salvador Ceballos
Tecnalia Energia
Spain

Jean-Frederic Charpentier
French Naval Academy
Research Institute
France

Gordon Dalton
Hydraulics & Maritime Research Centre
University College Cork
Ireland

Fernando Salcedo Fernandez
Tecnalia
Spain

Damian Flynn
University College Dublin
Ireland

Olav Fosso
NTNU
Norway

Alejandro Garces
Universidad Tecnológica de Pereira
Spain

David Kavanagh
ESB Networks
Ireland

Andrew Keane
University College Dublin
Ireland

Peter Kracht
IWES Fraunhofer
Germany

Joe MacEnri
ESB-International
Ireland

Joseba Lopez Mendia
Tecnalia
Spain

Marta Molinas
Tecnalia
Spain

Darren Mollaghan
Hydraulics & Maritime Research Centre
University College Cork
Ireland

Bernard Multon
SATIE Laboratory, ENS Cachan
France

Dónal Murray
Department of Electrical Engineering
University College Cork
Ireland

Dara O'Sullivan
Analog Devices (formerly of HMRC)
Ireland

Pierpaolo Ricci
Global Maritime Consultancy Ltd
(formerly of Tecnalia)

John Ringwood
National University of Ireland, Maynooth
Ireland

Eider Robles
Tecnalia Energia
Spain

Maider Santos-Mugica
Tecnalia Energia
Spain

Fergus Sharkey
ESB-International
Ireland

Elisabetta Tedeschi
Tecnalia Energia
Spain

Acknowledgements

The editors would like to thank all the authors for their contributions to this book. Their efforts have helped us to create the first book of its type in the field and what we hope will become a work of reference.

The editors would also like to thank The Hydraulics and Maritime Research Centre (HMRC) and University College Cork (UCC) for their support during the preparation of this publication.

Particular thanks to Darren Mollaghan and Dr Anne Blavette from HMRC for their tireless work in the preparation of the manuscript and the chasing of wayward authors.

We would also thank the IET for giving us the opportunity to produce this publication.

We would also like to acknowledge the Science Foundation Ireland Charles Parsons Energy Research Award, the Sustainable Energy Authority of Ireland, the Marine Institute, Enterprise Ireland, the International Energy Agency Ocean Energy Systems Implementing Agreement, the International Electrotechnical committee and the EU's Framework 7 Research Programme.

Foreword

The expected increase in the world's population to 10 billion by the end of the next century yields the 'grand challenges' of society. Amongst these are the provision of food and energy, both of which can be sourced from the ocean. The future sustainable, secure delivery of energy will be essential to the well-being of all of the world's nations. Renewable energy is expected to play a major part in future energy supplies both to reduce the impact on the world climate and also to make up the potential shortfall in conventional energy sources. Land-based renewable energy resources are limited by certain socio-economic factors which are not so prevalent in the ocean. We will have to look to the oceans in future to ensure that the energy supply for the planet is produced in a sustainable way. A recent report by IPCC, looking into the potential for renewable energy to mitigate climate change, stated that 'the world's oceans could supply many times the current electricity demand of the world'. The IEA Ocean Energy Systems Committee has identified that 'ocean energy could have up to 750 GW installed by 2050 with 160,000 direct jobs created by 2030'.

Ocean energy is an emerging industry sector and there are a number of promising developments under way. Significant commercial deployments in the gigawatt range are envisaged over the next 10 to 20 years in Europe, United States, Asia and South America.

I welcome this publication; it comes at an important time in the development of the nascent ocean renewable industry. I envisage that it will become the reference work for electrical systems in this field and will serve as an excellent text for educational and industrial applications.

The range of topics that are covered here will give the reader an excellent understanding of the topics required in relation to all aspects of the electrical systems. These topics are presented within the framework of practical applications. The contributing authors have wide experience both in research and industrial applications relevant to the electrical systems for ocean energy.

I hope that the reader will be able to utilise the information to assist in the realisation of the ocean energy industry as it endeavours to make a major contribution to the world energy supply and to mitigating climate change.

Tony Lewis
Professor – Energy Engineering
University College Cork
Cork
Ireland

Chapter 1

Introduction

I first started in the field of ocean energy in the mid-1990s as an undergraduate electrical engineer. At this time you could count the number of ocean energy electrical engineers on one hand.

Back then, as device concepts were emerging the focus was more on hydrodynamics and fundamental performance. This was rightly so. The thinking was that the electrical system came out of a box and could be added at the end. The electrical design aspects of the projects always took second billing and this was often the case because many of the groups developing the technology did not have in-house electrical engineering expertise. After I finished my PhD I moved to Australia to work on the design, build and commissioning of a full scale grid connected wave energy device. What I learned from this experience is that the electrical system and its control are fully integrated into the fundamental operation and performance of the device. I also learned that there were no how-to guides, ready references, guidelines or standards and that everything had to be adapted from other areas. Mind you, it was exciting to be carving a new path.

Thankfully much has changed in the past 10 years as we are now seeing full scale grid connected wave and tidal devices with plans afoot for hundreds of MW of arrays within the next few years. Of course this will present the engineer with a new set of challenges but I hope this book will become a reference for both those working and studying in the field and will help them to overcome these challenges.

We have been collaborating internationally to help catch up with guidelines, best practice and standards for the industry. Both the co-editor, Dr Dara O'Sullivan and myself have taken part in a number of international initiatives in this arena. The International Energy Agency Ocean Energy Systems (IEA-OES) Implementing Agreement (www.iea-oceans.org) has over the past number of years coordinated a number of technical annexes with participants from 20 countries. These annexes have provided guidelines for many aspects of ocean energy devices. One of the technical annexes in which we participated focused on the grid integration of wave and tidal systems.

More recently, and building on the IEA work, the International Electrotechnical Committee (IEC) has set up a technical committee TC114 to draft international standards for marine energy conversion systems. Again, several of these standards are directly focused on power quality and electrical performance.

The authors of the various chapters of this book have worked with us in the past on IEA and IEC tasks, as well as other collaborative projects including EU Framework 7 funded research. They are all experts in their respective areas and I am very proud that they have agreed to contribute to this publication.

The scope of *Electrical Design for Ocean Wave and Tidal Energy Systems* is quite wide. Each part of the electrical system is discussed as a subcomponent. These subcomponents are brought together to form a controlled system that can then be integrated into the grid. We give several worked examples throughout the book to highlight this.

The second chapter deals with the selection of generators and their interaction with power electronics. This is not trivial and there are subtle nuances in comparison to wind turbines. One aspect specific to wave energy devices is that the peak to average power ratio can vary significantly. The third chapter deals with power cables, connectors and umbilicals, an area where there is still much development to come. Chapters 4 to 6 deal with grid integration and power quality issues and a number of case studies are presented. Energy storage is discussed in Chapter 7 and the benefits of even small amounts of storage are clearly demonstrated. The implementation of control systems in ocean energy devices is presented in Chapter 8. Some overview and background as well as specific wave and tidal examples are given. The chapter on modelling and simulation brings all of the pieces together and also describes some codes and methods under development. To conclude we detail the relative costings of various systems and why the electrical design can have a significant influence on overall project lifetime cost. This is important to understand as large scale uptake of wave and tidal technology will not happen until the economics are right and that the industry can be competitive. It is also in keeping with the notion that the industry will build offshore power stations and it is the electricity that these power stations deliver which is the market commodity.

What will become clear through reading the book is that the electrical system and in particular the control system cannot be seen in isolation from the rest of the power conversion system. Unlike our previous thinking that the electrical system can be added at the end, we now know that we must include it right at the beginning of the development. A subtle change in the electrical system could result in significant costs or gains in performance, survival, device mass or project cost and make the difference between success and failure.

This is why this *Electrical Design for Ocean Wave and Tidal Energy Systems* is so timely in its publication as we embark on the build out of a new industry. This is the first book of its kind and brings a wealth of international knowledge and experience into the one publication.

Raymond Alcorn

Chapter 2
Electrical generators in ocean energy converters

Nomenclature

CP	Power coefficient of a turbine
DFIG	Double-fed induction generator
DTC	Direct torque control
FOC	Field-oriented control
MMF	Magneto-motive force
OEC	Ocean energy converter
O&M	Operation and maintenance
OTS	Off-the-shelf
OWC	Oscillating water column
PMSG	Permanent magnet synchronous generator
PTO	Power take-off
SCIG	Squirrel cage induction generator
SG	Synchronous generator
SVM	Space vector modulation
TEC	Tidal energy converter
TSR	Dimensionless tip speed ratio of a turbine blade
VF-DPC	Virtual flux direct power control
VFOC	Virtual flux-oriented control
VOC	Voltage-oriented control
WEC	Wave energy converter

2.1 Introduction

D. O'Sullivan

On-line synchronous generators form the backbone of traditional fossil fuel generation power systems. These run at effectively constant speed, are synchronized to the electricity grid, and are optimized for the speed at which they run. Power, torque and fault current ratings are well defined and standardized. However, in the

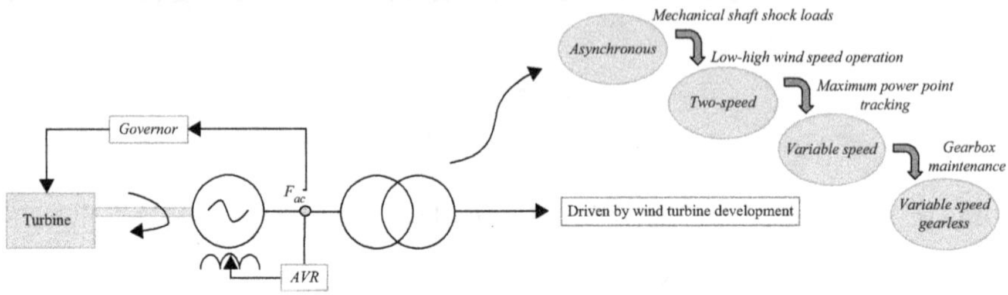

Figure 2.1 Development of generator technology

field of renewable power generation, generator selection, rating and design are more complex processes which are inextricably linked to the overall operation of the entire conversion system; speed variation in the turbo-generator control is often vital in order to maximize the primary power take-off (PTO) efficiency from the renewable source, which by its nature is usually highly variable in time. For example, in wind power generators the available PTO increases substantially if the turbine rotational speed is controlled to increase as a defined function of the wind speed. This control strategy is known as maximum power point tracking and has been responsible for the gradual transition of wind power generator technology from fixed or dual speed to variable speed machines. Likewise, under heavy gust or swell conditions, a fixed speed generator will experience severe shock loads on the generator shaft, whereas if the speed is allowed to increase, the inertia of the system will absorb some of the extra power input. This mechanical consideration initially led to the adoption of asynchronous generators in wind turbines where the slip range was utilized to provide a small measure of speed variation and speed compliance. Extensions to the speed range were also provided for by pole changing or rotor resistance variation. However, in recent years, the improvement in cost and performance levels of high power switching transistors has led to the adoption of fully variable speed controlled generators. These have typically taken the form of either gear-coupled double fed induction generators (DFIGs) with power electronics control of the rotor voltage and frequency or direct-coupled synchronous generators (SGs) with power electronics control of the stator voltage and frequency, and either brushless field excitation or permanent magnet excitation. This development process is illustrated in Figure 2.1.

This chapter details the generator drive train requirements relevant to ocean energy applications, both wave and tidal. The specific impact of the wave and tidal energy device technologies on the generator design and selection is also highlighted. Finally, the design of power electronic controllers associated with variable speed generator control is described.

2.2 Overview of generator drive train options

An overall classification of generator drive train topologies can be made between fixed speed and variable speed solutions.

2.2.1 Fixed speed solutions

In fixed speed solutions the generator is linked directly to the supply grid (Figure 2.2). Therefore the angular speed of the rotor is practically fixed and determined by the supply grid frequency, the gear ratio and the number of pole pairs of the generator. Generally speaking, fixed speed solutions have simple and reliable construction of the electrical parts, require less maintenance and are more economic. Squirrel-cage induction generators are the preferred solution for this topology but wound rotor synchronous generators can also be considered.

Even though fixed speed solutions are simpler than variable speed solutions, there are several drawbacks associated with them, of which some drawbacks are as follows:

- *Low energy capture*: The power extraction is usually a function of the rotational speed of the prime mover and the ocean wave or tidal stream conditions. For each condition there is an optimum rotational speed. In general terms an increase in efficiency of between 5% and 20% can be achieved when variable speed solutions are employed [1].
- *Mechanical stress*: Since the speed of the prime mover is fixed, power fluctuations are directly transmitted as torque pulsations, causing mechanical stress. Solutions using squirrel-cage induction generators are less mechanically demanding than their wound rotor counterparts. This is due to the compliance effect of the slip during transients.
- *Power quality*: Mechanical power fluctuations are not only transmitted to the drive train but also transferred to the grid. Therefore appreciable electrical power fluctuations that degrade power quality are produced when fixed speed solutions are used especially with low inertia drive trains.
- *Reactive energy compensation*: Squirrel-cage induction generators consume reactive energy. Usually reactive compensation elements to compensate for this consumption are needed.

2.2.2 Variable speed solutions

The problems related to fixed speed can be overcome by means of variable speed solutions. Variable speed is achieved using power converters as an interface

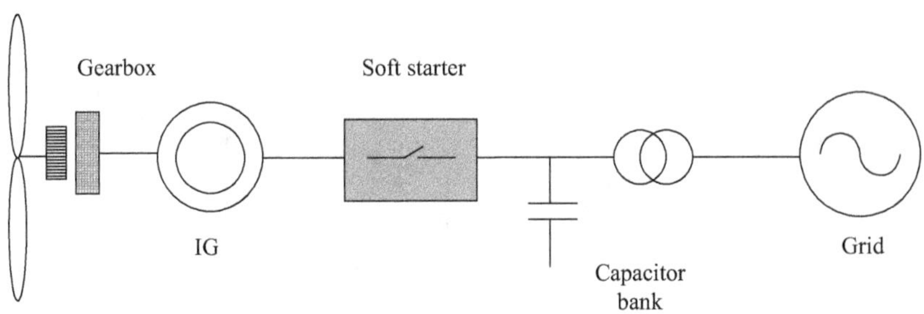

Figure 2.2 Fixed speed topology (assuming turbine-like prime mover)

between the electrical generator and the grid. The main advantages of variable speed solutions with regard to fixed speed solutions are the following:

- Power electronics decouples the grid from the generator.
- Enhanced PTO is feasible due to better matching of the resource and the prime mover speed.
- The rotor and the drive train can act as a flywheel storing or delivering energy. Therefore electrical power fluctuations are reduced.
- It is less mechanically demanding due to the inherent mechanical compliance in a variable speed system
- Active and reactive power can be fully controlled.

The main disadvantage of variable speed solutions is the increase of the cost due to the use of power electronics. There are many variable speed topologies depending on the generator type and the power electronics. Perhaps the most common nowadays are those using a fully controllable back-to-back converter (Figure 2.3) in a double fed induction generator configuration (Figure 2.4) or in a full-converter configuration (Figure 2.5).

In the DFIG configuration the rotor winding is connected to the grid through a bidirectional power converter. The converter size is usually about 30–40% of the rated power. It allows variable speed operation and active and reactive power control within certain limits. An important drawback of this topology is the necessity of using slip-rings, which increases the maintenance requirements. This is

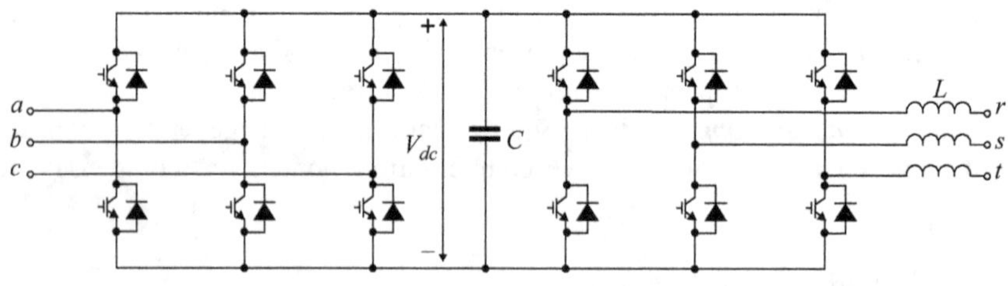

Figure 2.3 Back-to-back converter

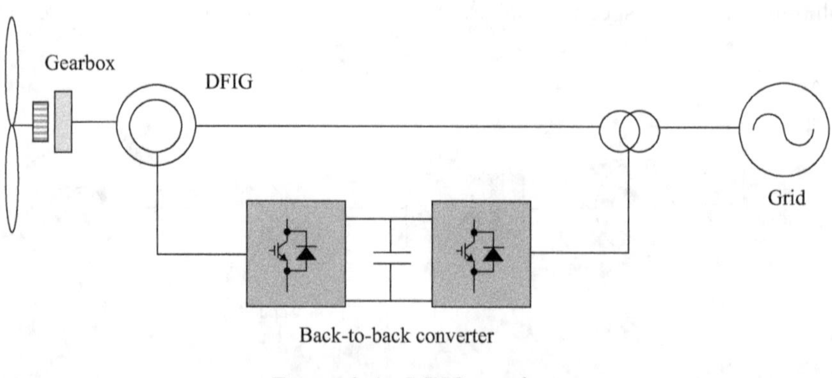

Figure 2.4 DFIG topology

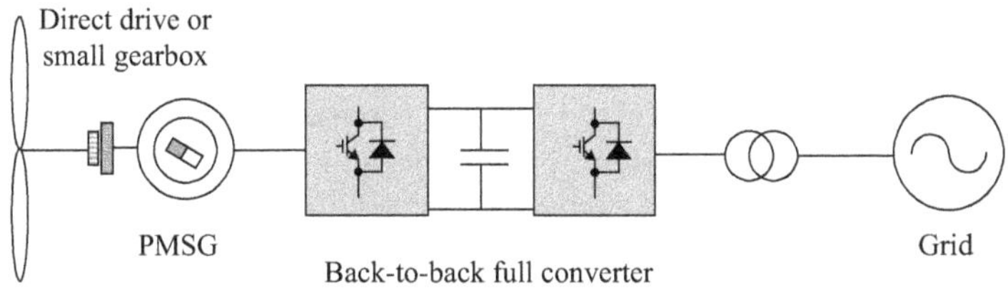

Figure 2.5 Full converter topology

an important issue especially for offshore applications where the time windows for maintenance are limited due to weather and sea state conditions. This aspect is discussed in more detail in section 2.4.1.

In the full-converter configuration the stator is connected to the grid through a full-power bidirectional converter as shown in Figure 2.5. Permanent magnet synchronous generators (PMSGs) and also wound rotor synchronous machines or squirrel-cage induction generators are technically feasible options for this topology. None of these options require slip-rings and so their maintenance requirements are decreased. In addition to a better control over the speed, active and reactive power are achieved since these magnitudes can be fully controlled over the whole range (from 0% to at least 100% of their nominal values). Depending on the generator technology choice different aspects regarding cost, efficiency, manufacturability, suitability for offshore applications etc. can dominate considerations. However the grid connection performance of all the options is very similar.

2.3 Overview of generator functionality in ocean energy converters

The approach to selection of the generator type and size in an ocean energy converter (OEC), whether wave or tidal, is first dependent on the role of the generator within the overall converter system. This functionality can vary significantly even within the same device. In this section the different operational functionalities of the generator are categorized, and their influence on the generator design parameters is discussed.

2.3.1 Power conversion

The most basic functionality requirement of a generator in a wave energy converter (WEC) is that of mechanical to electrical power conversion – in similar manner to the functionality of a generator in a fossil fuel power plant. In this case, the generator does not participate in controlling the PTO action, but simply converts the incoming mechanical power to an electrical output. Some examples of this would be:

- Fixed speed tidal turbines.
- Point absorbers with hydraulic PTO, accumulator smoothing, and constant speed hydraulic motor and generator.
- Overtopping devices with multiple constant speed hydro turbines.

The torque range of the generator in this case will be completely dependent on the primary power capture and prime mover mechanisms, and any inherent energy storage. It is likely that the generator will have to operate over a wide load range continuously, and so part-load efficiency becomes an important variable. Typically, the generator will operate at constant speed and will be directly grid connected.

2.3.2 Prime mover efficiency optimization

This is probably the most common requirement of a generator in ocean energy devices, and corresponds to the generator functionality of the majority of variable speed wind turbines. The generator speed and/or torque are specifically controlled to optimize the performance and efficiency of the prime mover [2]. Some examples of this are:

- OWCs with variable speed air turbines.
- Variable speed tidal turbines.
- Overtopping or pump devices with variable discharge or variable jet hydro turbines.

In these cases, the generator control is the means of prime mover efficiency optimization. It is important for the generator to have a wide speed and torque control range in order to optimize this efficiency. Generator torque ratings can be eased due to the wider speed range and consequent absorption of some of the input power in the system inertia.

2.3.3 Power smoothing

The generator control can allow for smoothing of the electrical output power, in conjunction with the inertia of the rotating system, which can be enhanced by additional flywheel inertia. This is beneficial to output power quality, and may overlap in terms of control strategy with prime mover efficiency optimization, but may also compromise it somewhat due to the limitation in response time if there is significant added inertia.

The use of a generator in this mode of operation can be beneficial to its rating specification, as significant torque pulsations can be absorbed by the inertia. The generator and power converter peak power rating can then be relatively close to the maximum mean power rating of the OEC. This aspect is discussed in more detail in later chapters.

2.3.4 Device damping control

In this case, the generator is controlled to directly influence device motions. This is the most effective approach to enhancing overall OEC efficiency, and also places the most stringent demands on the generator ratings. The generator control is utilized to adjust the reaction force to the device motion in, for example:

- Point absorbers with hydraulic PTO.
- Tidal current oscillating hydrofoils.
- Point absorbers with direct PTO.

Directly controlling the motions of the device can optimize the primary PTO efficiency, which is the most significant optimization from a system efficiency point of view. However such control schemes can lead to significant over-rating of the generator and associated power electronics to meet the peak torque requirements of such a design [3]. This is discussed in further detail in Chapter 8.

2.4 Generators in wave energy systems

Suitability to sustained and reliable operation of the generator in the harsh environmental conditions of the offshore marine environment is clearly a significant and important requirement. This is examined in terms of the mechanical and environmental wear sustained by machines in such an environment. One important issue to be tackled is the feasibility of the use of brushed machines such as the DFIG and the brushed SG in such an environment, and the consequent maintenance requirements.

2.4.1 Brush operation

Brush wear in brushed machines is the result of mechanical friction and electrical erosion. Friction produces carbon dust, while the result of electrical erosion is the vaporization of carbon with little physical residue. In order to achieve a good coefficient of friction between the carbon brush and the slip-ring, it is necessary to establish a good carbon composite film [4]. A good film layer can reduce the coefficient of friction to 10% of the original bare coefficient. In order to maintain a good working film, brushes should ideally operate close to the rated load current. Operation in over-current causes slip-ring blackening and reduced brush life, whereas protracted light load operation results in film removal and increased brush friction and wear. The differing input resource profiles of typical wave and wind energy converters are depicted in Figure 2.6 for a time series of 40 s. While the resource input to the wind device is oscillatory, its variation around the average level is significantly less severe than for the wave device, where the resource input fluctuates around zero every half-wave cycle. The high pulsating nature of wave energy power flows is thus clearly not well matched to the desired electrical operating point of a brush–slip-ring arrangement.

It is evident from Figure 2.6 that without significant inherent energy storage, WECs require a high peak to average rating, and so will operate close to their peak or overload current ratings for a significant proportion of the time. This represents a further complication in the employment of brushed machines. In a publication sponsored by Vestas Wind Systems [5], it was discovered that under high current operation, brush–slip-ring systems can periodically enter film instability modes that are characterized by severe brush wear and high brush temperature, increasing the wear rate by approximately 40% more than the expected wear rate. This is less of an issue for tidal energy converters which will have an input power profile more akin to a wind energy device.

The other factors that inhibit good brush film formation are high humidity and the presence of chemical contaminants in the air. While these environmental factors

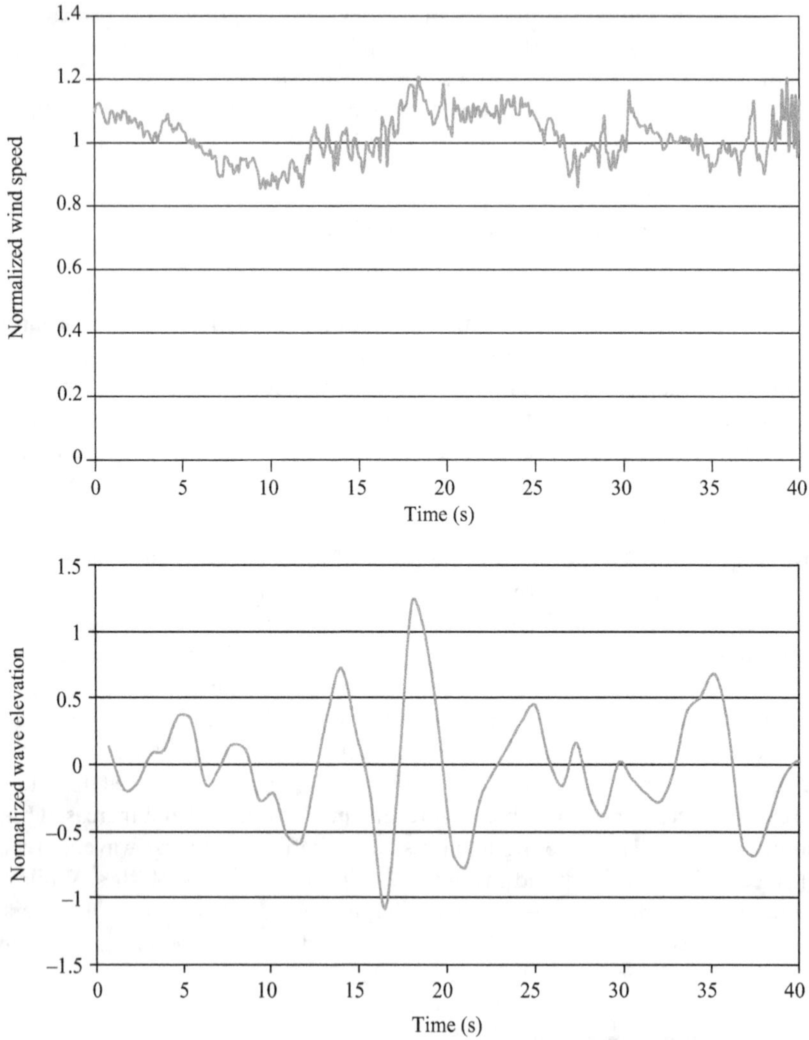

Figure 2.6 Normalized resource time series for (top) wind turbine and (bottom) WEC over a 40 s time slot

are issues for offshore wind turbines also, humidity and air vapour control are being built into modern offshore wind turbines for these reasons [6]. The inclusion of such air quality control in the generator enclosure of WECs is usually less feasible since the generator is often situated in an inaccessible location.

The minimum brush life of a general purpose, slip-ring machine is typically 3,500–8,750 h. The effective operating period of an OEC at a good site is around 5,000 h annually. Thus, in order to avoid costly outages, or even generator damage, the brushes ought to be changed at least twice annually, which corresponds with best practice in the wind industry, where brush changes are carried out typically every 6 months and potentially every 3 months.

2.4.2 Operation and maintenance

The option of using the DFIG machine or the brushed SG machine requires the presence of brushes in the system which, as previously mentioned, must be maintained and replaced on a regular basis, typically twice a year. Depending on the site specifics this is potentially a more serious issue for OECs than for wind energy, and even offshore wind. In the case of WECs, if on-board maintenance is considered safe up to a 1 m swell, then statistically this only allows a probable 7 days in the year for maintenance, in a typical North Atlantic offshore location. If on-board maintenance is allowed up to a 1.5 m swell, there will be on average about 55 days in the year when such maintenance is possible. This problem became apparent for the first time in the Bockstigeen wind farm approximately 12 km of the Swedish west coast [7]. Docking of the maintenance boat proved to be extremely difficult even at a wave height of a little over 1 m. Subsequent to this experience, alternative approaches are being explored in accessing offshore wind turbines, including movable docking rails, submarine vehicle and diver access points, as well as helicopter pads located on the nacelle [7, 8]. Moreover, once access is possible, the actual maintenance procedure can proceed in a relatively stable and protected environment. Clearly the situation is even less straightforward for floating offshore WECs. Access is likely to be only by boat, and the working environment itself may not be stable. These factors and the consensus of the industry and research community appear to strongly support long lifetime, low or even zero on-board maintenance designs being a requirement. These considerations would appear to rule out the use of the DFIG (as well as the brushed SG) in offshore WECs, despite its clear advantage in terms of size, cost and efficiency.

Some WEC technologies such as the Aquamarine Oyster™ locate the generator onshore, alleviating the issue of access and maintenance, thus opening up the option of brushed machines.

2.4.3 Corrosive environment

All machine types will be protected to a high degree from the worst effects of the environment. However, if the generator is located offshore in the marine environment, it is vulnerable to the effects of saline air or moisture. This will have the most detrimental effects on a PMSG machine since NdFeB, which is the material of choice for high-performance permanent magnet machines, is very sensitive to corrosion [9]. These materials can be destroyed within days if certain forms of corrosion take hold. This fact represents a disadvantage for the PMSG in this regard, although epoxy coatings such as VACCOAT™ have been developed for the protection of such rare earth materials in saline environments [10].

2.4.4 Mechanical issues

In full-scale wave energy devices, motions up to 6 m in amplitude can occur in time periods of 3 s, along with pitching motions leading to angular accelerations greater than 7 deg/s^2 [22]. It is evident that such conditions apply severe mechanical stress on the system components, and bearings and couplings will have to be rated to absorb these shock loads. Moreover the pitching motions of the device will induce gyroscopic loadings on the bearings.

Initial analysis of the impact of these motions on bearing design indicates that the static loading of the bearings for a horizontal axis rotary generator, within a typical full-scale device layout, increases at least twofold.

Generators with a higher power-to-weight ratio will have an advantage in this regard, as the shock loading and bearing ratings will not be as severe. In the power range of interest, the 4 pole SCIG has a ratio of 4–4.5 kg/kW, as compared to 3–3.8 kg/kW for the 4 pole SG and PMSG [11]. Hence the SCIG has a weight penalty of 20–30%. The surface magnet PMSG, however, has an additional problem in that the permanent magnets are brittle and can be cracked under mechanical shock unless precautions are taken.

2.4.5 Requirements by WEC category

As outlined previously, there exists a profusion of wave energy devices under development. It is not easy to come to any clear conclusions regarding which category or categories of device will end up being commercially successful. In this section, a range of some of the major device categories are examined, and the specific generator machine characteristics assessed for each category. The device categories to be considered are:

- Oscillating water columns (OWC) with air turbines.
- Point absorbers with hydraulic PTO.
- Overtopping or pump devices with hydro PTO.
- Point absorbers with direct PTO.

2.4.5.1 OWC [12, 13]

In an OWC device, wave action is converted to an oscillating air flow by means of hydrodynamic pressure variations across a water column inside the device which in turn pressurizes and depressurizes the air in an air chamber. The oscillating airflow enters and exits the air chamber through a ducting arrangement in which is usually inserted a bidirectional air turbine, such as a Wells or impulse turbine [14]. The presence of the duct acts effectively as a gearing mechanism converting low-velocity airflow in the chamber to high-velocity flow across the turbine blades. The consequences of this, for the generator, are that the turbine shaft can usually be gearless and still operate in a relatively high speed range. In Wells turbine designs off-the-shelf (OTS) two- or four-pole machines are possible. Impulse turbines tend to operate at a somewhat lower speed, but can still operate within the range of a six- or eight-pole machine. A wide operating speed range is generally optimal in these situations, since the applied airflow profile is typically highly variable, and the efficiency of the air turbine is optimized by varying the speed to match an instantaneous or averaged operating point of maximum efficiency yield [2]. These devices do not tend to have any significant inherent energy storage, apart from the inertia in the mechanical system. Thus, generator machines with high-peak-to-average torque ratios are suitable in order to keep the machine rating reasonable. However, if there is full rated power electronics in the system, this will have to be rated close to the peak generator power.

2.4.5.2 Point absorbers with hydraulic PTO [15, 16]

Point absorber devices are generally axi-symmetric about a vertical axis and are characterized by being small in comparison to the incident wave length. A buoyant body, termed a displacer which can be either surface piercing or submerged, moves with the wave motion against a stationary or slow moving reactor. The relative motion of the two drives a hydraulic ram. This slow moving pumping action is converted to a higher speed rotating motion in a hydraulic circuit which may involve one or more accumulators and one or more hydraulic motors. The hydraulic motor is generally directly shaft connected to the main generator. The speed and torque characteristics of the generator in this case are greatly influenced by the hydraulic circuit design. If significant accumulator energy buffering is present in the hydraulic circuit, the rotational speed of the generator can be constant [15] or have a narrow range of variation, although variable speed designs offer some advantages in terms of efficiency [16]. Rotational speeds can be designed to match OTS generator designs. If there is sufficient flexibility in the hydraulic circuit design, the torque and power rating of generators in these devices can be relatively benign, and be selected close to the device rated power.

2.4.5.3 Overtopping/pump devices with hydro PTO [17, 18]

Overtopping devices extract energy from the sea by allowing waves to impinge on a structure such that they force water up over that structure into a reservoir, thus raising its potential energy. Pump type devices use the direct forces of the waves on a hinged device structure to pump water into a raised reservoir in a similar manner. The potential energy of the water is converted to kinetic energy using a conventional hydro turbine. After exiting the turbine, the water is then returned to the sea. Due to the presence of the large reservoir, the turbine–generator combination can run at fixed speed; however, allowing some variation in the speed range enhances the hydro turbine efficiency. Variable discharge operation is then possible with fixed guide vanes and fixed runner blades [17]. As with conventional hydro turbines, speed levels tend to be quite low, of the order of 100–300 rpm, generally necessitating a gearbox. The presence of the large reservoir buffers the generator from the inherently large peak-to-average power ratio of the resource, and allows for total generator power rating close to the device-rated power level.

2.4.5.4 Point absorbers with direct PTO [19]

Instead of driving a hydraulic ram or pump, the motion of a point absorber can be directly connected to the electrical generator via a mechanical linkage of some sort. One example of this would be a direct-drive linear generator [20], where the motion of the point absorber is directly coupled to a linear translator, which acts as the moving element in a linear generator. Another example would be a direct-drive rotary generator, where the bidirectional linear motion of the point absorber is converted to bidirectional or unidirectional rotary motion of a generator via mechanical couplings such as pulleys, clutches, belts or gears.

Linear generators are most definitely a non-OTS design, and represent a very difficult generator design problem due to their low velocity. This results in very large torque and current requirements, which in turn are limited by magnetic saturation and thermal limits in the generator windings. A further issue is the attractive forces between stator and translator. This can be alleviated by utilizing a non-ferromagnetic core in either the translator or stator. However, these generators are still very much at the development stage. They require converter connection to the grid as they are inherently variable speed. PMSGs are realistically the only feasible machine design in this case.

Direct-drive rotary generators can be OTS designs, given appropriate gearing arrangements. These require converter connection to the grid also as they are inherently variable speed. If a clutching mechanism is present, the generator can be unidirectional, and the presence of inertial storage will improve the peak-to-average rating requirements of the machine. Clutching mechanisms can present maintenance issues, so bidirectional generator operation can also be considered. High part load efficiency is very important in this case [21], as the operating point of the generator is constantly changing making PMSGs the most attractive option again.

In both cases, the generator system must absorb high torque pulsations, since there is no inherent energy storage. This can be absorbed somewhat by inertial speed changes, but it is likely that the peak power rating of both generator and power electronics will be significantly greater than the average device power rating.

2.4.5.5 Summary by WEC category

Table 2.1 provides an overview of the speed and torque ranges as well as the typical generator types suited to the different categories of WEC. These can help to provide initial guidelines to the type of generator that may be required [22].

2.5 Generators in tidal energy systems

J. Bard and *P. Kracht*

The choice and design of generator systems for TEC are influenced by a variety of factors. The requirements on the generator system derive from the role within the overall tidal energy converter (section 2.2), economic and ecological aspects, environmental conditions and also device-specific characteristics like the structural design and the turbine technology. On the one hand, many of the resulting requirements can be considered standard as they also have to be accounted in fields like industrial applications and conventional electricity production. These are for instance a suitable torque-speed range, efficiency, viability etc. On the other hand, particular requirements like submerged operation and highly extended maintenance intervals can arise which, if at all, are only found in special fields like the oil and

Table 2.1 Summary of generator torque and speed ranges for each WEC category

OEC	Prime mover speed range	Generator peak torque range (expressed in per unit, pu)	Generator type
A	600–1500 rpm	2–4 pu; reduces with added inertia	• VS SCIG/SG/PMSG
B	1000–3000 rpm	Close to 1 pu with high accumulator storage, up to 4 pu as storage reduces	• FS SG/SCIG-SVC for high-storage designs • VS SG/PMSG/SCIG for low-storage designs
C	100–250 rpm	Close to 1 pu	• GBC: FS SG/SCIG-SVC or VS PMSG/SCIG • Low-speed VS PMSG
D	0–3 m/s 0–400 rpm	2–5 pu; reduces with added inertia	• Linear: PMSG (custom) • Rotary: GBC VS SG/PMSG

VS = variable speed; FS = fixed speed; GBC = gearbox coupled; SVC = solution requires static VAR compensator for reactive power requirements

gas industry. In the context of developing different TEC concepts from down-scaled prototypes to first pre-commercial full-scale demonstration devices several approaches have been used to meet the requirements of the generator systems. This section describes the characteristics of applied generator technologies and uses selected examples for illustration. In order to ease the understanding of the interaction between the generator system and the overall TEC, brief descriptions of the current resource, the turbines applied and different TEC concepts are given.

2.5.1 Tidal stream and tidal energy converter characteristics

2.5.1.1 Tidal stream characteristics

Tidal currents provide a highly reliable source of renewable energies. If one considers average time intervals of some minutes the current velocity is highly predictable and is designated as mean velocity hereafter (Figure 2.7). The mean velocity shows roughly four peaks per day in almost all locations in the world. As the current velocity in the open sea is too low to allow for efficient energy conversion, only sites where the currents are accelerated by narrows or other geographical conditions qualify for the installation of tidal energy converters. One can assume that this requires peak currents above 2 m/s, which occur in many parts of the world. Some places even exhibit peak currents more than 8 m/s [23]. Typical examples of suitable sites are the Bay of Fundy, Canada [24], and the tidal test site at EMEC in the Fall of Warness, UK [25], where peak current velocities around 4 m/s can be found.

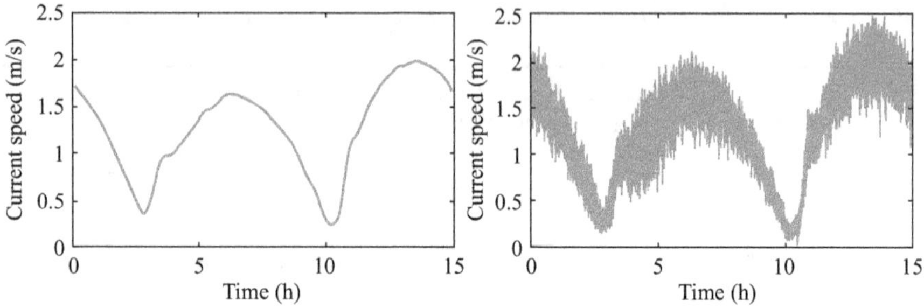

Figure 2.7 Example of marine current velocity data, measured in the Bristol Channel, England: (left) time-averaged curve; (right) high-resolution curve showing turbulences

The mean current velocity is superimposed by speed fluctuations at smaller scales, which can be referred to as turbulence (Figure 2.7). On the one hand, this comes from the effect of waves and wind at the free surface and on the other hand, from bottom and side friction with the associated shear stress in the water column [26]. Although in-depth understanding of the physical phenomena and corresponding quantification are topics of ongoing research, it is observed that those fluctuations are typically 10% of the mean speed, or more [27, 28].

Forming the basis for the design process of a TEC project and thus also affecting the applied generator system a profound knowledge of the current resource in terms of both the mean values and the expectable turbulences at the given site is required. This can only be achieved by a thorough site assessment, including long-term measurements of currents and turbulences.

In the majority of TEC concepts either horizontal or vertical axis current turbines are used as prime movers to convert the power in the marine currents to a rotating mechanical power. The power harvested by the turbine can be expressed by the power coefficient, C_P, which gives the ratio between the incident marine current power on the rotor swept area and the mechanical output power of the turbine. C_P is a function of the dimensionless tip speed ratio, TSR, which is defined as the ratio between the blade tip speed and the current velocity. Figure 2.8 depicts a typical C_P-TSR curve, which is quite similar for both horizontal and vertical axis turbines. Another common aspect of both turbine types is the rather low rotational speed. Considering turbines rated in the power range from 100 kW to several megawatts, rated speeds in the range of around 10 rpm will be found.

Differences between the two types of turbines are that the vertical axis one shows an oscillating torque, a slightly lower C_P and an even lower rated speed. These effects are caused by the constantly changing angle between the rotor blades and the current stream and also the fact that each blade is operated in the lee of the other blades during significant parts of each revolution. Despite these negative aspects vertical axis turbines are applied in some TEC concepts, which is due to some significant advantages mainly in terms of structural integration as described below. A further advantage of vertical axis turbines is that the power conversion is

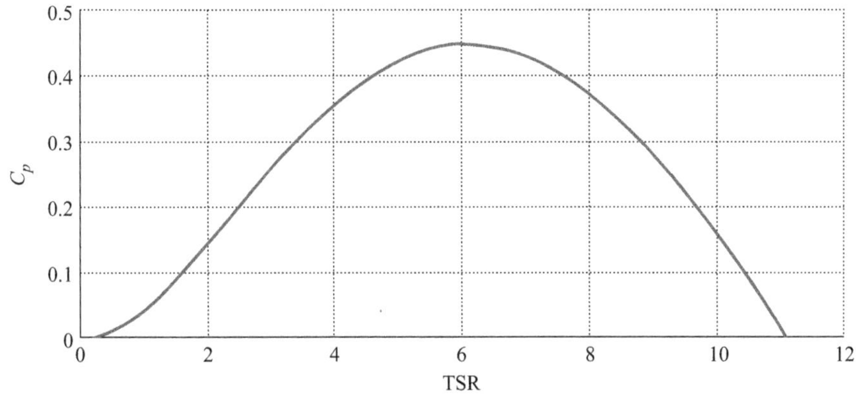

Figure 2.8 Illustrative C_P-TSR curve

not affected by the current direction. Considering horizontal axis turbines mainly two approaches are used to deal with the reverting flow. Some concepts use fixed symmetrical blades, which result in some disadvantages in terms of efficiency and controllability. Another approach is the application of pitched blades, allowing for spinning the blades according to the flow direction. The blade angle can be adjusted for power reduction of the turbine, which is, if required, a further advantage of pitched blades. But obviously this needs to be balanced against, e.g., increased system complexity. Further options in the turbine design – not detailed here - are for instance tubed turbines [29] and turbines using rather the drag than the lift effect [30].

Alternative prime mover concepts

Besides the types of prime movers discussed above, several other operating principles are studied by different device developers. Examples are for instance oscillating hydrofoils [29] and tidal sails [31]. Obviously these operating principles can result in completely different requirements on the generator systems. A detailed consideration of these technologies is not the scope of this chapter.

Operation modes of tidal turbines

Tidal turbines have to deal with a wide range of current velocities. The full range is typically covered by up to three different turbine operation modes (Figure 2.9). At low current speeds the turbine is stopped (region 1), as the combination of energy production, system losses, component wear etc. does not allow for an economical operation. In region 2 the turbine output power is below the rated power of the device. In this region the control objective usually is to maximize the energy capture by operating the turbine at the peak of the C_P-TSR curve (Figure 2.8), which means the turbine is operated with variable rotor speed. The trade-off between increased costs due to loads on the structure, rated power of the components etc. and the energy harvested can make it appear reasonable to limit the turbine output power at some point (region 3). According to the C_P-TSR curve this can either be

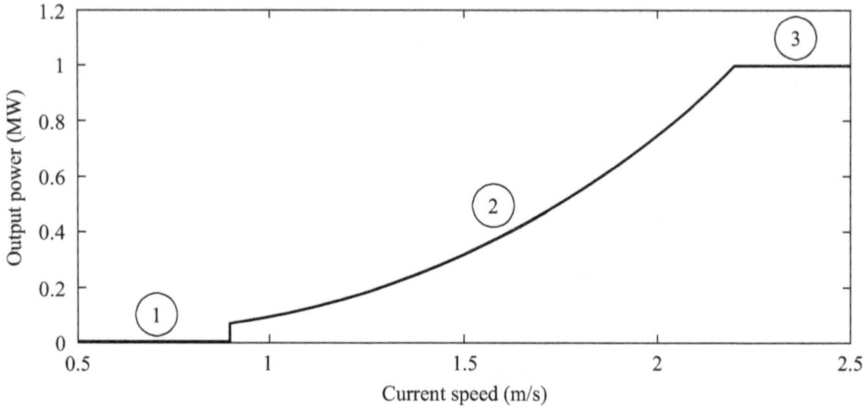

Figure 2.9 Illustrative steady-state power curve showing the different turbine operation regions

achieved by operating the turbine above or below the optimum TSR, both of which can turn out to be problematic. Operating the turbine below the optimum TSR, using the so-called stall effect [7], is likely to result in high forces on the turbine and the generator system needs to be designed for high-peak torques in case of turbulences. In case the turbine is operated above the optimum TSR cavitation might become a serious issue and additionally all components of the turbine–generator system should be designed for the occurring high rotational speeds. As mentioned above blade pitching gives a further way to control the turbine, avoiding high torque variation in the stall region and also high rotational speeds.

In conclusion it can be said that both turbine and current characteristics and the applied control schemes are quite similar to what is found in wind energy.

Structural integration of the turbine

In wind energy the structural layout of the devices has converged towards the well-known design of standard horizontal axis wind energy converters. Looking at tidal energy converters a different situation can be found. Quite a number of different concepts are still being studied. In the following paragraphs three examples of successfully tested TECs are presented.

The supporting structure of the SeaGen device [32] consists of a surface piercing mono-pile set in a hole drilled in the seabed. The two turbines of the TEC are mounted on a cross-beam, which for maintenance and repair can be lifted above sea-level by means of hydraulic rams (Figure 2.10). Load shedding at high current velocities and adaption to the reversing flow direction of the two horizontal axis turbines are achieved by pitched blades. Another example of a TEC utilizing a horizontal axis turbine with pitched blades (not depicted here) is the HS1000 device [33]. The generator of the HS1000 is housed in a nacelle mounted on a seabed-fixed tripod and has to be lifted by a floating crane for maintenance and repair.

Figure 2.10 SeaGen device, turbines lifted above sea-level for inspection and maintenance (reproduced by permission of Marine Current Turbines Ltd)

The Kobold turbine [34] follows a completely different concept. The Kobold turbine is a vertical axis turbine mounted below a floating platform (Figure 2.11).

The examples given here reflect only an excerpt of the multitude of different concepts currently under development; for a more complete overview of technologies the authors refer to [29]. Nonetheless these three different concepts show that the structural layout highly affects the requirements on the generator system as discussed in the following sections.

2.5.2 Generator system specifications

2.5.2.1 Requirements on the generator system

The basic requirement on the generator system is to match the turbine characteristics – basically this results in the need for a variable speed generator system, adapted to the low speed range of the tidal turbines. A detailed specification of the generator system can only be achieved by consideration of the overall system. Requirements like maximum speed and torque of the generator system derive from factors like the control schemes and mechanisms – pitched/unpitched turbine – and the expected turbulences at the site. Independently from the generator technology applied, it should be taken into account that apart from turbulences and the influence of any wave action, the tidal currents are deterministic. From this deterministic behaviour load profiles for the generator system rating can be derived, which can be used for the generator rating, i.e. the generator rated power does not have to match the maximum rated power of the TEC. Instead the generator can be operated in overload condition during peak currents, using the thermal time constants of the generator.

Figure 2.11 *Kobold turbine: (left) scheme of turbine rotor and electrical generator; (right) prototype device, driving a pump to dissipate the generated power (reproduced by permission of Ponte di Archimede International Spa)*

By means of load profiles also the impact of full- and part-load efficiency of the generator system on the overall system efficiency can be determined. The importance of part-load efficiency decreases with the proportion of time the turbine is operated in region 3 (Figure 2.9). Further aspects to be accounted are for instance operation and maintenance (O&M) costs, reliability, availability and low environmental impact of the generator system components.

Some further crucial requirements derive from the specific TEC concept. Here some examples shall be given which are derived from the aforementioned TEC concepts. Both in the SeaGen and the HS1000 device and also in almost all other TECs applying horizontal axis turbines the generator system is submerged, which highly influences the generator system design. First, the generator system must either be protected against or be designed to operate in direct contact with the seawater. Second, as access to the system can be extremely cost and time intensive the system must be designed for high robustness and extended maintenance intervals. A positive aspect of the submerged operation is the relatively low ambient temperature, which allows efficient cooling of the components. Another aspect combined with the application of horizontal axis turbines are the high thrust loads, which needs to be accounted for in the bearing design of the turbine–generator system. Considering the example of the vertical axis turbine a completely different situation is found. Here the generator system is located in a machinery room above sea-level. Obviously also in this case robustness and long maintenance intervals are key factors for commercial success, but the relatively enhanced accessibility of the generator system definitely gives some more scope. Considering the bearings, instead of thrust loads, high bending torques from the turbine now need to be absorbed. Additionally as in this example the support structure of the turbine is a floating platform; further loads from the platform motions have to be absorbed by the bearings.

2.5.2.2 Comparison of generator technologies

In wind energy currently four different generator technologies are used for large wind turbines [35], namely PMSG, DFIG, SCIG, SG, which on a first glance are also candidate systems for TECs. However due to some drawbacks of DFIGs and SGs – maintenance requirements due to slip-ring wear (DFIG) and complex rotor structure (SG) – these types of generator systems do not appear to be appropriate technologies for TECs and are rarely – if at all – used in this field. In wind energy SCIGs have been the standard generators utilized in the 1990s, but are used less and less due to some improved characteristics of other generator technologies such as the PMSG as discussed below, as well as the availability of lower cost high-power converters. On the contrary SCIGs are widely used in TEC projects, mainly as a result of two considerations. First, the rotor structure of an SCIG is very simple and thus this type of generator system shows a very high robustness. Second, SCIGs operated as motors are the quasi-standard electrical machine in industrial applications. Therefore SCIGs are also highly available in standard designs for marine and subsea application (see below). Also PMSGs, a rather new technology, are often used in TEC designs. Compared with SCIGs, PMSGs show a lower weight and size and an increased efficiency both at full and part load [7]. These advantages need to be balanced against the drawbacks of PMSGs. These are mainly uncertainties in both the reliability – corrosion and cracking magnets, which is a topic of ongoing research – and the investment costs. The price for the raw materials of the permanent magnets accounting for a significant portion of the costs of a PMSG (ca. 20–30%) is highly fluctuating [36]. Regardless of the type of generator a full back-to-back converter is typically used for the grid connection allowing variable operation over the full speed range (see next section). However, the speed range of standard electrical machines is well above the low rated speeds of tidal turbines. The most common method for speed adaption, which is also applied in many TEC projects, is the application of gearboxes, which has some major disadvantages. Due to the additional losses the overall system efficiency is decreased. Gearboxes furthermore require maintenance on a regular basis, and they can have a significant effect on the system reliability – the highest downtimes in wind energy converters are caused by gearbox failures [37].

To overcome the disadvantages associated with gearboxes so-called direct-drive systems were proposed for wind energy converters and in the 1990s the first wind turbines equipped with SG direct-drive systems were successfully introduced. Obviously a direct-drive generator system eliminates the drawbacks combined with the application of a gearbox, but the direct-drive technology also has some drawbacks. In a direct-drive system the generator must be designed to meet the low speed–high torque characteristics of the turbine, which results in large-diameter generators. In wind energy so-called rim generators are applied in direct-drive systems. In this machine design the generator rotor comprises a ring-shaped active part, which is connected to a central shaft by means of rotor arms. This means that contrary to standard electric machines the rotor is not solid. Although this results in a weight reduction, direct-drive generator systems show a high weight penalty compared to a drive train, comprising a standard high-speed generator and a gearbox [35]. Obviously with the high weight come additional costs for material, installation, transportation etc. Additionally low-speed electrical machines tend to show a lower efficiency than high-speed electrical machines [39], which partly neutralizes the efficiency benefit of direct-drive systems. These disadvantages of direct-drive systems can be partly compensated, if PMSGs are used. PMSGs can be built in a more compact fashion, resulting in a lower weight [35], and as no excitation is needed show a higher efficiency. Despite the still significant weight and cost penalty, PMSG-based direct-drive systems are considered a very promising generator technology for wind energy converters, which is reflected in the fact that many wind turbine manufactures have introduced PMSG-based direct-drive system in recent years [40]. Also for TECs direct-drive PMSG systems are promising and are applied in some designs. In principle the drawbacks of the technologies, high weight, costs etc., are also problematic in TEC projects, but the large diameter of the direct-drive system has been accounted for by some developers by placing the turbine in the centre of the generator (Figure 2.12). This appears to be a feasible solution as the diameter of tidal turbines is significantly smaller than that of a wind turbine rated at equal power. One common device utilizing a PMSG direct-drive system is the Open-Centre Turbine [38].

Besides the principal benefits and disadvantages of direct-drive generator systems one crucial factor should be considered in the decision making – this is the fact that a direct-drive rim generator is definitely a non-OTS design, which brings additional risks and costs in a development project. The experiences from wind energy show that for instance a significant improvement in terms of reliability cannot be achieved without significant and ongoing research and development effort [37]. Additionally both the industry and research community are studying advanced concepts, aiming at further weight and size reduction and increased

Figure 2.12 Integration of a rim-generator in a TEC

efficiency [41, 42]. This research also includes studies on improved PMSG designs especially for TEC applications [43].

Also the question on how to deal with the maritime environment and the possibly submerged operation is closely linked to the weighing of benefits and disadvantages of designing the generator systems from standard components compared to the application of specifically developed components. When looking at a design like the Kobold turbine (Figure 2.11) it appears that the requirements on the generator system are quite similar to requirements on electrical machines installed on marine vessels. For this application OTS components from various manufactures are available, which provide required features like enhanced bearing arrangements and corrosion protection, are certified for marine applications and have proven track records. In 2002 a first prototype of the Kobold turbine was installed in the Strait of Messina, Italy. Initially the generator system comprised an SG and a gearbox for speed adaption. In the course of grid connecting the device in 2005 the generator system was redesigned to better match the available shaft power, which was found lower than expected due to lower current speeds. The actual generator system consists of an SCIG and a gearbox (personal communication, March 2013). If one looks at the examples of horizontal axis turbines the generator is operated submerged, resulting in requirements common for instance in the oil and gas industry. Commonly two different approaches are used to realize a submerged operation of electrical equipment including electrical drives. The first is to install the equipment in a watertight nacelle. On the one hand, this approach requires some effort for the construction of the nacelle. Especially the dynamic seals for the rotor connection need to be carefully designed. On the other hand, the generator system can be designed from standard components [44], which benefits in terms of costs and availability of the electrical equipment. One example for a device relying on this technology is the HS1000. Additional to the generator system, comprising an SCIG and a gearbox, the nacelle houses electrical equipment such as control and sensor hardware. For maintenance and repair the nacelle needs to be recovered from the seabed, which is a cost and time intensive procedure. Therefore the generator system is designed for long maintenance intervals of 5 years [33]. The design of the SeaGen device follows the second approach to deal with the submerged operation, which is to protect the single components of the generator system against being flooded by seawater. This means an increased number of dynamic seals are required. In case of the SeaGen device the SCIG, the gearbox and the brake have to be separately protected. On the one hand, this increases the risk of a seal-failure. On the other hand, the design and construction of a watertight nacelle is avoided and components like electrical machines for subsea operation are available OTS. In [45] it is described in detail how the generator system for the Seaflow device, a predecessor version of the SeaGen device, was designed from standard components used in wind energy and the oil and gas industry. As can be seen from the aforementioned examples, both conventional approaches to deal with the submerged operation are successfully used in TECs and have proven their feasibility. Major advantages of these approaches are the availability, reliability and the reduction of development time and risk in a project. This has to be weighed against the drawbacks of these technologies. These are on the one hand high investment costs for either subsea components or the design and construction of a watertight nacelle and on the other hand the risk of water contamination due to leaking oil from gearboxes, bearings or oil-filled generators, and serious component damages in case of a seal-failure.

Actually the risk of failing dynamic seals seems to be one major concern in TEC projects, which is accounted by some developers by introducing new concepts in their TEC designs. Obviously direct-drive generator systems eliminate the potential risks combined with a gearbox. An approach to improve the reliability of the generator itself is simply to flood the air-gap of the generator. The idea behind this technology is to avoid dynamic seals and use only static ones, which directly protect the machine windings, instead of the whole air-gap of the machine. One example of a device – among others – utilizing this technology is the Open-Centre Turbine. Additionally magnetic bearings are utilized in the design, resulting in a lubricant-free device. Other developers apply seawater lubricated bearings [46, 47], a technology well-known for instance from applications like ship propulsion and hydropower.

Generally speaking, any avoidance of submerged equipment – especially movable parts – will result in an increased reliability and ease of maintenance. Some developers follow the consequent approach to fully avoid the subsea installation of dynamic seals [46]. This means that an unpitched turbine is applied and no brake system is incorporated in the TEC. This approach reduces the risk of a system failure simply by reducing the number of critical components. But special attention has to be paid to particular scenarios such as the event of a power network or power electronics fault. In this case the turbine will accelerate up to a current velocity depending on run-away speed. This operation point has to be addressed in the design of the generator and the bearings, which both must withstand the increased mechanical stress in run-away conditions. Additionally in case a PMSG is used high induced voltages will occur. Therefore enhanced generator winding insulations and circuit breakers, to protect the power electronics, might be required.

2.6 Power electronics for generator control in ocean energy converters

S. Ceballos and *E. Robles*

Previous sections have clearly established the potential requirement for variable speed control of the main generator in the power conversion drive train. This section details typical modern power electronic controller techniques capable of implementing fully variable speed control.

First, a general overview of different control methods will be presented. Field- and voltage-oriented controllers will be introduced and PI (proportional-integral) tuning procedures will be discussed. Finally some issues regarding reliability of power converters are analysed.

2.6.1 Control of back-to-back converters in full-converter variable speed configurations

Section 2.1 described how variable speed solutions with back-to-back converters can improve the performance of renewable generation power take-off systems. This section describes a control algorithm for a variable speed solution with full-converter configuration and a PMSG. A field-oriented control (FOC) algorithm for the generator-side converter and a voltage-oriented control (VOC) for the grid-connected converter will be developed. A comprehensive description of these and other control strategies for power converters interfacing synchronous and asynchronous machines can be found in [48].

2.6.2 Generator-side power converter control

The aim of the generator-side power converter controller is to control the speed and/or torque of the electric generator according to certain predefined references calculated by a higher level controller.

Figure 2.13 shows a general classification of the variable frequency control methods [49]. These control methods can be divided into scalar and vector controls. In the scalar control, which is based on a relation valid for steady state, only the magnitude and frequency of the voltage, current and flux linkage vectors are controlled [50]. Thus, the control system does not act on the space vector position during transients. Therefore this control is used when high dynamics are not demanded. On the contrary, in vector control, which is based on a relation valid for transients, also the instantaneous position of the voltage, current and flux space vectors are controlled. The control system adjusts the position of the space vectors and guarantees their correct orientation for both steady state and transients.

Among the vector control methods, the most widely used technique is the FOC, which uses current control loops in a synchronous reference frame [51] and

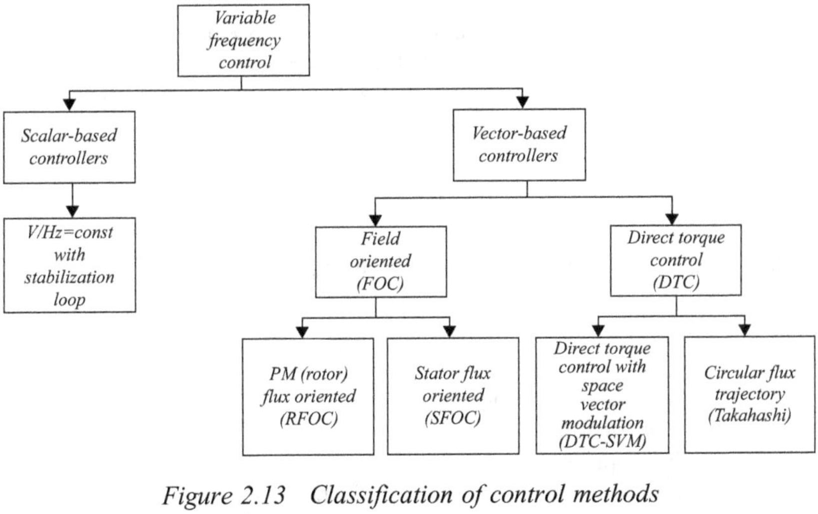

Figure 2.13 Classification of control methods

pulse width or space vector modulators. Also a very extended control scheme is the direct torque control (DTC) developed in [52]. The DTC is a vector control scheme with closed torque and flux loops. Using bang-bang hysteresis controllers made this control concept very fast and not complicated. However, it has several drawbacks such as the requirement of a fast sampling time, a variable switching frequency and a high torque pulsation, which can be overcome using DTC-SVM [49]. This method can be considered as a mixture between FOC and DTC. A detailed description of these control algorithms can also be found in [48].

This section describes the implementation of a FOC for PMSGs. A dynamic model of the PMSG is developed, the block diagram of the controllers is introduced and some general rules to tune the controllers are given.

2.6.2.1 Generator dynamic model

The electrical dynamic equations of a PMSG are derived using the power conservative stationary *dq* reference frame, and assuming that the *q* axis is synchronized with the magnetic flux:

$$v_{1d} = R_s i_{1d} + L_d \frac{di_{1d}}{dt} - L_q \omega i_{1q} \tag{2.1}$$

$$v_{1q} = R_s i_{1q} + L_q \frac{di_{1q}}{dt} + L_d \omega i_{1d} + \sqrt{\frac{3}{2}} \psi_m \omega \tag{2.2}$$

$$T_e = \sqrt{\frac{3}{2}} p \left[\psi_m i_{1q} + i_{1d} i_{1q} (L_d - L_q) \right] \tag{2.3}$$

where v_{1d} and v_{1q} are the d and q components of the voltages at the generator terminals respectively, i_{1d} and i_{1q} are the transformed stator currents, T_e is the electromagnetic torque, R_s is the stator resistance, L_d and L_q are the dq axis stator inductances, p is the number of pole pairs, ψ_m is the magnetic flux and ω is the electrical rotational speed.

Assuming a machine with low saliency, the d and q stator inductances are very similar so, from now on, it is assumed that $L_d \approx L_q = L$. In addition, the reference of the stator magnetizing current I_{1d}^* is set to zero. With these assumptions, (2.3) is reduced to:

$$T_e = \sqrt{\frac{3}{2}} p \psi_m i_{1q} \tag{2.4}$$

The mechanical dynamic equation of a generator is given by

$$T_m - T_e = J \frac{d\omega_m}{dt} + D\omega_m \tag{2.5}$$

where T_m is the mechanical torque, ω_m is the mechanical rotational speed, J is the inertia and D is the friction.

Applying the Laplace transformation to these equations the model of the generator can be represented by the block diagram shown in Figure 2.14. This block diagram represents a multiple input multiple output system with two control variables v_{1d} and v_{1q} and three controlled variables i_{1d}, i_{1q} and ω_m. It is also important to point out the existence of two coupling terms between the currents. Due to these coupling terms any variation in a single variable affects all the rest.

2.6.2.2 Generator-side converter controller

A FOC to regulate the controlled variables i_{1d}, i_{1q} and ω_m is described here. Figure 2.15 displays the block diagram of the controller. It consists in two nested control loops; an external PI controller regulates the mechanical rotational speed of the generator while two internal PI controllers regulate the dq stator currents. In addition, decoupling terms to make the PI regulators immune to the effects of the coupling terms of the system are also included.

Assuming that the internal control loops are much faster (at least five times faster) than the external control loops, the tuning of the speed and current

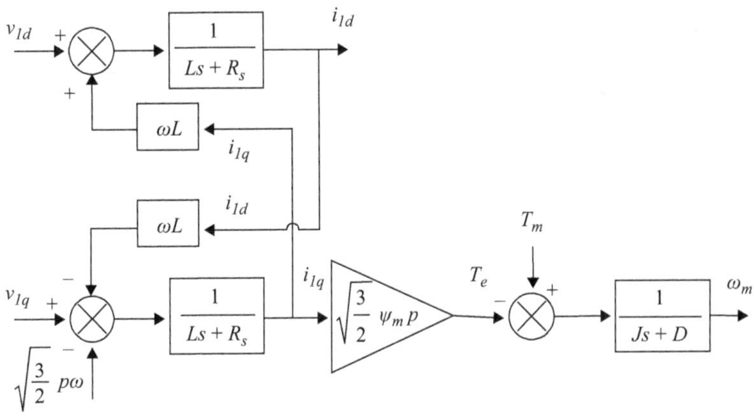

Figure 2.14 Block diagram representation of PMSG model

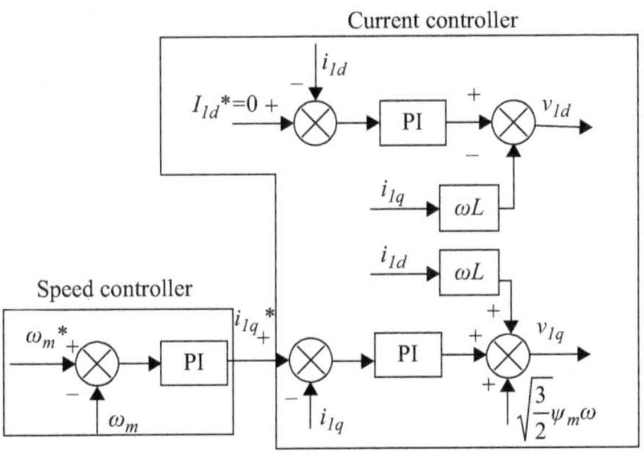

Figure 2.15 Block diagram representation of generator-side converter controller

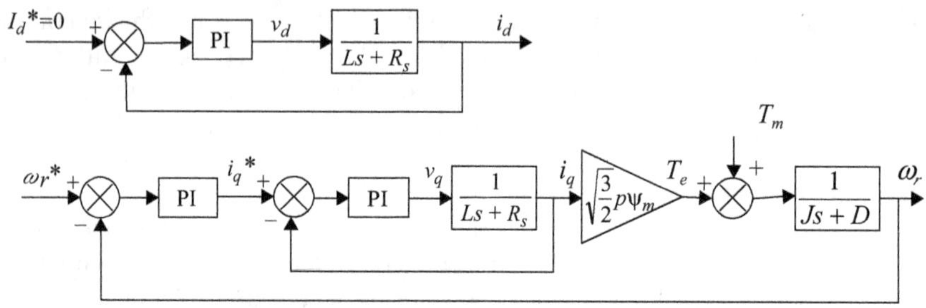

Figure 2.16 Speed and current control loops once the coupled terms have been compensated

controllers can be done independently. In order to tune the current regulators, the zero-pole cancellation method described in [53] has been used. After decoupling the current loops the transfer functions affecting the active and reactive components of the current are simplified to a first-order system as shown in Figure 2.16:

$$G(s) = \frac{1}{Ls + R_s} \tag{2.6}$$

The PI current controllers are mathematically represented by

$$G_c(s) = K_{1pc} + \frac{K_{1ic}}{s} = K_{1ic}\frac{1 + \frac{K_{1pc}}{K_{1ic}}s}{s} \tag{2.7}$$

The closed-loop transfer functions for the dq components of the current are thus:

$$G_{cl}(s) = \frac{i_{1d}(s)}{I_d^*(s)} = \frac{i_{1q}(s)}{I_q^*(s)} = \frac{G(s)G_c(s)}{1 + G(s)G_c(s)} = \frac{\frac{K_{1ic}}{R_s}\frac{\left(1 + \frac{K_{1pc}}{K_{1ic}}s\right)}{s\left(1 + \frac{L}{R_s}s\right)}}{1 + \frac{K_{1ic}}{R_s}\frac{\left(1 + \frac{K_{1pc}}{K_{1ic}}s\right)}{s\left(1 + \frac{L}{R_s}s\right)}} \tag{2.8}$$

If the zero introduced by the PI controller is placed on the pole introduced by the system, that is:

$$\frac{K_{1pc}}{K_{1ic}} = \frac{L}{R_s} \tag{2.9}$$

The closed-loop transfer function of the system becomes a first-order transfer function:

$$G_{cl}(s) = \frac{\frac{K_{1ic}}{R_s s}}{1 + \frac{K_{1ic}}{R_s}s} \tag{2.10}$$

Setting a closed-loop settling time for the *dq* stator currents equal to T_{1c}, the integral constant of the current controllers is given by

$$K_{1ic} = \frac{4R_s}{T_{1c}} \tag{2.11}$$

Substituting (2.11) in (2.9) the proportional constant can be expressed as

$$K_{1pc} = \frac{4L}{T_{1c}} \tag{2.12}$$

Expressions (2.11) and (2.12) allow tuning in an analytical way the PI *dq* current controllers. The proportional and integral constants are a function of the electrical parameters of the generator and the settling time.

The settling time has to be selected in such a way that:

- T_{1c} should be small to provide quick dynamic to the system.
- T_{1c} should not be too small to avoid saturation of the voltages given by the converter (overmodulation).

Figure 2.17 shows an experimental test that analyses the performance of the current controller in the presence of sudden reference changes. The reference of the active component of the current of a PMSG is a square waveform with a 200 ms period. Figure 2.17 displays the actual active component of the current. The results show the good dynamic behaviour of the current controller.

Once the inner current controllers have been tuned the focus is now put on the speed controller. If it is assumed that the inner current control loops are at least five times faster than the external speed control loops, their effect in the speed dynamics can be considered negligible. Therefore the simplified system block diagram affecting the speed can be represented as shown in Figure 2.18.

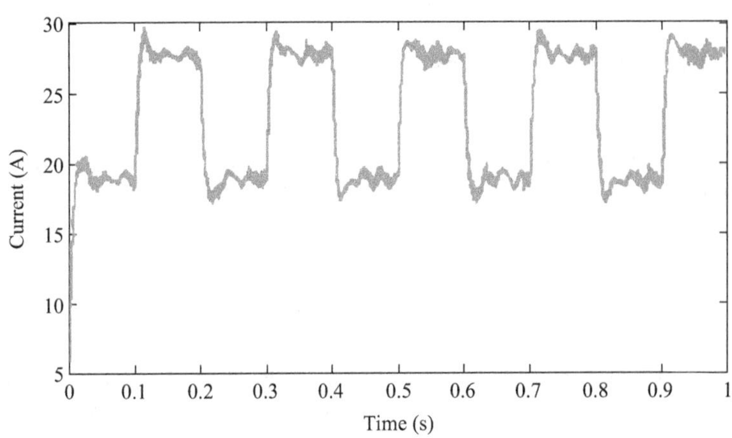

Figure 2.17 Performance of the current controller with zero-pole cancellation method, square waveform reference with a 200 ms period

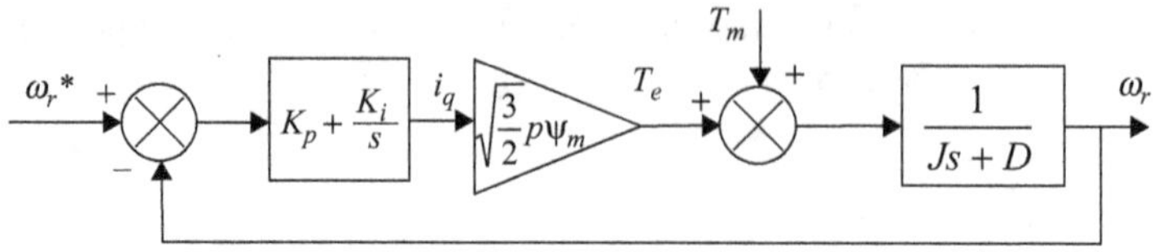

Figure 2.18 Simplified speed control loop

The zero-pole cancellation method can also be used to tune the speed controller. Applying this method the following expressions for the proportional and integral constants of the speed controller are obtained:

$$K_{Ps} = \frac{4J}{T_s\sqrt{\frac{3}{2}}p\psi_m} \qquad K_{Is} = \frac{4D}{T_s\sqrt{\frac{3}{2}}p\psi_m} \tag{2.13}$$

where T_s is the closed-loop settling time for the rotational speed.

This tuning procedure is simple and gives good results when T_s is small or the inertia of the drive train is high enough. Otherwise speed fluctuations due to mechanical torque variations (T_m) can have a considerable influence in the response of the system. In that case, a tuning procedure based on the analysis of the root locus can be used to improve the dynamic response of the system. Using this strategy the closed-loop complex poles of the system can be placed to obtain a desired dynamic response usually characterized by a defined settling time and damping factor. In addition a pre-filter can be used to minimize the overshoot caused by the zero introduced by the PI controller [54]. Figure 2.19(a) shows the position of the complex poles and zeros of a speed controller developed for a low-speed PMSG when this method is used.

Figure 2.19(b) displays simulated and experimental speed responses. When the generator speed is stable at 40 rpm, a speed reference change up to 60 rpm is produced. Due to the action of the controller this speed reference is followed. Once the actual speed of the generator is 60 rpm, at $t = 20$ s, an increase of the mechanical torque T_m up to 100% of its nominal value is applied. Again, after a transient period, the speed is controlled properly. Obviously the duration and speed waveform during the transient can be modified, changing the position of the poles and the zero in the root locus.

2.6.3 Grid-side power converter control

The control of a grid-connected converter can be considered as a parallel problem to the vector control of an electrical machine. In this case the controller aims to control the DC-bus voltage and the reactive component of the current injected into the grid. Alternatively, it is also possible to control the active and reactive power injected into the grid.

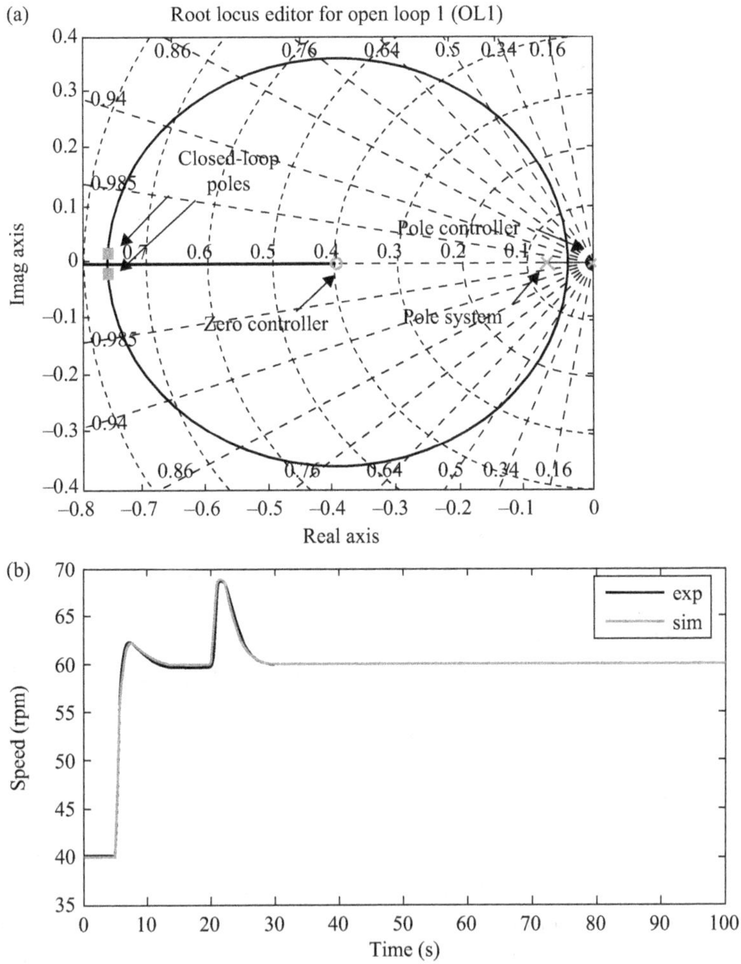

Figure 2.19 *(a) Speed root locus; (b) simulated and experimental speed response (speed reference step at t = 5 s and torque step at t = 20 s)*

Figure 2.20 shows a general classification of control strategies for grid-connected power converters. They are the following:

- Voltage-oriented control (VOC).
- Virtual flux-oriented control (VFOC).
- Direct power control (DPC).
- Virtual flux direct power control (VF-DPC).

The VOC of a grid-connected converter is very similar to the FOC of a converter connected to an electrical machine, while the DPC is very similar to the DTC of a machine. Both the VOC and the VFOC use internal current control loops in a rotating *dq* reference system [55, 56] and pulse width or space vector modulators to

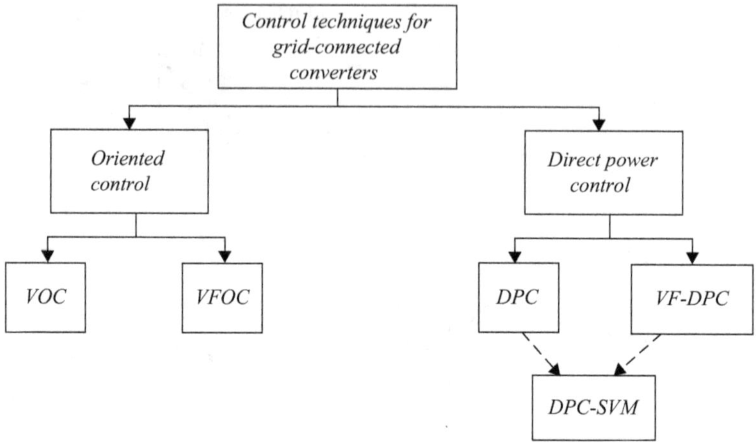

Figure 2.20 Classification of control methods for grid-connected converters

guarantee a good dynamic and static performance. The main difference between VOC and VFOC is the way they estimate the angular position of the grid voltage vector. While the VOC gets the voltage vector directly measuring the grid voltage, the VFOC uses the concept of virtual flux [48].

The DPC is based on the instantaneous active and reactive power control loops [57, 58]. There are no internal current control loops nor pulse width modulator blocks, because the switching states of the converter are selected by a switching table which is based on the instantaneous errors between the commanded and estimated values of active and reactive power. Therefore a correct and fast estimation of the active and reactive power is needed. The main difference with the VF-DPC is the way the instantaneous active and reactive power are obtained. The former obtains it from the grid voltage, while the VF-DPC obtains from the grid voltage flux. The control principles of the DPC are very similar to those of the DTC schemes for electrical machines, and share the same drawbacks [59]. To overcome them, an SVM (space vector modulation) strategy was introduced to the DPC structure obtaining the DPC-SVM control scheme [60]. The DPC-SVM scheme includes the advantages of the SVM (i.e. constant switching frequency and unipolar voltage pulses), and preserves the advantages of the DPC (i.e. simple and robust structure, lack of internal current control loops, good dynamics etc.). A detailed description of these control methods can be found in [48].

This section describes the implementation of a VOC. A dynamic model of the grid-side converter, including the grid filters and the DC bus, and the block diagram of the controllers are introduced. Some general rules to tune the regulators are given.

2.6.3.1 Grid side converter dynamic model

The dynamic equations modeling the AC side of the grid-side converter using the power conservative *dq* transformation are given by

$$v_{2d} = R_f i_{2d} + L_f \frac{di_{2d}}{dt} - L_f \omega i_{2q} + e_d \tag{2.14}$$

$$v_{2q} = R_f i_{2q} + L_f \frac{di_{2q}}{dt} + L_f \omega i_{2d} + e_q \tag{2.15}$$

where v_{2d} and v_{2q} are the d and q components of the voltage at the grid converter terminals, i_{2d} and i_{2q} are the transformed grid currents, e_d and e_q are the transformed grid voltages, R_f is the grid filter resistance, L_f is the grid filter inductance and ω is the grid angular frequency.

The behaviour of the DC bus, assuming that the converters have an efficiency of 100%, is described by means of expression:

$$Cv_{dc}\frac{dv_{dc}}{dt} = P_{gen} - v_{2d}i_{2d} - v_{2q}i_{2q} \tag{2.16}$$

where v_{dc} is the DC-bus voltage, P_{gen} is the power injected into the DC bus by the converter connected to the generator and C is the DC-bus capacitance.

Equation (2.14)–(2.16) constitute the dynamic model of the grid-side converter. It is important to highlight that there is a non-linearity in the system introduced by (2.16). In order to develop a linear controller, this equation should be simplified. Assuming that the filter losses are negligible and that the d axis is synchronized with the grid voltage, i.e. $e_q = 0$, (2.16) can be rewritten as follows:

$$\frac{1}{2}C\frac{dv_{dc}^2}{dt} \approx P_{gen} - e_d i_{2d} \tag{2.17}$$

(2.14), (2.15) and (2.17) represent a simplified model of the grid-side converter. The block diagram representation of this model is shown in Figure 2.21.

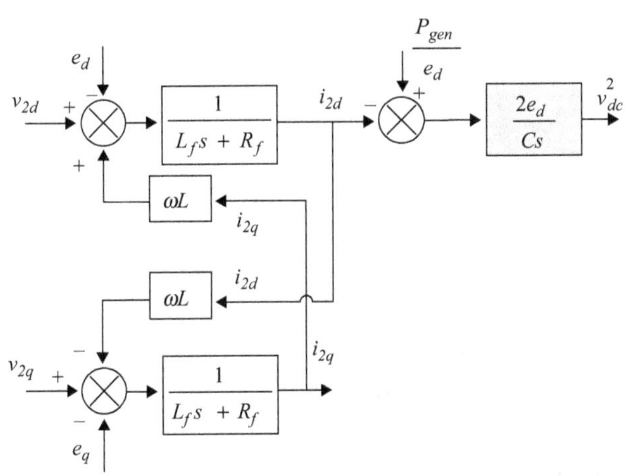

Figure 2.21 Block diagram of the linearized model of the grid-side converter

2.6.3.2 Grid-side converter controller

A VOC to regulate the variables i_{2d}, i_{2q} and v_{dc}^2 has been used. The block diagram of the grid-side converter controller is shown in Figure 2.22. Again two nested control loops, an inner one for the grid current components and an outer one for v_{dc}^2, have been employed. Following a similar approach to that used to tune the generator current controllers, the proportional and integral constants of the PI current regulators are given by

$$K_{2pc} = \frac{4L_f}{T_{2c}} \quad \text{and} \quad K_{2ic} = \frac{4R_S}{T_{2c}} \tag{2.18}$$

where T_{2c} is the closed-loop settling time for the dq grid currents.

Figure 2.23 shows the simulated behaviour of a grid-connected converter with the proposed controller. The current references for the d and q components of the current are 20 A and 0 A respectively. At $t = 25$ ms the references change to 45 A and 5 A respectively. A 4 ms settling time is assumed. These results confirm the good dynamics that can be achieved when this control method is used.

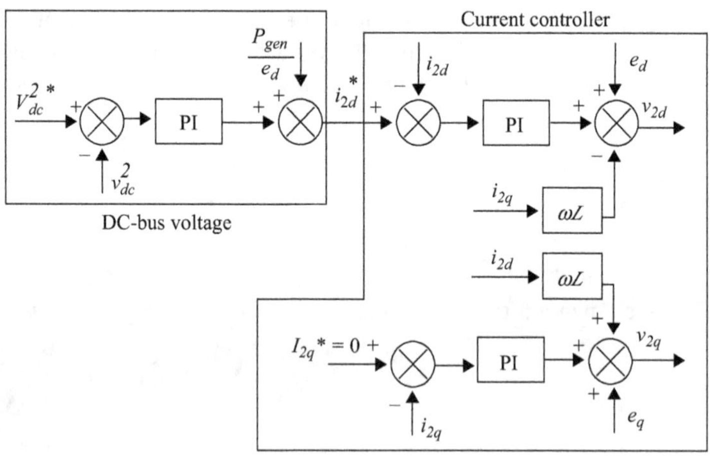

Figure 2.22 Block diagram of the grid-side converter controller

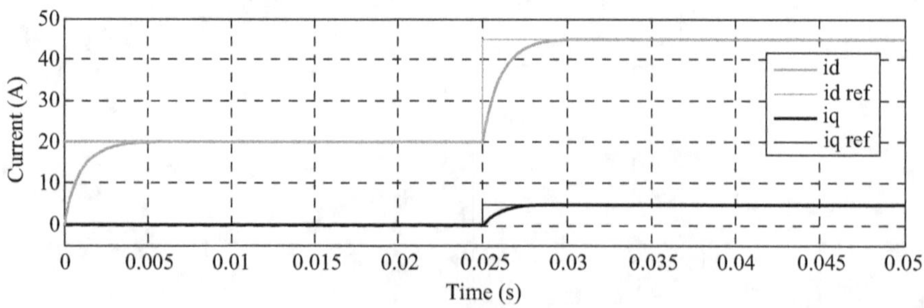

Figure 2.23 Performance of the grid-side current controller with zero-pole cancellation

The voltage controller is composed of a PI regulator and a feed-forward term comprising the power injected into the DC bus by the generator-side converter P_{gen}. The inclusion of this factor improves drastically the dynamics of the DC-bus voltage, given that any change in the generator power is rapidly compensated by the feed-forward factor. It is also noteworthy that the DC-bus dynamics are mainly governed by the grey coloured block of Figure 2.21, provided that the current control loops are significantly faster than the voltage control loop. This transfer function has a pole at the origin and, therefore, there is no need of including an integral controller to remove the stationary voltage error. This assumption is only valid for the simplified model represented in Figure 2.21, when the feed-forward term has been introduced on the controller and when a 100% efficiency of the converters is assumed. It is not fully correct for the non-simplified model given by (2.16). Nevertheless, it is still a good approximation when the proportional constant of the controller is high enough. Therefore based on this assumption, the voltage regulator proportional constant is given by

$$K_{Pvdc} = -\frac{2C}{e_d T_{vdc}} \tag{2.19}$$

where T_{vdc} is the closed-loop settling time for v_{dc}^2.

If a high-precision control of the DC-bus voltage is required, an integral controller can be introduced and tuned according to the desired dynamics by means of the root-locus method, similarly as it was done previously for the generator speed controller.

2.6.4 Fault operation of variable speed drives

Reliability of variable speed drives is an important issue in numerous applications. In particular, it is of great interest in offshore systems since the periods of time for maintenance are constrained due to weather conditions. A number of studies have addressed this area and several solutions for variable speed drives have been proposed [61–65]. Figure 2.24(a,b) shows two attempts to endow a variable speed drive with fault tolerant capabilities. The solution shown in Figure 2.24(a) [61] requires the neutral of the generator to be available.

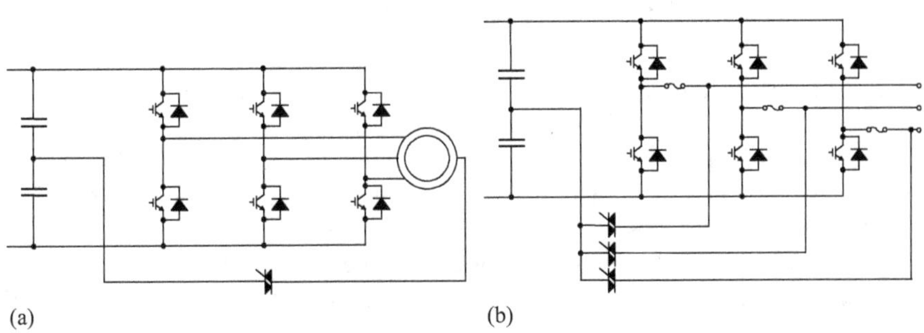

Figure 2.24 (a) Fault tolerant solution for open circuits; (b) fault tolerant solution for both open and short-circuits

This solution is able to accommodate open circuit faults. In case one of the switches of the converter fails in open circuit, the pair of thyristors connected to the neutral of the generator are activated connecting this point to the midpoint of the DC bus. In order to maintain the rotating MMF after the fault, thus guaranteeing a disturbance-free operation, the currents of the healthy phases should be increased by a factor of $\sqrt{3}$ and shifted away 30 degrees from their balanced position. Therefore a non-zero common mode current will flow through the generator neutral point. In this solution the torque and the speed of the electric machine can be maintained after the fault but the drive has to be over-rated to accommodate higher currents. Otherwise a reduction in the torque by a factor of $\sqrt{3}$ is obtained. More flexible solutions based on the same principle can be found in [62, 63].

Figure 2.24(b) shows a solution able to deal with single short-circuit or open circuit faults [64]. In case of a short-circuit, the fuse of the faulty leg is blown and the faulty leg is isolated. Then, the generator terminal that was initially connected to the faulty leg is now connected to the midpoint of the DC bus activating the associated pair of thyristors. Once this is done a set of voltages with a phase difference of 60 degrees is generated. For instance if phase *a* fails, and hence the terminal *a* of the generator is connected continuously to the neutral point, the following set of voltages taking as a reference the DC-bus midpoint has to be generated:

$$\left.\begin{aligned} v_{a0} &= 0 \\ v_{b0} &= A\sin\left(\omega t - \frac{5\pi}{6}\right) \\ v_{c0} &= A\sin\left(\omega t + \frac{5\pi}{6}\right) \end{aligned}\right\} \tag{2.20}$$

Under this assumption, a balanced set of line-to-line voltages is obtained as follows:

$$\left.\begin{aligned} v_{ab} &= A\sin\left(\omega t + \frac{\pi}{6}\right) \\ v_{bc} &= A\sin\left(\omega t - \frac{\pi}{2}\right) \\ v_{ca} &= A\sin\left(\omega t + \frac{5\pi}{6}\right) \end{aligned}\right\} \tag{2.21}$$

It is important to highlight that a reduction on the voltage by a factor of 2 with regard to the non-fault operation is obtained when this solution is applied. Therefore the torque of the machine can be maintained but the rotational speed is reduced to a half after the fault. In order to overcome this drawback four-leg fault tolerant solutions have been proposed [62].

It is also important to highlight that special attention should be put during the design of these drives in order to avoid undesired turn on of the thyristors due to high *dv*/*dt*.

More details on these and other fault tolerant solutions for variable speed drives can be found in [65].

2.7 Summary

This chapter has provided an outline of the issues confronting the designer of an OEC when selecting or designing the electrical generator and its associated control system. It is clear that, especially for WECs, this is a complex problem due to the high variance in device PTO type, and indeed the functionality of the generator itself within the power chain. In this regard, no single generator technology selection can be highlighted as being optimal. However, some overall guidelines for the generator selection have been provided for the different WEC categories, and generic guidelines related to offshore O&M minimization.

Specification and design of the generator system in TECs have been reviewed and shown to be highly dependent on the device, turbine and current characteristics. Both SCIGs and PMSGs are promising generator technologies for application in TECs, showing individual advantages and disadvantages. It can be assumed that some of the drawbacks of PMSGs, for instance the reliability issues, will be solved in the context of further research and development. Thus it is not unlikely that in the long-term PMSGs will turn out to be the superior technology.

As important as the choice of the generator technology are the concepts to deal with the special requirements of TEC devices like marine environment, submerged operation and extended maintenance intervals. Various approaches to address these requirements have been successfully tested in TEC projects. Ongoing sea trials and also new installations will bring valuable knowledge about the long-term reliability of these concepts in the field.

A consistent requirement in both WEC and TEC system generators is the need for variable speed control and operation. In this context, the power converter technology and controller algorithms related to this need have been introduced. Grid-side and machine-side control loop performance has been outlined with useful design guidelines for these control loops. Finally, the provision of fault tolerant power converter solutions – clearly important in the offshore environment – has been introduced.

2.8 References

[1] R. Datta and V. T. Ranganathan, 'Variable-speed wind power generation using a doubly fed wound rotor induction machine: A comparison with alternative schemes', *Power Engineering Review, IEEE*, vol. 22, pp. 52–62, 2002.

[2] P. A. P. Justino and A. F. d. O. Falcao, 'Rotational speed control of an OWC wave power plant', *Journal of Offshore Mechanics and Arctic Engineering*, vol. 121, pp. 65–70, 1999.

[3] J. K. H. Shek, D. E. Macpherson, M. A. Mueller, J. Xiang, 'Reaction force control of a linear electrical generator for direct drive wave energy conversion', *Renewable Power Generation, IET*, vol. 1, pp. 17–24, 2007.

[4] R. J. Hamilton, 'DC motor brush life', *IEEE Transactions on Industry Applications*, vol. 36, pp. 1682–1687, 2000.

[5] M. V. R. S. Jensen, 'Long-term high resolution wear studies of high current density electrical brushes', in *Proceedings of the Fifty-First IEEE Holm Conference on Electrical Contacts, 2005*, IEEE, Chicago, USA. 2005, pp. 304–311.

[6] Techwise A/S 2002, *Elsam. Offshore wind farm. Horns review annual status report for the environmental monitoring programme*, 1 January 2001–31 December 2001. Available at: http://www.offshore-wind.de/page/fileadmin/offshore/documents/Umweltmonitoring/HornsRev_2001_Annual_Status_Report_for_the_Environmental_Monitoring_Programme.pdf. Accessed September 2010.

[7] E. Hau, *Wind Turbines: Fundamentals, Technologies, Application, Economics,* 2nd edn. Springer, 2005.

[8] Sustainable Energy Ireland, *Offshore Wind Energy and Industrial Development in the Republic of Ireland*, Sustainable Energy Ireland, 2004.

[9] J. Puranen, *Induction Motor Versus Permanent Magnet Synchronous Motor In Motion Control Applications: A Comparative Study*. D.Sc. dissertation, University of Lappeenranta, Finland, 2006.

[10] L. Zapf, 'High end corrosion protection of rare-earth permanent magnets', Vacuumschmelze GmbH KG 2009. Available at: http://www.electrochem.org/meetings/scheduler/abstracts/216/1830.pdf. Accessed September 2010.

[11] L. H. Hansen, P. H. Madsen, F. Blaabjerg, H. C. Christensen, U. Lindhard and K. Eskildsen, 'Generators and power electronics technology for wind turbines', in *The 27th Annual Conference of the IEEE Industrial Electronics Society, 2001, IECON '01*. 2001, vol. 3, pp. 2000–2005.

[12] A. F. d. O. Falcão, 'Control of an oscillating-water-column wave power plant for maximum energy production', *Applied Ocean Research*, vol. 24, pp. 73–82, 2002.

[13] D. V. Evans, 'The oscillating water column wave-energy device', *IMA Journal of Applied Mathematics*, vol. 22, pp. 423–433, 1978.

[14] T. Setoguchi and M. Takao, 'Current status of self-rectifying air turbines for wave energy conversion', *Energy Conversion and Management*, vol. 47, pp. 2382–2396, 2006.

[15] A. R. Plummer and M. Schlotter, 'Investigating the performance of a hydraulic power take-off', in *European Wave and Tidal Energy Conference*, Uppsala, Sweden, 2009.

[16] P. Ricci, J. Lopez, M. Santos, J. L. Villate, P. Ruiz-Minguela, F. Salcedo and A. F. d. O. Falcão, 'Control strategies for a simple point-absorber connected to a hydraulic power take-off', in *European Wave and Tidal Energy Conference,* Uppsala, Sweden, 2009.

[17] M. Jasinski, W. Knapp, M. Faust and E. Fris-Madsen, 'The power takeoff system of the multi-MW wave dragon wave energy converter', in *European Wave and Tidal Energy Conference*, Porto, Portugal, 2007.

[18] L. Margheritini, D. Vicinanza and P. Frigaard, 'SSG wave energy converter: Design, reliability and hydraulic performance of an innovative overtopping device', *Renewable Energy*, vol. 34, pp. 1371–1380, 2009.

[19] H. Lendenmann, K.-C. Strømsem, M. D. Pre, W. Arshad, A. Leirbukt, G. Tjensvoll and T. Gulli, 'Direct generation wave energy converters for

optimized electrical power production', in *European Wave and Tidal Energy Conference,* Porto, Portugal, 2007.

[20] M. A. Mueller, 'Electrical generators for direct drive wave energy converters', *IEE Proceedings – Generation, Transmission and Distribution*, vol. 149, pp. 446–456, 2002.

[21] T. K. A. Brekken, H. Hapke and J. Prudell, 'Drives Comparison for Reciprocating and Renewable Energy Applications', in *Twenty-Fourth Annual IEEE Applied Power Electronics Conference and Exposition, 2009. APEC 2009,* Washington DC, USA, 2009, pp. 732–738.

[22] D. O'Sullivan and A. W. Lewis, 'Generator Requirements and Functionality for Ocean Energy Converters', in *International Conference on Electrical Machines Rome*, Italy, 2010.

[23] I. Bryden and G. T. Melville, 'Choosing and evaluating sites for tidal current development', *Proceedings of the Institution of Mechanical Engineers, Part A: Journal of Power and Energy*, vol. 218, no. 8, pp. 567–577, December 2004.

[24] J. Culina and R. Karsten, 'Comparison of different resolution models and observed current profiles in the Bay of Fundy, Canada using turbine-relevant metrics', *Proceedings of the 9th European Wave and Tidal Energy Conference*, CD-ROM, Paper no. 94, Southampton, UK, September 2011.

[25] http://www.emec.org.uk/facilities/tidal-test-site/. Accessed October 2012.

[26] E. Osalusi, J. Side and R. Harris, 'Structure of turbulent flow in EMEC's tidal energy test site', *International Communications in Heat and Mass Transfer*, vol. 36, no. 5, pp. 422–431, May 2009.

[27] I. A. Milne, R. N. Sharma, R. G. J. Flay and S. Bickerton, 'Characteristics of the onset flow turbulences at a tidal-stream power site', *Proceedings of the 9th European Wave and Tidal Energy Conference*, CD-ROM, Paper no. 97, Southampton, UK, September 2011.

[28] J. Thomson, B. Polagye, V. Durgesh and M. C. Richmond, 'Measurements of turbulence at two tidal energy sites in Puget Sound, WA', *IEEE Journal of Oceanic Engineering*, vol. 37, no. 3, pp. 363–374, July 2012.

[29] S.E. Ben Elghali, M. E. H. Benbouzid and J. F. Charpentier, 'Marine tidal current electric power generation technology: State of the art and current status', in *IEEE International Electric Machines & Drives Conference, IEMDC'07*, vol. 2, pp. 1407–1412, Antalya, Turkey, May 2007.

[30] http://www.hales-turbine.co.uk/technology.html. Accessed March 2013.

[31] http://www.tidalsails.com/technology. Accessed March 2013.

[32] http://www.marineturbines.com/Seagen-Technology. Accessed March 2013.

[33] http://www.hammerfeststrom.com/products. Accessed March 2013.

[34] http://www.pontediarchemede.it. Accessed October 2012.

[35] Y. Duan and R. G. Harley, 'Present and future trends in wind turbine generator designs', *IEEE Conference on Power Electronics and Machines in Wind Applications, 2009. PEMWA 2009*. pp. 1–6, Lincoln, Nebraska, USA, June 2009.

[36] A. Vath, 'Antriebskonzepte für Offshore-Windenergieanlagen', *Konstruktion, special Antriebstechnik*, S1/2012, Springer-VDI-Verlag, Germany, 2012.

[37] F. Spinato, P.J. Tavner, G.J.W. van Bussel and E. Koutoulakos, 'Reliability of wind turbine subassemblies', *Renewable Power Generation, IET* , vol. 3, no. 4, pp. 387–401, December 2009.

[38] http://www.openhydro.com/technology.html. Accessed March 2013.

[39] H. Polinder, F. F. A. van der Pijl, G.-J. de Vilder and P. J. Tavner, 'Comparison of direct-drive and geared generator concepts for wind turbines', *IEEE Transactions on Energy Conversion*, vol. 21, no. 3, pp. 725–733, September 2006.

[40] H. Polinder, 'Overview of and trends in wind turbine generator systems', *Power and Energy Society General Meeting, IEEE*, Detroit, MI, USA, July 2011.

[41] U.S. Department of Energy, *Advanced wind turbine drive train concept: Workshop report*, June 2010. Available at: http://www.osti.gov/bridge/. Accessed March 2013.

[42] D. Bang, H. Polinder, G. Shrestha and J.A. Ferreira, 'Promising direct-drive generator system for large wind turbines', *Wind Power to the Grid – EPE Wind Energy Chapter 1st Seminar, EPE-WECS 2008*, Delft, Netherlands, March 2008.

[43] O. Keysan, A. S. McDonald and M. Mueller, 'A direct drive permanent magnet generator design for a tidal current turbine (SeaGen)', *IEEE International Electric Machines & Drives Conference (IEMDC)*, Niagara Falls, Ontario, Canada, May 2011.

[44] T. Hazel, H. Baerd, J. J. Bremnes and J. Legeay, 'Subsea high-voltage power distribution', *Petroleum and Chemical Industry Conference (PCIC)*, Toronto, Canada, September 2011.

[45] P.L. Fraenkel, 'Marine current turbines: Pioneering the development of marine kinetic energy converters'. *Proceedings of the Institution of Mechanical Engineers, Part M: A Journal of Power and Energy*, vol. 221, pp. 159–169, 2007.

[46] R. Arlitt and K. Argyriadis, 'Development and certification of the Voith Hydro HyTide 110 tidal turbine', *3rd International Conference on Ocean Energy*, Bilbao, Spain, October 2010.

[47] B. V. Davis, 'Low head tidal power: A major source of energy from the world's oceans', *Energy Conversion Engineering Conference, IECEC-97*, Honolulu, HI, USA, July–August 1997.

[48] M. P. Kazmierkowski, R. Krishnan and F. Blaabjerg, *Control in Power Electronics – Selected Problems,* Academic Press, 2002.

[49] D. Swierczynski, *Direct Torque Control with Space Vector Modulation (DTC-SVM) of Inverter-Fed Permanent Magnet Synchronous Motor Drive.* Ph.D. thesis, Faculty of Electrical Engineering, Warsaw University of Technology, 2005.

[50] M. P. Kazmierkowski and H. Tunia, *Automatic Control of Converter-Fed drives*, Elsevier, 1994.

[51] F. Blaschke, 'A new method for the structural decoupling of AC induction machines', in *Proc. Conf. Rec. IFAC*, International Federation of Automatic Control, Duesseldorf, Germany, October 1971.

[52] L. Takahashi and T. Noguchi, 'A new quick response and high efficiency strategy of induction motor', *IEEE Transactions on Industry Applications*, vol. 22, no. 5, pp. 495–502, September–October 1985.

[53] M. Malinowski and S. Bernet, 'Simple control scheme of three-level PWM converter connecting wind turbine with grid', in *International Conference on Renewable Energies and Power Quality ICREP'04*, Barcelona, Spain, 2004.

[54] J. Zaragoza, J. Pou, A. Arias, C. Spiteri, E. Robles and S. Ceballos, 'Study and experimental verification of control tuning strategies in a variable speed wind energy conversion system', *Renewable Energy*, vol. 36, no. 5, pp. 1421–1430, May 2011.

[55] M. Malinowski, *Sensorless Control Strategies for Three-Phase PWM Rectifiers*, Ph.D. thesis, Faculty of Electrical Engineering, Institute of Control and Industrial Electronics, Warsaw University of Technology, 2001.

[56] M. P. Kazmierkowski and L. Malesani, 'Current control techniques for three-phase voltage source PWM converters: A survey', *IEEE Transactions on Industrial Electronics*, vol. 45, no. 5, pp. 691–703, October 1998.

[57] T. Ohnishi, 'Three-phase PWM converter/inverter by means of instantaneous active and reactive power control', in *Proceedings of the Annual Conference on IEEE Industrial Electronics (IECON'91)*, Kobe, Japan, 1991.

[58] V. Manninen, 'Application of direct power control modulation technology to a line converter', in *Proceedings of the European Conference on Power Electronics and Applications (EPE'05)*, Sevilla, Spain, 2005.

[59] M. Jasinski, *Direct Power and Torque Control of AC/DC/AC Converter-Fed Induction Motor Drives*, Ph.D. thesis, Faculty of Electrical Engineering, Institute of Control and Industrial Electronics, Warsaw University of Technology, 2005.

[60] M. Malinowski, M. Jasinski and M. P. Kazmierkowski, 'Simple direct power control of three-phase PWM rectifier using space-vector modulation (DPC-SVM)', *IEEE Transactions on Industrial Electronics*, vol. 51, no. 2, pp. 447–454, April 2004.

[61] T. H. Liu, J. R. Fu and T. A. Lipo, 'A strategy for improving reliability of field-oriented controlled induction motor drives', *IEEE Transactions on Industrial Applications*, vol. 29, no. 5, pp. 910–918, September–October 1993.

[62] S. Bolognani, M. Zordan and M. Zigliotto, 'Experimental fault-tolerant control of a PMSM drive', *IEEE Transactions on Industrial Electronics*, vol. 47, no. 5, pp. 1134–1141. October 2000.

[63] J. R. Fu and T. A. Lipo, 'Disturbance-free operation of a multiphase current-regulated motor drive with an opened phase', *IEEE Transactions on Industry Applications*, vol. 30, no. 5, pp. 1267–1274. September–October 1994.

[64] J. R. Fu and T. A. Lipo, 'A strategy to isolate the switching device fault of a current regulated motor drive', in Proceedings of the *IEEE-IAS Annual Meeting*, vol. 2, pp. 1015–1020, 1993.

[65] B. A. Welchko, T. A. Lipo, T. M. Jahns and S. E. Schulz, 'Fault tolerant three-phase AC motor drive topologies: A comparison of features, cost, and limitations', *IEEE Transactions on Power Electronics*, vol. 19, no. 4, pp. 1108–1116, July 2004.

Chapter 3

Cabling umbilical and array layout

P. Ricci and J. L. Mendia

3.1 Definition of grid connection layout

3.1.1 Introduction

Since only a few marine energy devices have been operating while connected to the grid and always for limited time and at a small scale, the problem of properly designed electrical configuration equipment and infrastructure has not yet drawn much attention among the marine energy community.

Ongoing projects mostly concern single ocean converters to be deployed at short distance from shore and are principally aimed at demonstrating the technology rather than maximizing the power transmission. Deployment sites have been often chosen mainly for practical and economical reasons, depending on the location of suitable grid connection points at the coastline, and the requirement for additional electrical infrastructure has been minimal to avoid additional costs. The limited distance to shore allows a reasonably efficient power transmission at low or medium voltage and for this reason some devices have been actually working without carrying any kind of power converter.

However, it is widely acknowledged by stakeholders that the key to successful commercialization of marine renewables in the future is the deployment in large-scale farms possibly with an installed power of the order of hundreds of megawatts. As shown in [1, 2] and in Chapter 10, this is due to the very high costs associated with the installation and operation at sea so that only projects of large magnitude can be economically feasible.

Considering that most of the wave and tidal technologies developed up to date are characterized by relatively small rated power (up to 1 or 2 MW), farms of several tenths or even hundreds of individual devices are required to reach the total installed power required to break even.

Although recognized by the industry, the need for arrays of many devices has not yet been tackled in depth by the research community because much of the efforts in the previous decades was focussed on the proof of concept and on the design of the technology, often conceived as an isolated element with little regard on the effective components and infrastructure required for it to generate electricity and transmit it to the grid. The design of marine energy arrays and the related infrastructure, however, constitutes a fundamental step in the development of marine renewables particularly in light of the economic constraints set by the industry.

The definition of the most adequate array configuration is a complex task which has connections to different technical aspects and considerations, which will be briefly presented in sections 3.1.3–3.1.7. Many of the questions and challenges related to it are still unanswered and will probably be among the primary subjects of research and development in the next years.

Even if the disposition and configuration of the array and its elements are completely defined, the engineering of the components that form what can be called a grid connection infrastructure represents a very challenging job due to the harshness of the environment and the specificity of the technologies. It is likely that future large-scale farms will consider the example of offshore wind installation as primary reference for the design of electrical infrastructure but several solutions will need consistent adaptation to the case of marine energy technologies, as presented in section 3.2.

This is particularly the case of dynamic cables and subsea connectors for which the design requirements and specifications posed by marine energy applications are distinctly different from the ones typically applied in the offshore industry. Bespoke solutions are needed in this case as the design of the connection is largely dependent on the device dynamics and the deployment location. Section 3.3 will review applicable standards and procedures for cable components and highlight the most important issues associated with marine energy projects.

3.1.2 General requirements and constraining factors

The large majority of marine energy converters (MECs) being developed aim at producing electricity directly at sea. Alternative ways of generating and transmitting energy do exist (hydraulic transmission, desalination, hydrogen) and have been applied in a few cases with reasonable success [3], but most of the large-scale projects for future arrays are considering the generation of electricity to be carried out directly at sea, usually on board of the device.

One could think of large batteries for storage so as to reduce or even eliminate the need for grid connection, but this system would probably be inefficient and would require nevertheless periodical access to the devices so that, in most projects, a proper electrical interface has to be designed to convey the power generated by the MECs and feed it into the grid.

From a very general perspective, the configuration of this interface should satisfy the following requirements (see also [4]):

- Connect the offshore marine energy generators to the onshore electrical grid for transmission of the generated electrical power into the general grid system (export power).
- Where required, connect the offshore marine energy device to the onshore electrical grid for the supply of the marine energy device auxiliary electrical loads (import power).

- Where required, connect the offshore marine energy device to the onshore electrical grid for the supply of electrical power to energize and start up the marine energy device (peak import power).
- Ensure that all equipment are being operated within its voltage and current limits throughout the power range of the marine energy farm and over the range of operation of the grid.
- Ensure that all equipment operate stably and predictably during individual generator start-up and during minor grid disturbances.
- Ensure that the protection systems operate correctly, and safely isolate faulty equipment.
- Ensure that any adverse effects of the marine energy generators on the electrical grid are remedied according to the local grid connection requirements.
- Ensure that grid support is provided as defined by the local utility in their grid connection requirements.
- Minimize energy losses in transmission of electrical energy back to shore.
- Minimize the capital cost of the overall (onshore and offshore) connection to the grid.
- Facilitate economic operation and maintenance of the marine energy farm in terms of isolation and access for maintenance and connection/disconnection of devices.
- Maximize reliability of the grid connection.
- Minimize downtime of generators during routine operational procedures (e.g. device maintenance) and following failures of the electrical connection (e.g. due to grid disturbances).

These general requirements are applicable to any kind of electrical connection configuration of a marine energy device or array of devices. These can be very complex systems consisting of several components like the ones seen in the example of Figure 3.1.

For practical purposes, however, a specific connection infrastructure should be first identified by the layout of the electrical lines, the indication of the transmission voltage along each line and the definition of the rated generated power of each connected device.

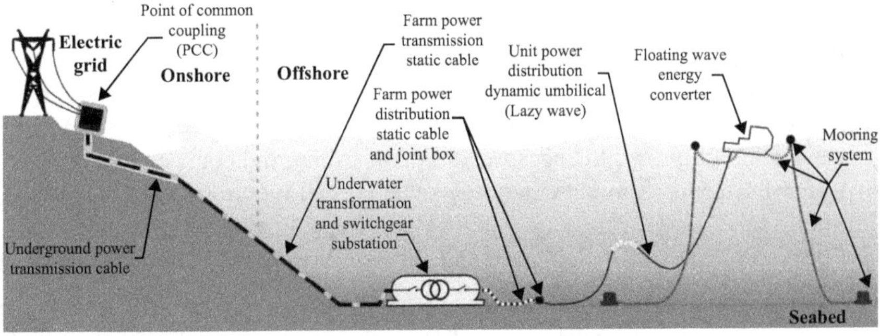

Figure 3.1 Example of electrical connection of wave energy converter

The engineering of the infrastructure is then dependent on the choice of these specific parameters as well as on other environmental constraints (water depth, metocean conditions etc.) in such a way that a set of practical design requirements might be defined depending on the layout established in the first place.

It is clear, therefore, that the choice of the transmission voltage is one of the first steps in the design of the connecting configuration and will serve as one of the basic requirements for the subsequent design of the components outlined in section 3.2.

However, before selecting the grid connection configuration, the main factors related to the proposed farm site as well as to the device characteristics should be carefully analysed, particularly when they might constitute additional constraints or requirements for some specific solutions (e.g. when considering accessibility for maintenance of the electrical equipment).

Site-specific environmental factors are:

- Distance to shore (hugely important on the assessment of layout and cable design).
- Energy resource (the performance of the device should be assessed against the resource to find out the capacity factor and other relevant quantities for the design of the cables and the connecting equipment).
- Bathymetry (knowledge of the bathymetry is crucial for any kind of work related to installation, operation and design of marine structures as well as laying of subsea cables).
- Grid capacity (the strength of the grid connection point would be very important for large-scale installations and might affect the design of components or the definition of specific control strategies).
- Seabed soil and sedimentation (cable routing and installation requires also knowledge of the soil of the seabed; sedimentation is crucial for both cabling and civil works).
- Environmental conditions (extreme parameters that drive the design of offshore structures, weather windows for operation and maintenance).

Device-specific features that might affect the design of the electrical connection are:

- Rated power (it determines the main requirement on the cables and connectors).
- Power converter and output voltage (this would largely influence the choice of the transmission equipment and at which point a substation or a connection hub is needed).
- Device mobility (it influences the design of cables and connectors).
- Control systems (it might determine some special requirements for the power line).
- Location of the connection point and accessibility.

Provided that all the devices are equipped with a power converter and a transformer, the selection of the type of technology has a major influence on the design of some items (such as the cable) but is less decisive on the definition of the global layout and, at some level, the design of the grid connection infrastructure might be neutral with respect to the type of device that is connected as indeed is the case of the existing marine energy testing sites [5–9].

However, the consideration of specific typologies of generation (such as, e.g., linear generators) might impose some choices in the transmission options presented in section 3.1.4, particularly when considering small-scale arrays or isolated devices.

3.1.3 Definition of array layout

When presented with the challenge of designing marine energy farms, there is little knowledge available to utilities and project developers: Practical experience in the development of marine energy arrays is almost inexistent and there are still huge uncertainties on the best options for the device configuration.

Clearly, the first step of a project would consist in the design of a basic layout, specifying the number and the disposition of the devices, the area they occupy (including moorings or foundations), the number and nature of their connections and the specification of their relative positions with respect to a common connection point or to the coastline.

As it will be discussed later, it is likely that future arrays will require several MECs connected to a common point where the energy produced is collected. From that point the energy is transmitted to shore by means of one or more subsea cables. The selection of the transmission voltage of the cable is largely dependent on the power installed and on the distance to shore, as made clear in section 3.1.6.

The definition of the inter-device electrical layout, however, might be different between two farms even if the transmission line presents the same characteristics and several options could be envisaged, as shown in Figure 3.2.

The choice between those options depends on a number of factors:

- Electric efficiency.
- Installation and operational requirements.
- Resource use (in terms of hydrodynamic performance).
- Power quality issues.
- Environmental impact.
- Moorings/foundations requirements.
- Availability and continuity of supply.

Those factors are related to a series of parameters which are not solely specific to the electrical connection design but are associated with other important aspects such as the hydrodynamics of the energy absorption or the requirements for accessibility of the devices.

Hence the design of the layout of marine energy has to be based on a balance between several criteria, as represented in Figure 3.3.

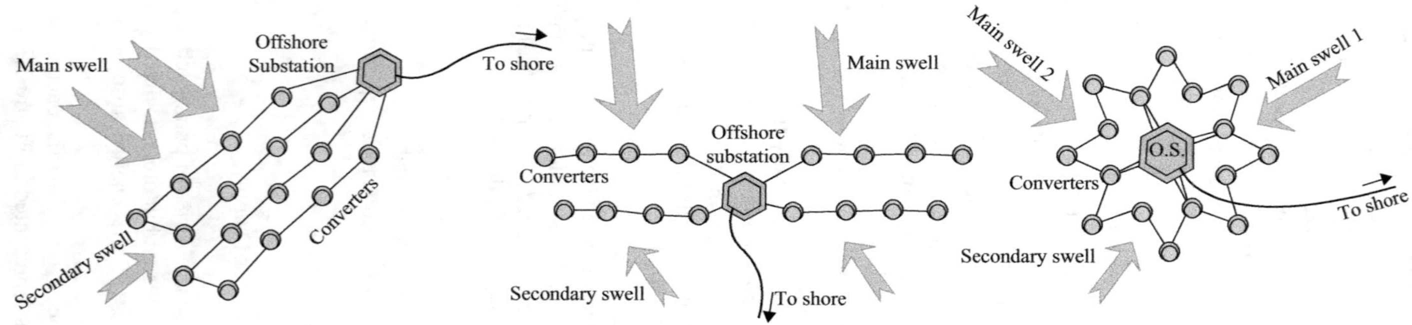

Figure 3.2 Examples of possible layouts

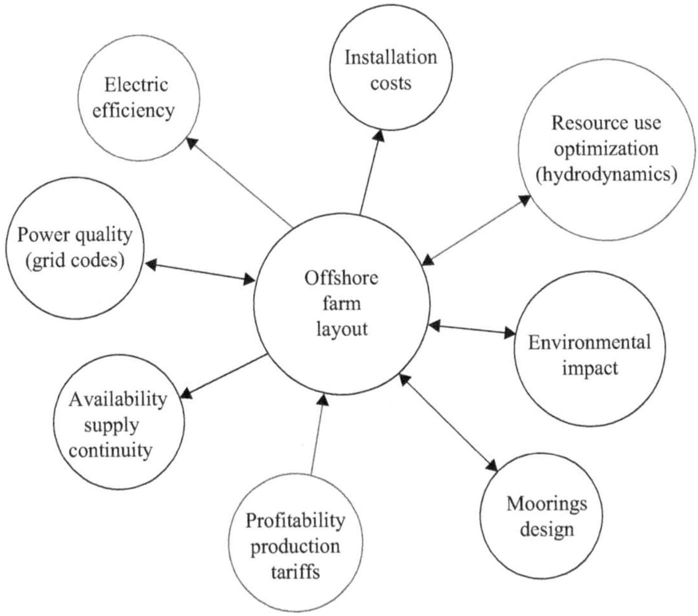

Figure 3.3 Offshore farm layout design main factors

3.1.3.1 Design constraints from hydrodynamic and moorings requirements

The assessment of the influence of the array geometric parameters on the power performance of marine energy farms has been the subject of numerous studies mostly based on numerical grounds with reference to small-scale experiments in a few cases. However, most of those were applied to specific device types or to relatively simplified cases with a limited number of combinations so that the subject is still debated and only very general conclusions can be drawn.

For instance, it appears clear that the disposition of the devices and their spacing have always some effect on the power absorption because the actual working principle of marine renewables relies on the hydrodynamic interaction of the converter with either the waves or the current.

For tidal devices, this influence is usually associated with the development of a wake downstream of the turbine and operates in the sense of a reduction [10] of the power extracted. Array configurations made of staggered rows placed in front of the main current direction are likely to be a solution though the inter-row distance should be sufficiently large as to avoid undesired wake interaction.

The effect of the hydrodynamic interaction in wave energy farms is even more complicated to ascertain: It has been shown by early analytical studies [11, 12] that constructive interference is possible and thus a larger power per device can be theoretically absorbed. However, this effect is typically limited to specific wave periods and recent studies carried out in irregular waves [13, 14] indicate that the actual gain in energy absorbed due to optimal configuration is of the order of a few percent whereas a decrease can be quantitatively relevant, especially in closely packed arrays.

Overall, it is nevertheless confirmed that, as far as the inter-device spacing becomes larger, the effect on the power absorption becomes negligible so that a very basic recommendation would be to maintain this parameter sufficiently large as to avoid largely negative interference.

Furthermore, apart from the considerations related to the hydrodynamic effects explained above, there are a number of technical requirements that might impose several constraints on the array configuration and on the inter-device distance.

First, all MECs will require station-keeping systems or foundations in order to maintain their position. Particularly, floating devices would probably adopt moorings with several anchors placed at a determined distance from the device and imposing a certain 'footprint'. If those devices were to be deployed and moored independently, the size of the 'footprint' might impose an important constraint on the inter-device distance. A possible alternative would be constituted by globally shared mooring arrangements for arrays in such a way that several converters might be moored and interconnected together. This would save infrastructures (in terms of anchors and chains) and might even prove to be beneficial for wave energy converters [15].

Another very important constraint will arise from the need for vessels and their equipment to operate at ease on the single devices for maintenance and repairs. This requirement will be often device- and site-specific, making it difficult to define guidelines to deal with it but it might as well be the most decisive one if the MECs are to be taken periodically onshore for maintenance.

3.1.3.2 Array inter-cabling configurations

Connecting individually to shore every device of the marine energy farm would enable a very flexible and reliable operation of the generator units. Nevertheless, in most cases this solution might lead to excessive cabling (and laying operations) costs even for small farms close to shore.

In addition, the number of devices connected in one circuit is limited as electrical barriers exist as a result of both the capacity of the collection cables and the voltage drop along their length. The maximum number of devices per circuit is therefore a function of the generator's rated capacity and the adequate spacing between the different units of the farm.

Therefore generating units are grouped into medium-voltage electrical collection subsystems within the marine farm. Those arrangements, so-called clusters, are then integrated together via offshore substations from where the transmission to shore is initiated. According to the specifications on voltage and power levels, some configurations might be more suitable than others.

Whether the distribution technology is AC or DC doesn't affect the cluster configuration since the interconnecting cable routes are similar, except in DC-series clusters as the devices are series-connected in order to raise the DC output voltage up at the node.

There are several types of clusters [16]:

- *String clustering without redundancy*: The devices are connected in parallel along a single collection cable. The main advantage is that if one arm of the system breaks the overall system continues working. This option is one of the most used in wind farms.
- *Star clustering*: The devices are connected independently to a cluster nodal platform. The advantage is that several parts are independent to each other so that in case of failure the overall infrastructure will continue working.
- *String without redundancy*: The system only has one transmission cable. If one of the MECs has any kind of failure this would affect the operation of the overall farm.
- *String with redundancy*: The devices are connected in parallel along a closed-loop collection cable and a switch controlling the power flow in the cluster.
- *DC-series clustering*: The devices are series-connected in various branches. This configuration is only used in DC clusters. Depending on power installed and distance to shore, DC transmission is becoming a more widely installed option in offshore wind energy.

It should be noticed that, as shown in Figures 3.4–3.8, the adoption of one specific clustering configuration does not depend, from a conceptual perspective, on the geometry of the array. Nevertheless, the design of the cable and the connecting equipment are certainly determined by the inter-device distance and, for

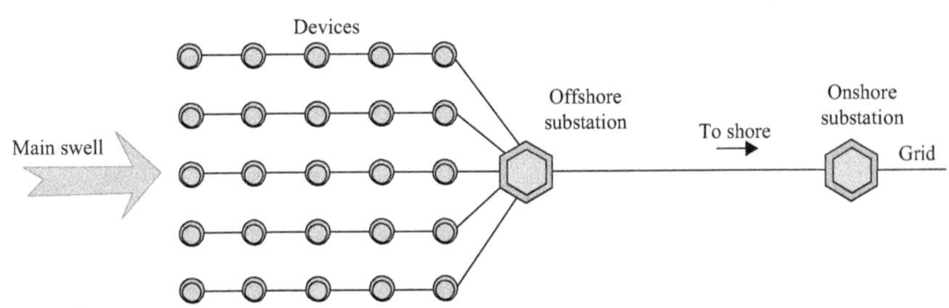

Figure 3.4 String series cluster in medium and large farms (AC and DC)

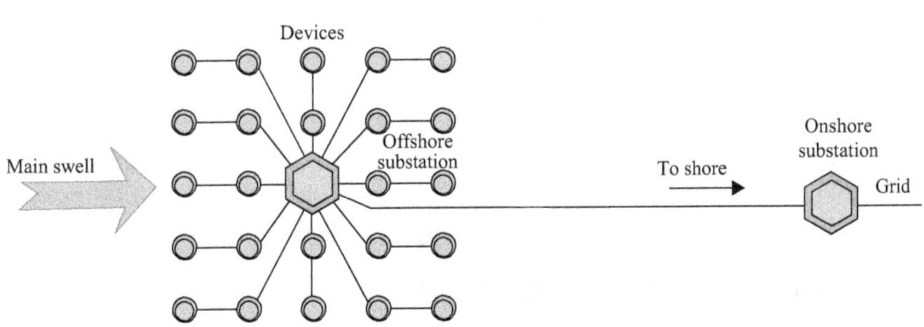

Figure 3.5 Star (radial) cluster

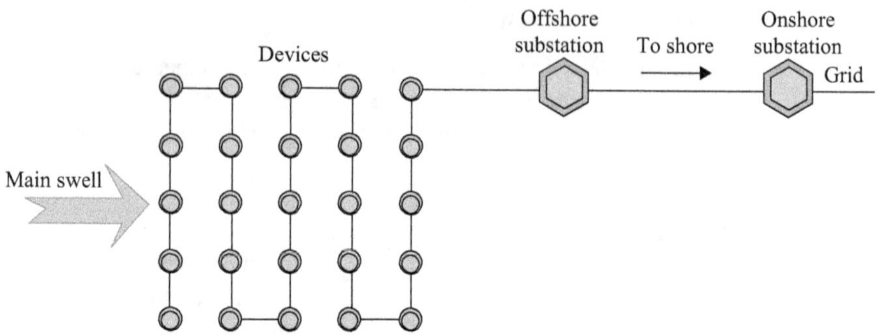

Figure 3.6 Full string cluster

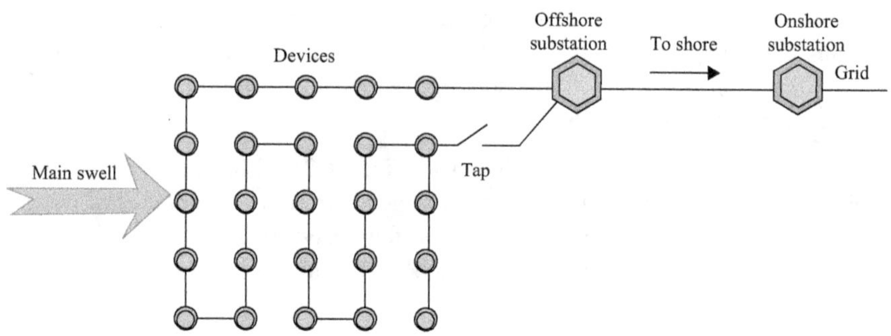

Figure 3.7 Redundant string cluster (AC and DC)

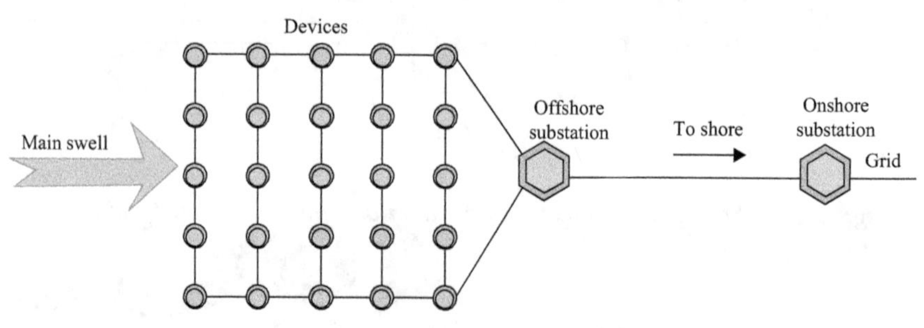

Figure 3.8 Series DC cluster

floating technologies, even by the dynamics of the MECs. In this case, the choice of the array spacing could be a trade-off between the need to avoid extreme mechanical loads on the one hand and excessive power losses on the other.

3.1.3.3 Optimization and future developments

Research on the design of marine energy array layout is still in its earliest stage and most of the studies carried out so far have focussed only on one of the issues presented before, not considering the influence on the performance of the change of the other factors due to a modification of the configuration.

However, it is clear from the reasons exposed previously that several aspects inform the decision on the best layout, and the optimization of the geometrical parameters of a marine energy farm should include all of them in order to be effectively useful for project developers and future operators.

In the short to medium term it is most likely that arrays will be composed of up to two rows arranged perpendicular to the prevailing wave or tidal direction. For single row arrays lateral device spacing has the potential to be optimized and ongoing studies are aiming at developing global tools capable of assessing quantitatively the balance between the different factors (hydrodynamic interaction, mooring costs, electrical infrastructure, installation and operational tasks etc.).

The best criterion for the comparison of different options will probably be driven by the economic performance, either in terms of net present value or cost of energy. As it will be shown in Chapter 10, techno-economic models [1, 2] might constitute the adequate tool for this purpose.

3.1.4 Power transmission options

From a theoretical point of view any marine energy device might be directly connected to the grid without any additional element, assuming that a proper power converter is installed on board. However, for efficiency and economical reasons, it is likely that power produced by arrays of converters will be collected and transformed before transmitting it to shore. This can be achieved in several ways and through several possible configurations. Some of these are introduced below.

At first distinction should be made based on the type of power transmission between offshore and onshore locations.

Electrical energy can be transported by alternating current (AC) or direct current (DC): For offshore plants the choice of whether to use a DC or AC transmission line is mainly determined by the distance to shore and the installed capacity [4, 17–19]. For projects located far from the grid connection point, or of several hundred megawatts in capacity, AC transmission becomes very expensive or, in some cases, practically impossible due to cable-generated reactive power using up much of the transmission capacity.

In such cases, high-voltage DC (HVDC) transmission is becoming an option. Such a system requires an AC/DC converter station both offshore and onshore; both stations are large installations whose building and operation might impose a number of engineering and economical challenges.

Currently most of the existing offshore transmission systems use HVAC for the transport of electrical power from the farm to the shore. Usually, an offshore substation is used to increase transmission voltage and therefore limit losses. Nevertheless, when the farm is not very distant from shore and the energy generator voltage transformer is high enough, offshore transforming substations may not be necessary.

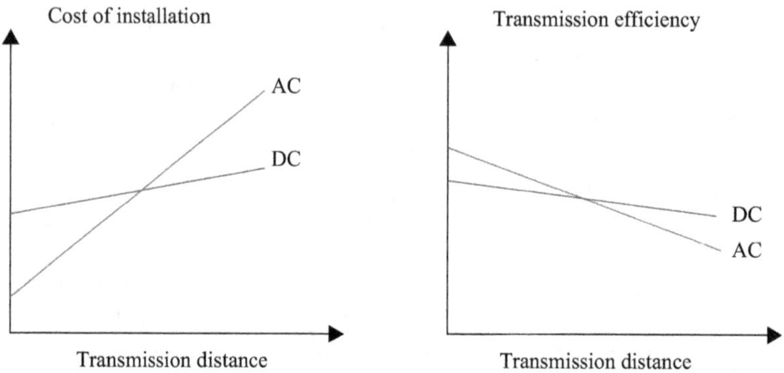

Figure 3.9 Installation cost and transmission efficiency for AC and DC transmission

There are two different preliminary issues to take into account for selection of the transmission options: capital costs associated with the initial investment and the efficiency of the transmission line (transmitted power divided by power produced).

The terminal costs of DC transmission systems are usually much higher than AC transmission costs. As we can see in Figure 3.9, as the distance to the shore increases, the investment for AC line increases more than HVDC in such a way that a break point, where HVDC and HVAC transmission costs are equal, can be identified.

The transmission efficiency is an important point to take into account to evaluate the transmission option for an offshore farm. When the distance to the shore is not large the efficiency of HVAC transmission is higher than HVDC transmission. However, similarly to the costs comparison, we can see that here also there is a break point over which the efficiency becomes higher in HVDC transmission option.

The placement of the power conversion equipment determines the kind of power signal the connection system of the device will have to deliver. It is expectable that future MECs will contain, apart from the electric generator, a power converter and a transformer on board in such a way that the output power will be transmitted through three-phase AC medium voltage lines (see Figure 3.10). However, alternative solutions are possible such as the case of wave energy devices operating with a linear generator whose power output is converted in a substation out of the device [20], as seen in Figure 3.11. In this case, the connecting cable from the device to the substation should be designed for low voltage transmission making unfeasible this option for very long line or high generated power (because of the losses that would be determined by possibly large currents).

Other possibilities include the rectification of the power signal on board in such a way that the device connection is performed through a DC cable, like in Figure 3.12. This option has been proposed for offshore wind because it would

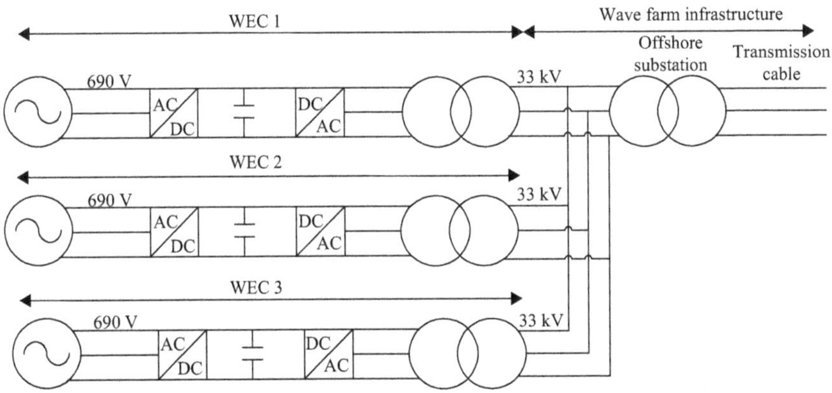

Figure 3.10 AC transmission: transformer located inside the MEC

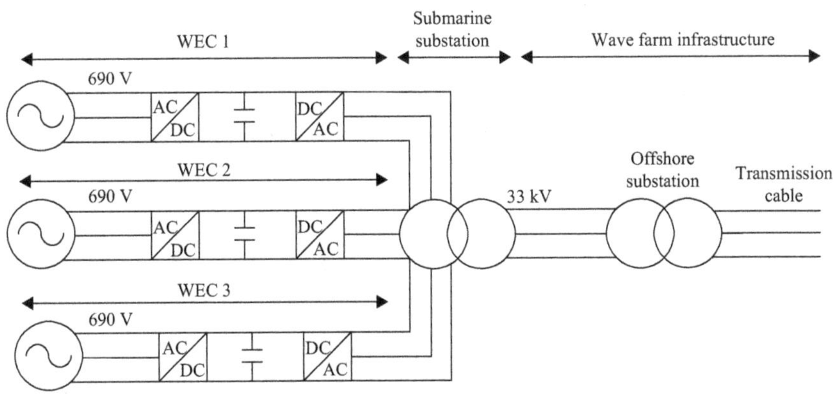

Figure 3.11 AC transmission: submarine substation with local transformer

allow increasing the transmission voltage by connecting in series the devices (avoiding in such a way the need of an offshore transformer) and might be preferable in case of large-scale plants far from shore when the high-voltage direct current transmission could be feasible. However, it would require nonetheless conversion equipment on board and possibly a transformer (in order to elevate the voltage from low to medium level).

3.1.5 Models for power transmission

For the modelling of AC transmission cables the distributed parameter model can be used. This model is commonly applied for the basic assessment of the power flow in simple electrical configurations [21]. Figure 3.13 illustrates schematically this approach: The impedance and admittance of any single line is applied to each infinitesimal section in order to correctly model the longitudinal variability of voltage and current.

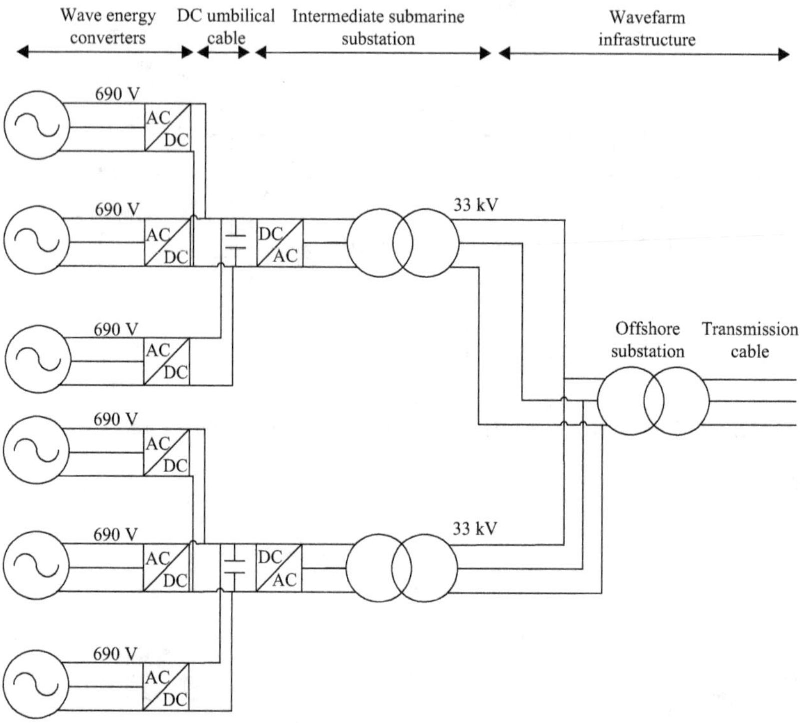

Figure 3.12 AC transmission Offshore substation with local transformer and inverter converting DC output from MECs

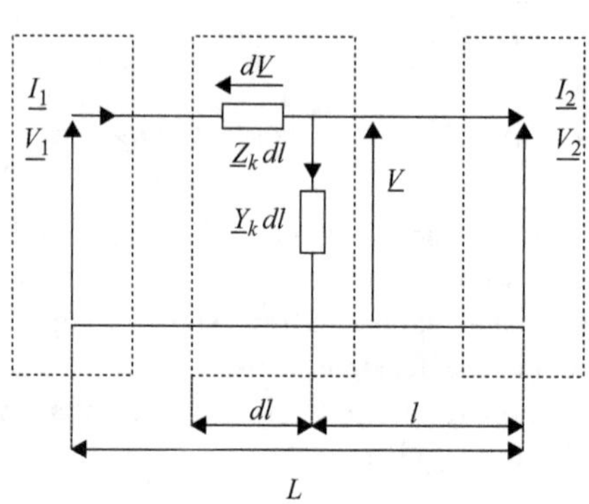

Figure 3.13 Scheme of distributed parameter model of the cable

The main equations of the model are

$$V_2 = V_1 \cosh\left(\sqrt{Z_l Y_l} L\right) - \sqrt{\frac{Z_l}{Y_l}} I_1 \sinh\left(\sqrt{Z_l Y_l} L\right) \tag{3.1}$$

$$I_2 = I_1 \cosh\left(\sqrt{Z_l Y_l} L\right) - \frac{V_1}{\sqrt{\frac{Z_l}{Y_l}}} \sinh\left(\sqrt{Z_l Y_l} L\right) \tag{3.2}$$

where V_1 and I_1 are voltage and current at the offshore terminal, V_2 and I_2 are voltage and current at the end of the line (onshore), Z_l is the impedance of the line per kilometre Y_l the admittance per kilometre and L is the length of the line in kilometres.

For the calculation of the parameters of the three-core cable line impedance and admittance, the first step is the design of the cable: The cross-section of one core is determined by taking into account transmission voltage and power capacity [22, 23].

For the thickness of the insulation and other layers, reference is made to international standards such as the IEC 60502 [24]. The expressions for the kilometric admittance and impedance of each core are described as

$$Z_l = R_{line} + jX_k \tag{3.3a}$$

$$Y_l = G_{line} + jB_k \tag{3.3b}$$

where R_{line} and X_k are respectively resistance and reactance per unit length of the line while G_{line} and B_k represent conductance and susceptance per unit length of the line.

Following [25] we can further write:

$$X_k = \omega L_{line} \tag{3.3c}$$

$$B_k = \omega C_{line} \tag{3.3d}$$

where L_{line} and $C_{line}$are inductance and capacitance of the line ω is the frequency of the grid. The resistance of the cable can be obtained through the relationship $R_{line} = \rho/S$ where S is the area of the cross-section of the conductor and ρ stands for the resistivity of the conductor material (in this case copper).

Cable losses in DC connection configurations are simpler to model than in the AC case. Under fair assumptions, only dissipative resistive losses are of importance in this case; therefore there is no need for complex models.

HVDC power transmission is performed through two typologies of conversion: the line commutated converter (LCC) and the voltage-source converter (VSC). More information on the structure and design issues associated with these configurations can be found in [4] and [19]. As for the results presented here, we considered different converter losses for the two cases (around 1–2% of the transmitted power for LCC and 4–5% for VSC; see [26] and [27]).

3.1.6 Efficiency of the power transmission

The model presented in the previous section can be applied to a real case to estimate the efficiency of the transmission [28].

In Figure 3.14 we can see the evolution of the power transmission efficiency against power and voltage transmission for a distance to shore of 20 km of an HVAC power transmission system.

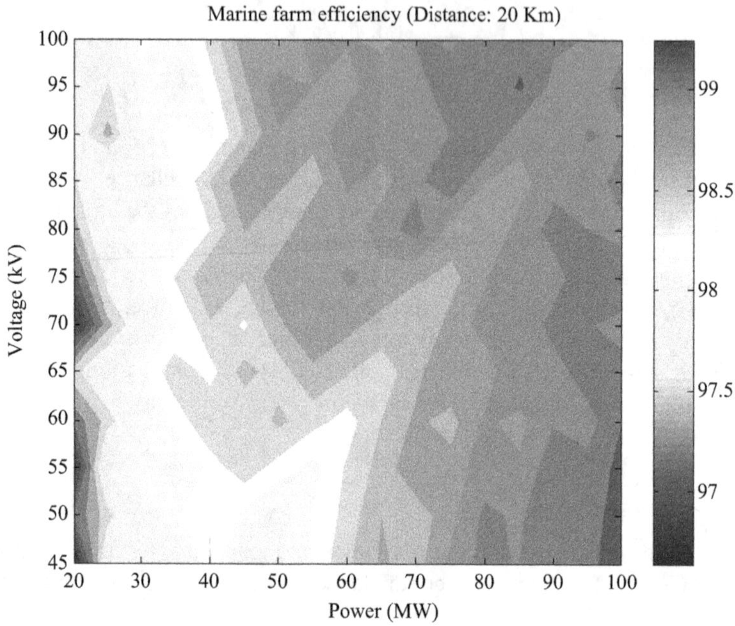

Figure 3.14 Marine farm efficiency for different AC transmission voltage and power installed (20 km)

When the distance to shore is relatively small, as in this case, the elevation of the transmission voltage has little effect upon the efficiency. Particularly in Figure 3.14 it can be seen how this parameter is very close to 1 for all the possible options. Generally better values can be observed for larger powers.

We can appreciate that for a power installed of 90 MW two different possibilities for high efficiency can be recognized: The first is characterized by relatively low transmission voltage. This corresponds to higher currents and possibly higher resistive losses but, on the other hand, also to very small reactive effects. On the contrary, the second area of high efficiency corresponds to very high transmission voltage, where reactive effects are predominant over Joule losses.

In [29] different connection options for offshore wind farms are presented. There are examples that can be compared and related to Figure 3.14. The first one is Kentish Flats (90 MW, 2005) which connects 8.5 km to shore with a 33 kV of transmission voltage without substation and the second one is Barrow in Furness (90 MW, 2006) that has a 33/132 kV substation and which connects at 7 km from shore. Even if these two cases are very similar in terms of distance to shore and transmitted power, it can be seen how the choice of the voltage is basically opposite, according to discussion relating to Figure 3.14.

In Figure 3.15 we can see the power transmission efficiency for a 100 km to shore case of AC transmission marine farm. In this example by contrast to the 20 km distance results, the best choice for efficiency is a relatively lower transmission

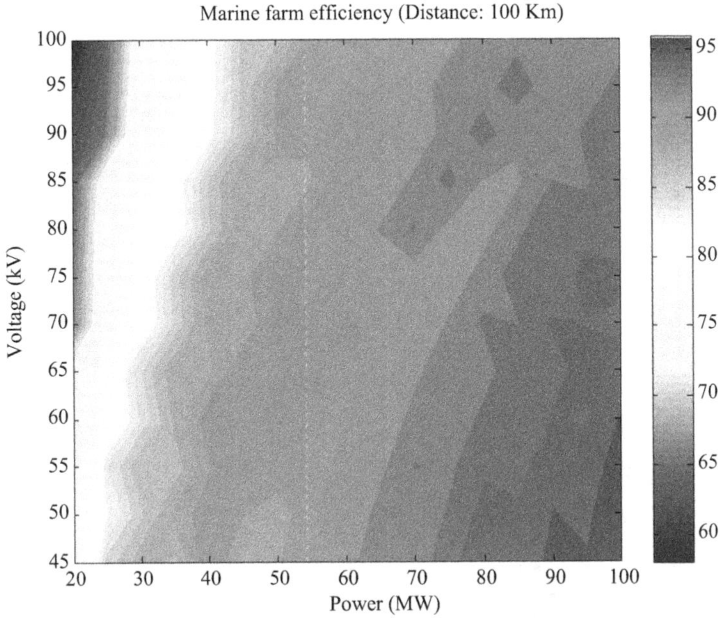

Figure 3.15 Marine farm efficiency for different AC transmission voltage and power installed (100 km)

voltage for lower powers, mainly because of the fact that at such long distances the capacitance effect of the submarine cables is very important.

To observe how AC and DC transmission efficiencies vary along the distance to shore, Figure 3.16 shows results for the case of an installed power of 20 MW and transmission voltage of 45 kV. As we can observe there is a 'break point' where the transmission efficiency is the same for AC and DC. Below this equilibrium distance, AC is generally more efficient whereas DC transmission is preferable above this limit.

Indeed, this 'break point' depends on the power installed and on the transmission voltage. The 'break point' distance is reduced as the transmission voltage is increased, particularly because DC Joule losses are consistently reduced for this case. Furthermore, the influence of the transmission voltage on this value is larger when the power installed is higher because of the increment of losses according with power installed.

The capacitance of HVAC insulated cable plays a major role in limiting the technically and economically feasible length of HVAC cable. Capacitance causes charging current to flow along the length of the cable and affects the reactive power generated by the cables, decreasing its capacity to transmit active power, especially over long distances. Because of this, it is necessary to provide reactive compensation at the cable's ends.

As we can observe in Figure 3.17, the reactive compensation effect increases with increasing distances of the cable to shore as the influence of the capacitance of the cable is larger.

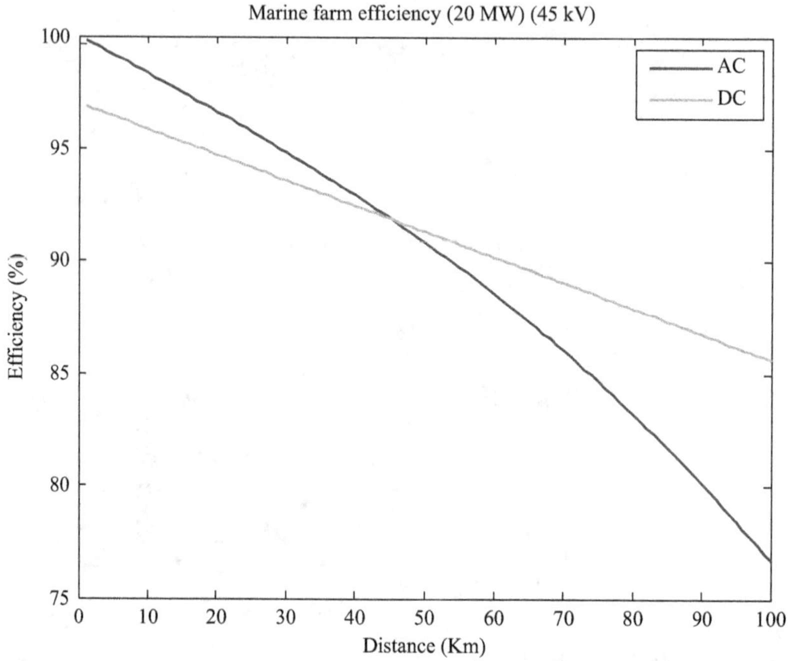

Figure 3.16 AC and DC efficiency comparison for 45 kV transmission voltage and power installed of 20 MW

The reactive compensation is defined with the aim of obtaining the minimum current intensity in the centre of the length at the transmission cable. In this way the current is the same at the beginning and at the end of the line.

3.1.7 Design of cable transmission

Preliminary grid connection configurations at a generic site should aim at managing a balance between the investment costs and transmission efficiency, since the latter would influence the future revenues of the installations. To perform such analysis, at least on an approximate basis, one should define first some basic inputs about the array such as distance to shore and power installed. Other parameters such as water depth and various environmental parameters might be pertinent, however, in a purely electrical analysis; these latter parameters would be much less decisive.

In absence of more specific information on the actual components that will constitute the electrical connection interface of the farm, the first step of the analysis is the determination of the voltage transmission and the required cable by carrying out some calculation of the efficiency with the models shown in the previous sections. This allows the selection of the most appropriate transmission voltage based on the location and the power installed.

Clearly, the adoption of a large voltage has a cost in terms of hardware which has to be taken into account in order to find the best solution so that the correct

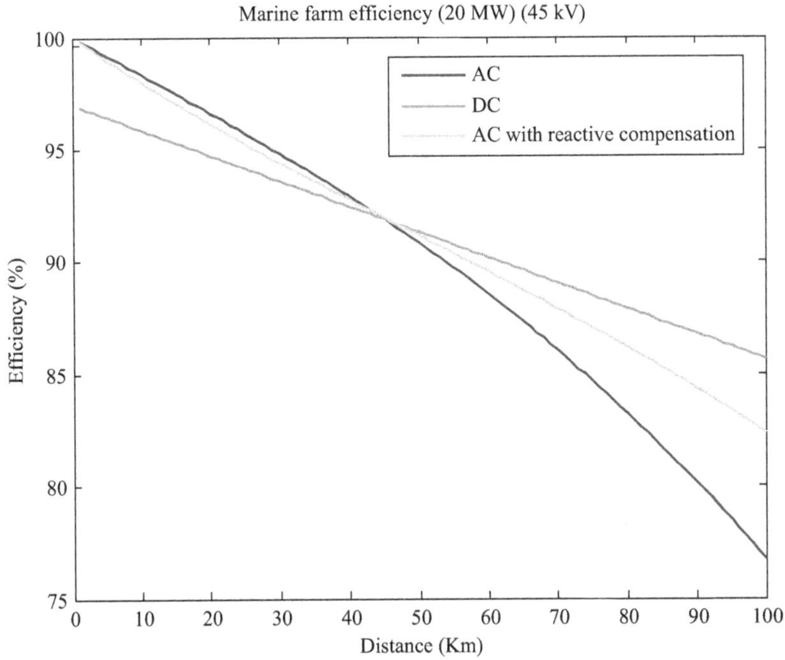

Figure 3.17 Marine farm efficiency against distance for AC transmission voltage of 45 kV, power installed of 20 MW, with and without reactive compensation and DC transmission

procedure for a preliminary design would involve a simple cost model that includes also the costs associated with the power losses encountered when applying not efficient configuration, as seen in [28] and analysed in more detail in Chapter 10.

The definition of the cable transmission is part of the whole design process of the array electrical configuration which, again, is the result of a techno-economic evaluation. In Figure 3.18 we present a brief description of the different steps to be taken into account while defining the layout of a wave farm.

3.2 Engineering of grid connection infrastructures

The problem of the design of grid connection infrastructures has been faced only marginally by device developers due to their relatively early stage of development, and the burden of defining adequate connection facilities for marine energy has been taken by utilities and public bodies involved in the design of testing sites. Even though the challenge is similar to the one experienced by the offshore wind sector, it has to be noticed that the increased depths (and possibly also longer distances to shore) at which wave energy devices are to be installed make some of the solutions currently being adopted by the offshore wind industry unfeasible. This might be true especially for future large-scale ocean energy arrays, which are likely to be placed at considerable distance to shore.

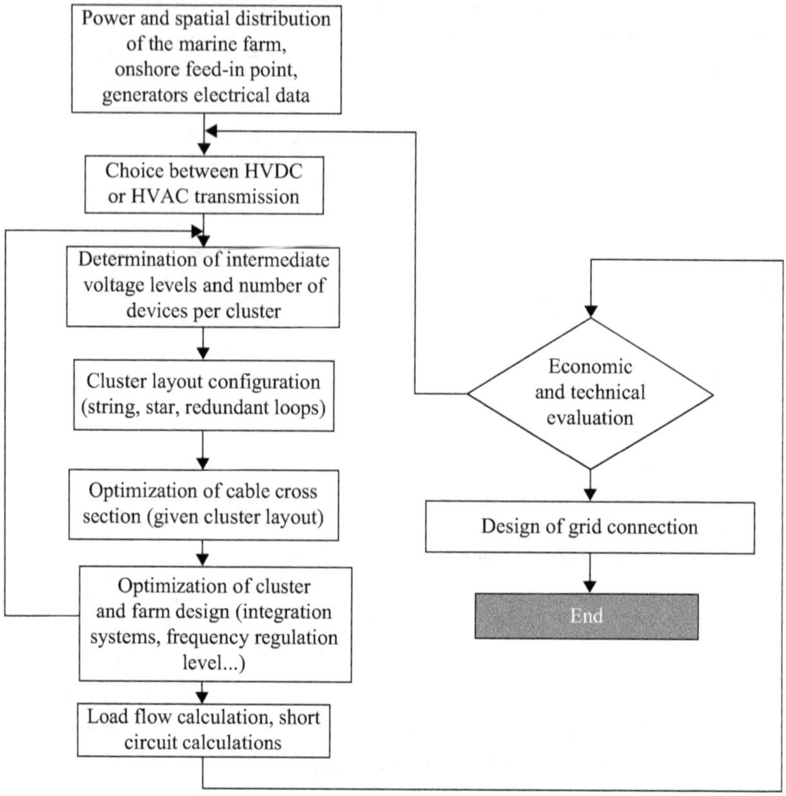

Figure 3.18 Design process of an array layout

However, the ongoing development of full-scale testing sites for wave and tidal technologies in several locations in Europe, especially when allowing the connection of several different devices, requires a detailed design and a thorough analysis of all the existing alternatives. The aim of this will be to identify the most suitable solution to the economical and technical constraints of the project. It is likely that this process will be even more important in the future, when very large scale commercial ocean energy deployments will take place offshore and the engineering of their electrical connection configuration will be crucial in assuring the profitability of the investment.

When the electrical characteristics of the connection are determined, the problem is the engineering of the structure required to support the connection. The grid connection electrical scheme will in general impose specific requirements on the design of the infrastructure. However, some general qualitative requirements, applicable to virtually any ocean energy deployment, can be formulated [30]:

- Most of the infrastructure can be considered permanent (i.e. it requires a relatively high design life) since the installation is supposed to be operational for a long time.

- The motions of the elements of the connecting infrastructures should be limited in order not to affect negatively the cable dynamics [31].
- The whole structure should be easy to maintain, operate and install.
- The components should have minimum volume and weight (low costs and ease of operation) and be commercially available (lower manufacturing costs).
- The environmental impact of the structure should be minimum during its life cycle.

These general requirements would turn into quantitative design constraints once the layout of the structure and the environmental conditions of the deployment site are determined.

Due to the limited experience in the operation of grid-connected marine energy devices and the variety of the existing concepts, there are virtually no commercial solutions available and the development of the ongoing demonstration projects is particularly difficult because of the frequent need of purposely designed systems.

As it will be presented in Chapter 10, however, these systems are capital intensive and require very large scale installations to ensure profit. Thus, it is likely that at a more advanced stage of development the marine energy sector will point to a reduced range of solutions and standardized procedures and components will be adopted.

3.2.1 Definition of connection points

The typical situation we are confronted with when designing a grid connection configuration is represented in Figure 3.19: One or more MECs located offshore have to be connected to the onshore grid through a transmission cable.

From a purely functional point of view, at least two connection points can be identified: one on the side of the converter, i.e. an offshore grid entry point, and the other one on the side of the grid.

3.2.1.1 Connection to the device

The design of the connecting procedure for bottom-fixed technologies does not constitute a novel challenge per se since the offshore wind industry has dealt with

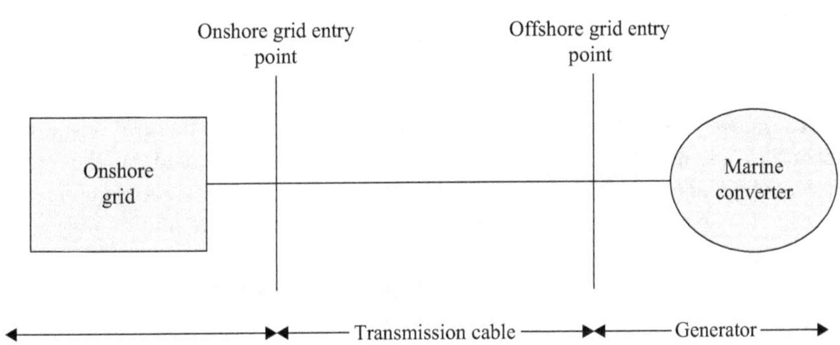

Figure 3.19 Basic representation of a grid connection scheme for a marine energy converter

this problem in the last decade. Offshore wind turbines have applied extensively steel J-tubes for the protection of the cable but due to the high failure rates that have been recorded, other options are being considered including the trenching of the cable from a well below the monopile to the substation or the use of free-hanging lines.

Marine energy characterized by fixed foundation can apply conventional static cable and allow the connection to be performed in a safe and dry environment. However, the effect of waves and current on the laid cable can be severe, particularly if considering the environment where tidal devices are going to be deployed.

Floating devices need to connect to the grid through a dynamic cable. Obviously this cable will have to connect to the power conversion equipment (or the generator in case of no power electronics hosted on board) inside the hull through a properly designed connecting point placed on the external surface of the MEC.

The definition of the connection point depends on a series of requirements:

- The motions in the connecting area should be of limited amplitude. This would lead to lower dynamical loads in the cable.
- Complete immersion of the cable should be guaranteed during operation and no oscillations outside the seawater should occur.
- The connected area should be close to the power conversion equipment (or the electrical generator).
- The connecting area should be relatively far from the mooring connection point or the extremities of the device in order to assure a low probability of interference with other systems and/or devices.
- The cable should be easy to maintain and inspect and allow simple access and disconnection from the MEC.

It is clear that the optimal choice satisfying all the requirements is dependent on the device. However, considering the majority of the floating technologies currently being developed, it can be said that in many cases a good choice would be to connect to the lower central part of the structure as close as possible to the centre of gravity (to avoid interference and large motions).

3.2.1.2 Connection point to the infrastructure

The dynamic umbilical cable coming from the device will connect to a grid infrastructure depending on the layout previously defined.

The most basic option would include direct connection to the grid onshore. In such cases an intermediate connection point should be provided on the seabed between the dynamic cable and an underground 'static' cable that would connect to the grid. This solution would probably require a long umbilical and might impose constraints on the choice of the voltage.

Large-scale arrays of MECs will require the definition of a specifically designed infrastructure to allow the connection of several devices to a common point linked to shore by a unique cable.

The design of these infrastructures is still a subject of research and some of the options that might be considered are summarized in section 3.2.3.

3.2.2 Components of grid connection infrastructures

As introduced in the previous section, the definition of the rated power of the farm and the distance to shore is the first step to determine a suitable configuration for the electrical connection. The assessment of different options could be based at a preliminary level on the evaluation of their efficiency and accordance to the grid requirements. However, electrical connection infrastructures in offshore locations represent often real challenges for offshore and civil engineers and usually require huge investments to assure reliable structures.

Generally the grid connection of a medium-large scale marine energy array would require the following physical elements:

- Device cabling and conversion equipment (power converter, transformer and umbilical cable).
- Cable connectors.
- Offshore substation.
- Subsea transmission cables.
- Onshore substation.

For small-size farms some elements might be different or even avoided.

Conversion equipment and umbilical cables for device inter-connection are currently device-specific and dependent on the generator type considered. It is expected that in the future most of the marine energy technologies will be provided with on-board converter and transformer while the design of the umbilical cable will probably be dictated by the deployment site. Umbilical cables for power transmission have been used in the offshore industry for decades but their application to marine energy technologies might require purposely designed solutions as dynamic loads due to motion of the devices (particularly wave energy floating technologies) are very different from the ones commonly experienced in offshore platforms.

On the other hand, the design of offshore substation, subsea cable and appropriate connectors is likely to be more related to the deployment site than to the device type. This is particularly true observing the fact that most of the testing infrastructures currently being proposed do not make reference to any particular device (apart from the distinction between wave and tidal technologies). Based on this rationale, previous experience of offshore wind industry is very useful to get an understanding of the technical issues related to the design of these elements.

3.2.2.1 Substations for marine renewable arrays

Offshore substations are used to reduce electrical losses by increasing the voltage and then exporting the power to shore. Generally a substation does not need to be installed if:

- The project is small (~100 MW or less).
- It is close to shore (~15 km or less).
- The connection to the grid is at collection voltage (e.g. 33 kV).

Early-stage marine energy projects are likely to satisfy all of these requirements; therefore building of properly designed offshore substation is not yet a primary need for ocean energy deployment. However, most future farms will be large and/or located far from shore, and they will require one or more offshore substations.

A number of offshore substations have been installed and operated for offshore wind energy farms, whose large size justified the high cost linked with their construction. Typically wind substations are fixed platforms based on concrete foundations and would probably not be suitable for deep-water deployments such that possibly required for wave energy devices. Offshore substation will typically comprise the following key components:

- Transformers.
- Electrical switchgear.
- Back-up electrical generator and batteries.

Future large-scale marine energy deployment would probably have to reconsider the design of purposely built substation since fixed structures with piled foundations would be too expensive for deep-water installations. For such cases there would be essentially two options:

- *Floating substations*: This option would allow the adoption of standard electrical equipment on board provided that watertight integrity is maintained and would be relatively easy to maintain and operate. Design of these structures would be however rather challenging because they should be capable of withstanding possibly very large wave loads and at the same time guaranteeing a very limited footprint (otherwise umbilical connection from devices might suffer severe damaging). Some solutions of this kind have been proposed but they are still at an early development stage and no concept has actually proven its feasibility. Experience from oil and gas industry is the key in many proposed designs such as moored semi-submersible platforms or tension-leg platforms. Some are actually being considered also as platform for floating wind turbines (e.g. work being carried out by researchers of the Massachusetts Institute of Technology).
- *Subsea substations*: Subsea installations would guarantee more safety in terms of load resistance and positioning but would require very expensive protection equipment for the electrical devices (most likely switchgear should include sealed compartments full of pressurized oil). Moreover maintenance would be very difficult or almost impossible in some cases since installations placed on the seabed would be operated only by remotely operated vehicles (ROVs), unless the water depth is low enough to allow divers' interventions. Disconnection of the cables would be extremely difficult and mostly require the adoption of wet-mate connectors (more expensive than the dry-mate ones). Subsea substations would probably be permanent structures and therefore would require very high reliability and redundancy to assure farm availability.

3.2.3 *Conceptual alternatives for connection units*

It is clear that one major challenge in the engineering of the infrastructure is how to design the junction point between the umbilical of the devices and the subsea cable going to shore.

If the voltage is the same for the device and the cable, as is the case of most demonstration projects, the junction box (hub) does not require housing any transformer and this gives the possibility to place this element on or below the surface without requiring special construction. Subsea substation solutions have been indeed proposed for marine energy [20] but they are designed for low power transmission and limited depths. In the present case four conceptual alternatives for the design of the structure supporting the hub are identified and presented. In all cases a main requirement is the solution to be modular, i.e. the possibility to make the junction box independent of its support structure in such a way that manufacturing, operation and installation are easy and the whole mooring or foundation system can be designed independently of the box design.

3.2.3.1 Subsea connection unit

The system is based on the use of a submerged junction box installed on the seabed and a series of connection points (Figure 3.20). The principal elements are:

- Gravity platform on which the hub is installed.
- Submerged hub.
- *Cable connectors*: They connect umbilical and dynamic cable. Since the only type of commercial connector available for the selected conditions is dry-mated, they require the connection to be performed above the surface and determine the adoption of intermediate dynamic cable.
- *Dynamic cable*: This is an intermediate cable that connects the umbilical of the device to the junction box.
- Marker buoy.

Both junction box and dynamic cables are permanent elements of the installation, and the connection and disconnection are performed by floating the connector at the end of the dynamic cable. The main advantages of this solution are:

- Mechanical stresses on the umbilical cables are reduced compared to other solutions.
- Components installed on the seabed are very little affected by stormy conditions.

Possible disadvantages are:

- The maintenance of the junction box is very complicated and switchgears could not be usable.
- Recovery of the connector might be difficult due to sediment dynamics.

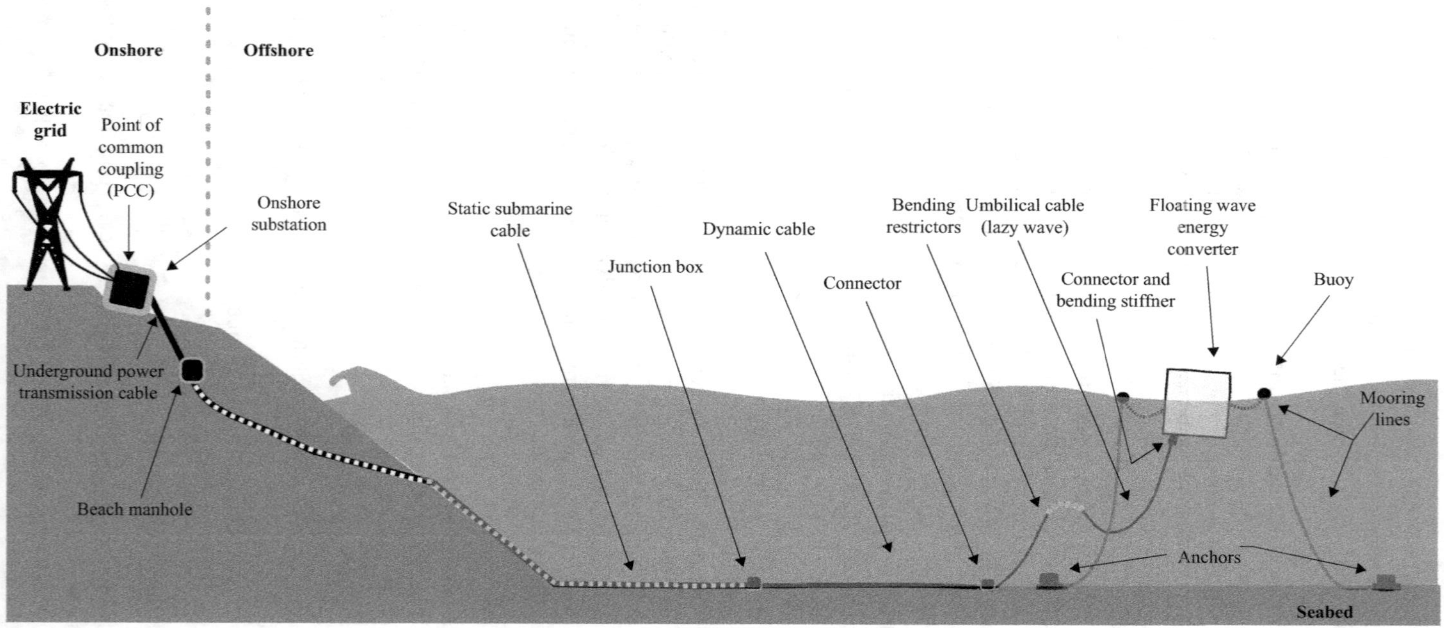

Figure 3.20 Configuration based on subsea connection unit

3.2.3.2 Connecting hub placed at intermediate depth

The system is based on the use of a submerged junction box installed on a platform located at 15–20 m below the surface. The principal elements are:

- Moored submerged platform hosting the hub and acting as a virtual seabed.
- Submerged hub.
- *Dynamic cable*: There might be the need for an intermediate cable connecting the junction box to the subsea cable since the platform motions might impose different requirements.
- *Cable connectors*: They connect umbilical cable of the device and junction box.
- Marker buoy.

In this case the umbilical cables of the converters might be directly connected to the hub. The connection might be performed by divers inside the platform in a watertight compartment or alternatively it might take place above the surface if the hub is allowed to be detached from the platform and made floatable. The main advantages of this solution are:

- Maintenance operations are realizable by divers without any additional equipment.
- If the hub is placed in a dry watertight compartment, standard connectors and switchgears could be adopted.
- Installation and decommissioning would be relatively easy.

Possible disadvantages are:

- The design of a platform with a watertight compartment could be difficult. On the other hand, also designing a system for detachment and refloating of the hub might be complicated.
- Dynamics of the platform might impose severe requirements on the umbilicals. Also the possibility of clashing of the cable should be checked.
- The solution is new and might be very expensive.

3.2.3.3 Bottom-fixed connection unit

The system is based on the use of a pile or a gravity structure as support for the junction box. The hub could be positioned above the surface but, in case the depth was too large, one could think of a submerged hub to work as a virtual seabed, similarly to what was proposed for the alternative presented in the previous section and in Figure 3.21.

The principal elements are:

- Fixed platform (pile or gravity structure).
- Hub (above or below the surface).
- *Cable*: In this case the cable might be placed inside the structure and might not require any intermediate section.
- *Cable connectors*: They connect umbilical cable of the device and junction box.
- Marker buoy.

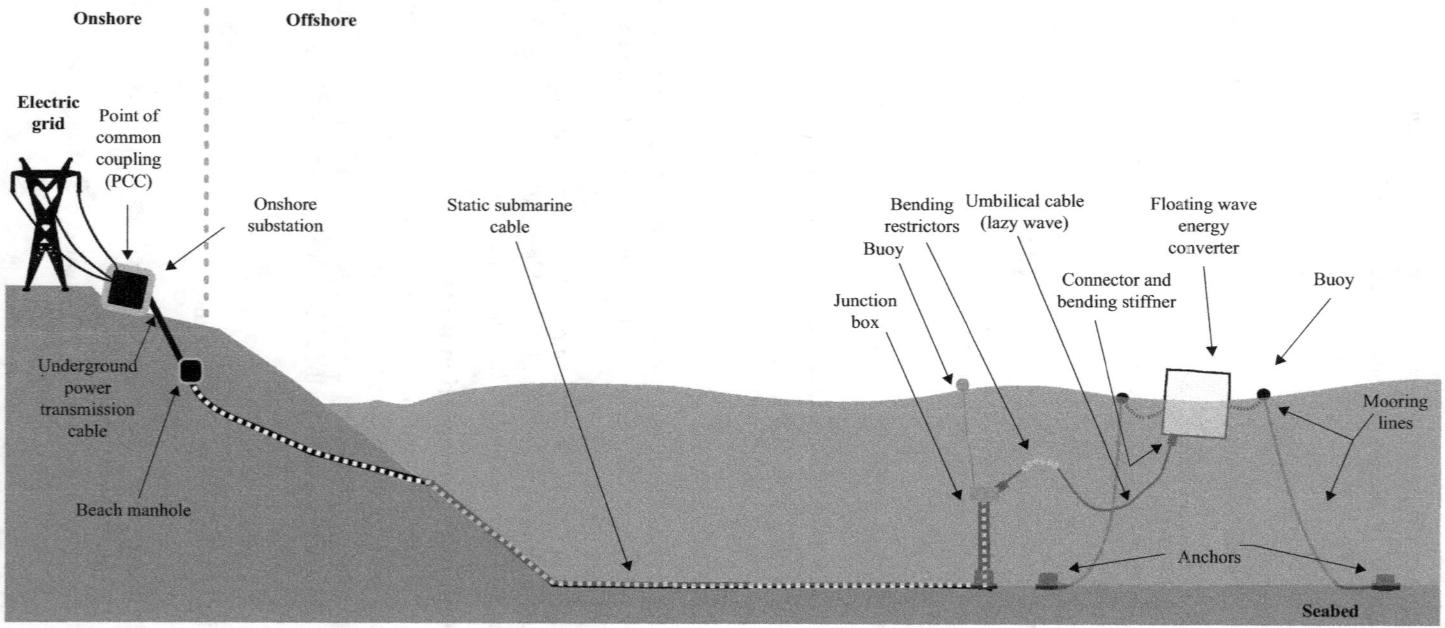

Figure 3.21 Configuration with connecting hub placed at intermediate depth

The connecting operations would be very easy when the hub is placed above the surface. If a submerged solution is considered, then there would be again the choice between watertight compartment and detachable hub (Figure 3.22).

The main advantages of this solution are:

- Ease of maintenance and operations. No need for any specific service to perform the connection.
- If the hub is placed above the surface or in a dry watertight compartment, standard connectors and switchgears could be adopted.

Possible disadvantages are:

- The construction and installation would probably be very expensive and might not even be possible depending on the soil.
- High environmental impact.
- Dynamics and possibility of clashing of the umbilical cables should be checked.

3.2.3.4 Floating hub

The system is based on a floating platform supporting the junction box (Figure 3.23).

The principal elements are:

- Moored floating platform.
- Hub.
- *Dynamic cable*: An intermediate cable connecting the junction box to the subsea cable has to be designed since the platform motions might impose different requirements.
- *Cable connectors*: They connect umbilical cable of the device and junction box.
- *Cable connector*: This would connect dynamic and subsea cable and might be permanent.
- Marker buoy.

The connecting operations would be very easy in this case because it could be performed above the surface.

The main advantages of this solution are:

- Ease of maintenance and operations. No need for any specific service to perform the connection.
- Ease of installation and decommissioning.
- Standard connectors and switchgears.

Possible disadvantages are:

- Dynamics of the umbilical cables might impose severe requirements on their design since they would link two floating structures.
- The design of the platform might be difficult due to the extreme environmental conditions and the requirement of very limited motions.

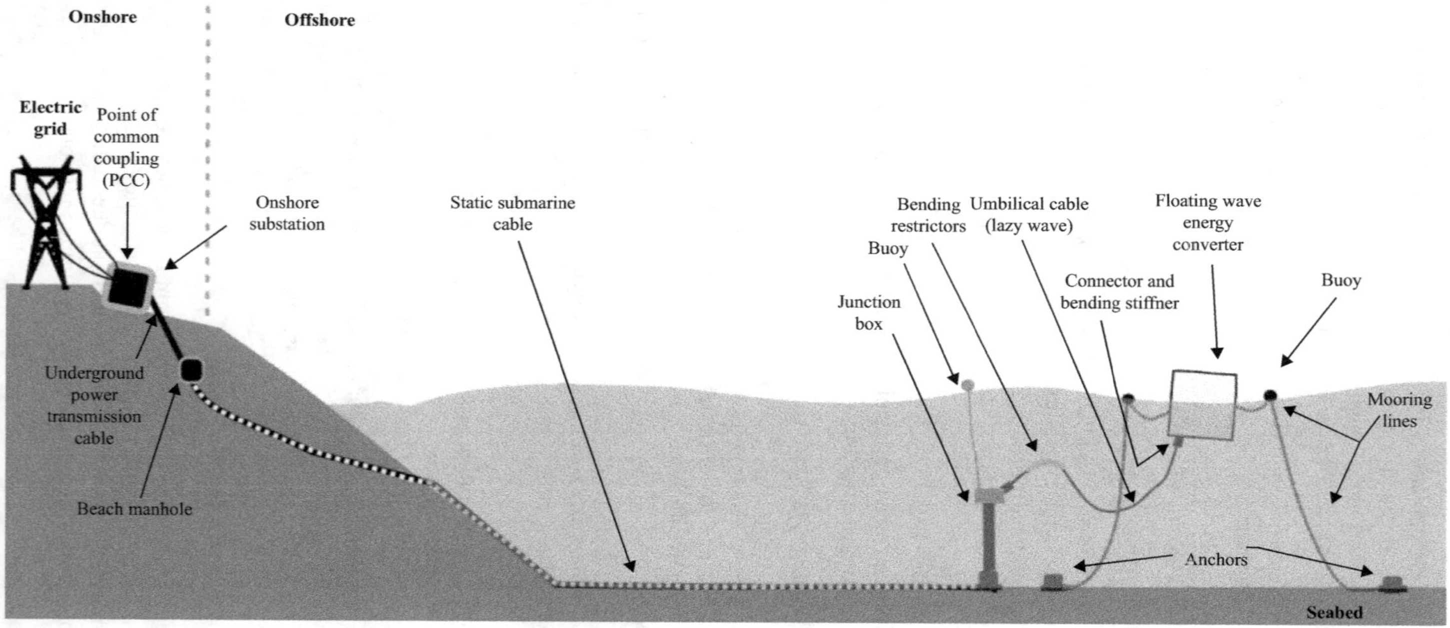

Figure 3.22 Configuration with bottom-fixed connection unit

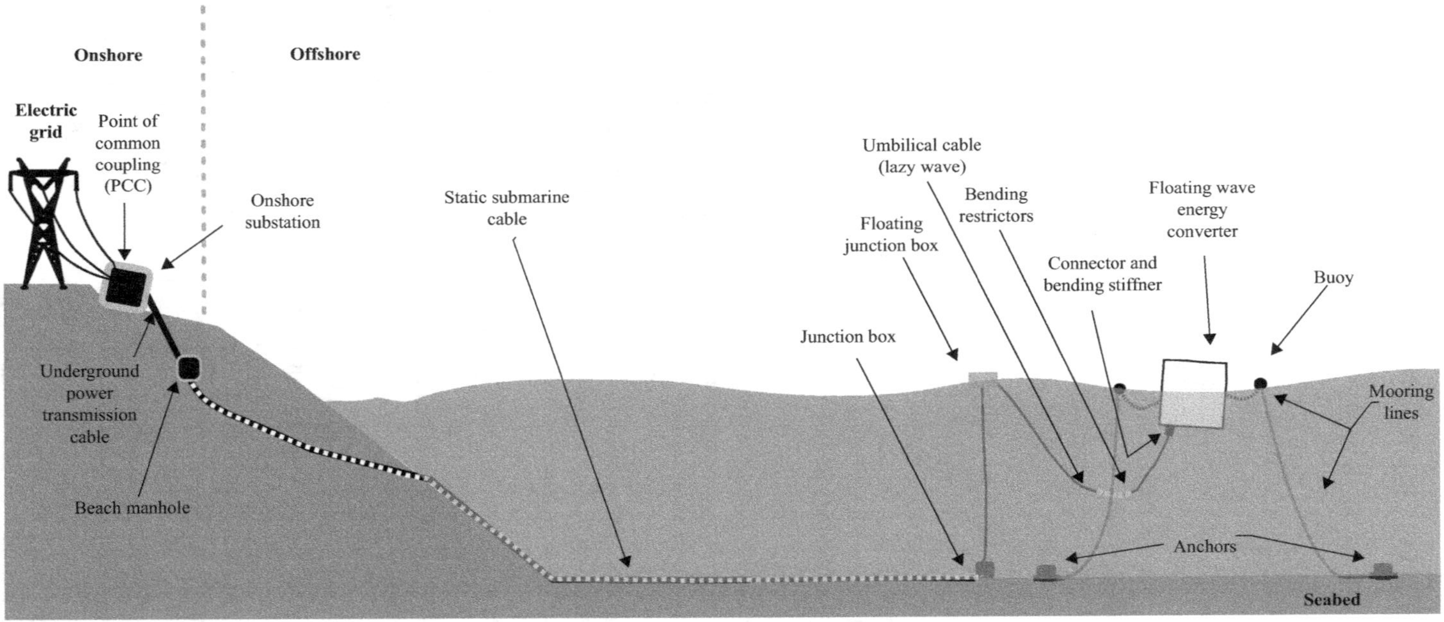

Figure 3.23 Configuration with floating hub

3.3 Dynamic cable and connector design

3.3.1 Reference standards and guidelines

It appears clear from the previous sections that subsea cables are fundamental elements of the energy chain of an offshore ocean energy convertor and that the profitability of future marine energy farms depends, in some measure, on the capability of designing cost-effective and reliable solutions for the cabling of the MECs.

Despite their acknowledged importance, relatively few information on their application to marine energy is available, partly due to the current lack of operational experience. Their design and manufacturing process can be extremely challenging, particularly when considering floating wave energy technologies. In this case, the cables applied for the connection of the device to the junction point are very similar to the umbilical cables encountered in the offshore oil and gas industry.

As for the case of mooring design and analysis, general methodologies and procedures for umbilical definition can be found in offshore engineering literature and standards referring to their specification, design, installation and manufacturing are applied by manufacturers and offshore technology industries:

- *ISO 13628 API Specification 17E: 'Specification for Subsea Production Control Umbilicals'* [32] specifies requirements and gives recommendations for the design, material selection, manufacture, design verification, testing, installation and operation of subsea control systems, service umbilicals and associated ancillary equipment for the petroleum and natural gas industries.
- *IEEE Std. 1120-2004: 'Guide for the Planning, Design, Installation, and Repair of Submarine Power Cable Systems'* [33] gives an exhaustive list of the numerous aspects to be taken into account for the planning, design, installation and repair of submarine power umbilicals.
- *DNV Guidelines on Design and Operation of Wave Energy Converters and DNV OSS-312* [34] is focused on wave energy devices but provides information and guidelines for power cable systems design and operation.
- *IEC 60502: 'Power Cables with Extruded Insulation and their Accessories for Rated Voltages from 1 kV (U_m = 1.2 kV) up to 30 kV (U_m = 36 kV)'* [24] specifies the construction, dimensions and test requirements of power cables with extruded solid insulation for rated voltages ranging from 1 kV ($U_m = 1.2$ kV) to 30 kV ($U_m = 36$ kV) for fixed installations such as distribution networks or industrial installations. Cables for special installation and service conditions are not included, in our case submarine use or shipboard application, hence necessary additional design considerations should be taken into account.
- *IEC 60092: 'Electrical Installations on Ships'* [35] gives the main system design features of electrical installations in ships, with special attention paid to cable construction, testing and installation. Detailed considerations related to shipboard and offshore application give useful guidelines for low-voltage submarine power cables' design adapted to MECs. As it will be commented

also later, however, care should be taken in applying the ratings specified in this reference since different operational and environmental conditions are to be expected in ocean energy applications.

- *IEC 60287: 'Electric Cables – Calculation of the Current Rating'* [36] specifies approaches and methodologies for calculating the current rating of electrical cables, providing therefore guidance for design and selection of the cross-section of the cable cores.

This amount of knowledge and experience can provide a first basis for the development of similar recommendations for the ocean energy sector but it is clear that purposely defined guidelines will need to be addressed in the future.

3.3.2 Composition of a typical subsea cable

Subsea cables for power transmission are generally made up of several layers, whose composition and relative importance are changing dependent on the application.

For MECs, since the primary function is the transmission of electrical power (and possibly communication or measurement signals) the cross-section is generally composed of one or more central cores containing the conductor.

Each core would require appropriate insulation and possibly a screening layer to keep the electrical field stress homogeneous. The whole ensemble is typically enveloped within a sheath.

Armouring is usually required for protection and structural support. Steel armours are practically always necessary for subsea dynamic cable configurations due to the high mechanical loads. An outer oversheath, generally polymeric, is also necessary for protection from corrosion of the armour (Figure 3.24).

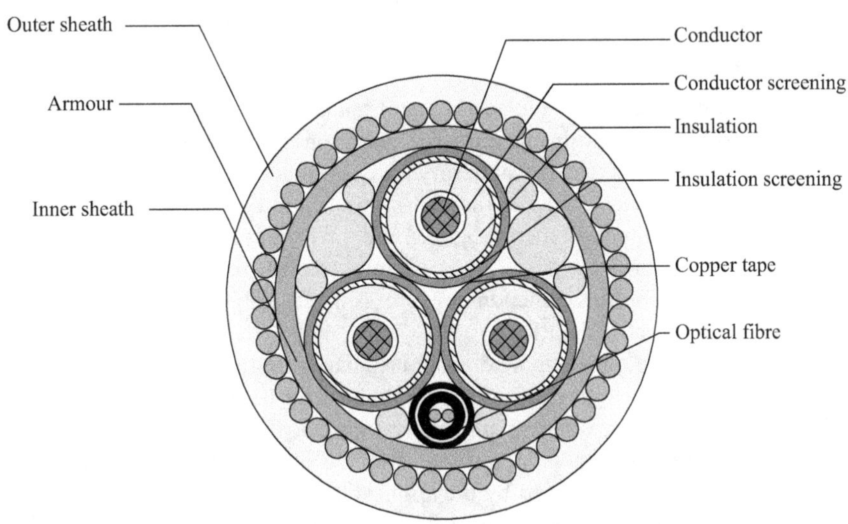

Figure 3.24 Example of a typical subsea cable cross-section

3.3.2.1 Conductor cores

The basic form of insulated three-core stranded conductor is used to carry three-phase AC. The conductor size must enable the cable to carry the desired load current without overheating and maintain voltage drop within desired limits.

The conductors of power cables are stranded (they contain several individual wires). The lay-up is usually ‘bunched’ which means that all wires are twisted in the same direction. Stranding assists cable flexibility and carriage of current as the latter distributes itself more uniformly across the cable cross-section than with a solid straight conductor.

Copper is generally preferred to aluminium for the conductors of all submarine cables as its use permits a higher current density, thereby reducing the overall diameter of the cable and its higher tensile strength is an advantage considering the dynamic stresses likely to occur during operation. In the event of the cable being damaged and seawater entering it, a copper conductor is much more resistant to corrosion than aluminium. However, it should be noticed that copper is usually more expensive and prone to price fluctuations than aluminium apart from the fact that its density is three times larger.

3.3.2.2 Insulation

Each conductor requires a proper thickness of insulation to meet its design voltage with the necessary degree of safety and reliability. An extra insulation of the conductive core is usually made in order to separate it from a potential metallic outer sheathing layer.

The required properties for subsea power cable insulation are the following:

- High insulation resistance to avoid leakage current.
- High dielectric strength to avoid electrical breakdown of cable.
- High mechanical strength to withstand handling and motion stresses.
- Non-water absorbing to maintain integrity in event of sheath damage.
- Alkali and acid resistant.

Synthetic polymers have replaced natural materials such as paper, mineral oil and natural rubber for the insulation of distribution power cables. The range of polymers available is extensive and variations in chemical compositions enable specific mechanical, electrical and thermal properties to be obtained.

Polymeric insulated submarine cables are in increasing demand for use in offshore oil fields and inter-island links up to a voltage of about 33 kV. Typical materials selected for this function are cross-linked polyethylene (XLPE) and ethylene-propylene rubber (EPR).

Considerable experience in land cable installations in many parts of the world has indicated that the service life of XLPE cables may be limited unless steps are taken to prevent moisture ingress into the insulation. It is therefore necessary to provide an impermeable moisture barrier over the insulation, and for submarine installations this is best achieved by the application of a lead alloy sheath.

Submarine cables insulated with EPR require no metal sheath and are installed as a ‘wet’ construction. This type of cable is suitable for use up to 33 kV provided selection is made of a special formulation of EPR and of a low-insulation design stress. Some three-core designs incorporate optical fibres or other communication cables in the interstices.

The immersed weight-to-diameter ratio of non-sheathed polymeric submarine cables is generally low and the cable is therefore liable to movement over the seabed when subjected to even modest tidal currents. The application of a metal (lead, aluminium or copper) sheath of appropriate thickness inhibits cable movement, apart from ensuring that the core insulation is maintained in a dry condition.

3.3.2.3 Screening

For most applications, polymer insulated cables have screened cores. Indeed, as the voltage levels rise above 3.8/6.6 kV, the tangential electrostatic stress acting along the insulation layers becomes higher than the radial stress acting across them. Leakage currents are set up leading to local heating and potential insulation breakdown, hence the need for screening (Figure 3.25).

The screen around the individual cores confines the stress to sound core insulation and also ensures that the flux is substantially radial.

The main functional element of the core screen is an earthed semiconducting layer in intimate contact with the insulation. This can comprise a semiconducting varnish in combination with a semiconducting tape but, nowadays, an extruded semiconducting layer appears to be a good solution. The combination of conductor screen, insulation and core screen are produced in one production process to ensure a composite extrusion without voids at the insulation surfaces.

As a result, the electric field in these cables is kept within a homogeneous dielectric and has a radial pattern. In order to provide a continuous earth reference for the semiconducting core screen and to supply the charging currents for all parts of the cable length, a layer of metallic tape or wires is usually applied over the different screens.

3.3.2.4 Optical fibres

Data exchange is necessary between the control and operation entities (monitoring for instance) of the marine energy farm. This function most readily will be carried

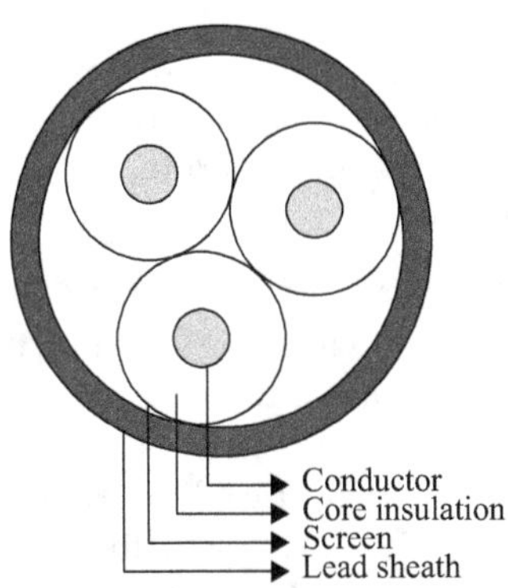

Figure 3.25 Composition of a screened power cable

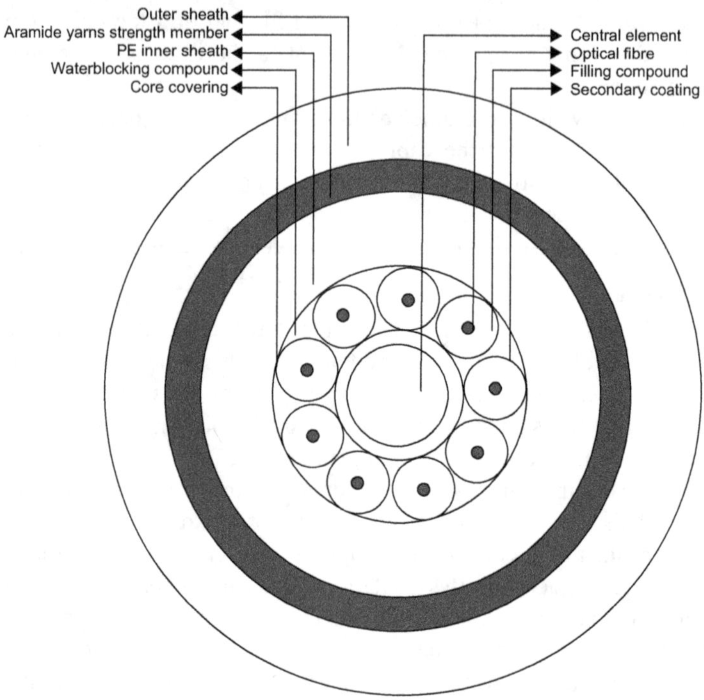

Figure 3.26 Typical structure of the cross-section of an optical signal cable

out by optical fibres incorporated inside the electrical power cable cores, granting the optical system the adequate mechanical protection (mainly bending radius and tensile strain limitations).

The construction and the integration of the optical fibre cables in the power umbilical are well-known processes widely used in offshore industry communication system to be adapted to ocean energy specific requirements (Figure 3.26).

3.3.2.5 Filler

Insulated conductors in a three-phase cable are usually symmetrically placed within a circular area, surrounded by a bedding sheath.

The core may be of two general types, free-flooding or jacketed:

- In a free-flooding type water is free to migrate through the internal parts of the core, filling the internal voids or interstices. A free-flooding cable is considered very reliable because each component is designed to be pressure-proof.
- In a jacketed core a pressure-restricted covering is applied on the outside surface. The function of the jacket is to form a pressure-restricted barrier against the intrusion of water or other media into the internal parts of the core and to act as an additional support layer for subsequent layers.

For the latter, it is often recommendable to fill the interstitial spaces with a soft material that could be depolymerized rubber, silicone rubber and/or cured urethane. These fillers can be used in umbilical cable cores to protect the most vulnerable components (electrical and optical wires) from high compressive loads and therefore allow them to move in the bundle during curvature variation of the umbilical. The fillers also aid in even distribution of handling and installation loads placed on the umbilical.

3.3.2.6 Armour

Some submarine cables may not require armouring, depending on the natural and manmade hazards and environmental conditions. In the case of power cables for marine energy applications, armouring becomes necessary for multiple purposes:

- Provide a partial path for return current.
- Carry tension during laying, operation and retrieval.
- Provide some physical protection against impacts.
- Control the bending radius to avoid kinking if laid over a sharp object.
- Provide abrasion resistance.
- Add weight to the cable to improve its dynamic stability.

A single layer of armour is common, but two layers of armouring may be used where there is a significant amount of abrasion or severe dynamic loads (as it is likely to occur in ocean energy applications). The pitch of armour wires may vary depending on the desired bending, stiffness, and coiling characteristics of the cable.

Since the cable torque concentrated in the suspended length between the vessel and the seabed can lead to kinking and knotting, double armour in opposite directions (double cross) is recommended for deep-water projects because it allows reaching a torque balance, reducing high tension elongations of the cable. This effect can also be neutralized by applying metal anti-twist tapes wound in the opposite direction to the armouring to produce a counter-torque beneath the armouring, avoiding an extra armour wires layer (handling and coiling improvement).

Round, flat or tape armouring may be used. Armour wires are made of various metals, including: aluminium, bronze and galvanized steel. Considering that magnetic armouring materials like galvanized steel might develop important electric losses (circuit eddy currents), in some cases the more expensive non-magnetic stainless steel could be required.

Material corrosion is also a main issue for armouring design, since it may have an effect on cable armour life. Sacrificial anodes and/or active cathodic protection systems may be used to prolong armour life.

3.3.2.7 Sheaths

Typical umbilical designs include sheaths for protection against corrosion and water entry. The most common cross-section configurations include a sheath layer covering the insulated cores and an outer oversheath for protection against abrasion of the armour and the cable as a whole.

The insulation must be sheathed to prevent the ingress of water and the sheathing in turn must be protected by mechanical armouring to withstand rough treatment particularly during laying.

Lead is commonly used in paper insulated high-voltage cables to provide a flexible, conductive, water-impervious barrier. Lead also adds weight to the cable, which helps hold it on the bottom in strong water currents, but could increase laying tensions. In polymeric-insulated cables an aluminium sheath is more commonly applied when water tightness is required, although this determines a careful design of the anticorrosion sheath.

The service life of a submarine cable will be dependent on the efficacy of the anticorrosion protection, as seawater is an aggressive environment in which to operate. Should the metal sheath of a non-pressure-assisted cable suffer corrosion damage, water would enter the cable and a voltage failure would ultimately occur.

Extruded polyethylene provides the most impermeable barrier to moisture ingress and is used practically universally for the anticorrosion sheath. It is applied directly above the metallic sheath. If protection against teredo (a marine worm present in shallow waters and notorious for boring holes in ships and other marine structures) is considered necessary, one or two layers of brass or copper tapes are applied over the polyethylene sheath.

Anticorrosion outer sheathes external to the armour wires are generally useless because of the impossibility of guaranteeing absence of damage to any thermoplastic or elastomeric layer during installation and operation. For such reasons, overall serving with textile tapes instead of an extruded oversheath is often preferred.

When a metallic sheath is applied over a single or three-core cable the external protection normally comprises a polyethylene anticorrosion sheath, armour bedding, one or two layers of armour wires and a bituminized textile serving overall.

3.3.3 Preliminary definition and design of umbilical components

The principal function of umbilical cables in wave energy technologies is the transmission of electrical power; therefore the preliminary definition and design of its components is based on the required electrical parameters.

Key variables for a preliminary design are the nominal electric power to be transmitted and the preferred operating voltage.

Subsequent steps of the design procedure of the electrical components are usually based on standard methodologies but some care is needed to take into account the specificities of the application.

3.3.3.1 Voltage designation and selection

Type of transmission

Voltage designation of subsea cables depends on the type of transmission considered. As mentioned before, it is most likely that power transmission will be performed through three-phase alternating current at least on the section connecting the MEC.

There are indeed some designs that consider the possibility to integrate a rectifier within the device [37]. In this case electricity would be transmitted through a direct current connection and the different units of a farm would be connected in parallel. For this kind of concept, two single core cables per device would be a suitable choice.

Apart from some different constants and some changes in the construction (the insulation layer would probably be different) the selection of DC cables is analogous to AC cables. In the remaining of section 3.3.3 we will refer to the case of three-core cables for three-phase AC service because most of the umbilical cables currently being manufactured for the offshore industry are of this type and future designs for MECs will probably take advantage from this experience.

However, it has to be noticed that the decision on the type of transmission depends on the kind of electric and electronic machinery to be placed inside the device. The Pelamis device [38], e.g., contains a transformer for elevation of the voltage up to 11 kV. It is clear that a sensible allocation of the space within a convertor for power electronics would represent an advantage for subsequent definition of the umbilical components and additional grid connection equipment.

Voltage designation of three-phase AC cables

Suitable insulation and cable construction can be specified for the required three-phase AC service performance. The design voltages for cables are expressed in the form U_0/U (formerly E_0/E). U_0 is the power frequency voltage between conductor and earth and U is the power frequency voltage between phase conductors for which the cable is designed, U_0 and U both being rms values.

Power cables in British Standards, e.g., are thus designated 600/1000 V, 1900/3300 V, 3800/6600 V, 6350/11000 V, 8700/15000 V, 12700/22000 V and 19000/33000 V. For transmission voltages above this it is normal to quote only the value of U and thus the higher standard voltages in the United Kingdom are 66, 132, 275 and 400 kV.

Standardization of system voltages has not been achieved worldwide although there is some move towards this. IEC has published voltage designations which are approaching universal acceptance particularly among European countries.

National standardization bodies also specify in some cases the maximum voltage for each nominal value. Table 3.1 shows normalized values for the typical medium voltage installations in Spain.

Table 3.1 Normalized voltage values for the Spanish three-phase electrical grid

Voltage class (kV)	Maximum voltage (kV)
1.8/3	3.6
3.6/6	7.2
6/10	12
8.7/15	17.5
12/20	24
15/25	30
18/30	36
26/45	52

DC system voltages, by which is meant DC voltages with not more than 3% ripple [25], are designated by the positive and negative values of the voltage above and below earth potential.

There is not any globally accepted definition of voltage classes as low voltage (LV), medium voltage (MV) and high voltage (HV); therefore the usual specification of the voltage rating is preferred. However, catalogues of the suppliers of power cables typically consider medium-voltage designs all the ones included between 3 and 20 kV.

Voltage class selection

As mentioned before, the selection of the appropriate voltage class for an umbilical cable specification depends on the kind of convertor and on the type of grid connection configuration.

Since to date only a few marine energy devices have undergone grid-connected sea trials and considering the different infrastructure and configuration that future arrays deployment might determine, it is not easy to define a reference value at this stage. However, taking into account field experiences and similar developments in the offshore wind energy industry, it could be reasonably assumed that umbilical connections will operate at medium voltages.

A sensible choice, especially for small-scale installations at limited distance to the connection point could be the utilization of standard values in order to reduce to a minimum the equipment for direct connection (for instance connection at the same voltage of the distribution grid would avoid the need for a site transformer).

Offshore wind turbines, e.g., are in many cases connected by 33 kV cables (or 30 kV according to IEC) to a substation or directly to the grid [39, 49].

Since early-stage marine energy installations will probably require minimum infrastructure for grid connection (at a preliminary stage it is better to host a converter on-board rather than building a substation for it), a transmission voltage of 11 or 33 kV on the umbilical side represents a suitable choice.

A precise value, however, will have to be defined according to national codes and detailed grid connection configuration.

3.3.3.2 Conductor core cross-section definition

The first step in designing or selecting an electrical cable consists in the definition of the value of the minimum conductor cross-section required. In a three-phase connection, key parameters for this determination are the apparent electrical power and the operating nominal voltage. Without precise indications on the kind of conversion system, one can consider the installed capacity as a reference for the real power and then apply a properly chosen power factor value since

$$S = \frac{P}{\cos\phi} \tag{3.4}$$

where S is the apparent power (in VAr), P the real power (in W) and $\cos\phi$ is the power factor (ϕ corresponds to the phase difference between voltage and current).

A reasonable and quite conservative value for the power factor could be taken equal to 0.8.

Once the voltage class of the cable is defined based on the criteria exposed in the previous sections, the required phase current can be found through

$$I = \frac{P}{\sqrt{3}U \cos \phi} \tag{3.5}$$

where I stands for current (in A) and U for the phase voltage (in V).

This value of the current will have to be less than the current-carrying capacity (sustained rating) determined by thermal verification.

For the calculation of the minimal section required for the conductor there are three criteria that should be considered simultaneously:

- *Heating of the conductor*: The density of current on the conductor must be limited to reduce the heating produced by the circulation of the current, since a maximum operational temperature is specified. This criterion determines a maximum value for the current.
- *Voltage drop*: The voltage drop (or equivalently the power loss) must be limited within acceptable values, particularly if the power is directly distributed to any kind of user. This criterion is mainly important for long lines, being its application to common umbilical not particularly restrictive.
- *Short-circuit current*: This value has to be limited to avoid overheating and extreme electro-dynamic stresses in case of occurrence of short-circuit. It might be a very important method for the determination of section of medium- and high-voltage lines.

3.3.3.3 Thermal design

Determination of the conductor optimal size under the thermal criterion is based on the choice of the minimum value of the cross-section capable of carrying continuously the current that is required by design parameters.

The maximum current that a specified conductor cross-section is capable of transmitting (current rating) depends on the construction of the cable and the type of insulation.

During service operation, cables suffer electrical losses which appear as heat in the conductor, insulation and metallic components. The current rating is dependent on the way this heat is transmitted to the cable surface and then dissipated to the surroundings. Temperature is clearly an important factor and is expressed as a conductor temperature to establish a datum for the cable itself. A maximum temperature, which is commonly the limit for the insulation material, without undue ageing, for a reasonable maximum life, is fixed.

Two of the important parameters in establishing ratings for standard operating conditions for particular installations are the ambient temperature and the permissible temperature rise.

Generally the maximum conductor temperature is determined by the insulation material and is specified by existing standards such as IEC 60287 (Table 3.2).

Table 3.2 Maximum conductor temperatures for different type of polymeric insulation [36]

Insulation	Maximum conductor temperature (°C)
Polyvinyl chloride	70
Polyethylene	70
Butyl rubber	85
Ethylene-propylene rubber (EPR)	90
Cross-linked polyethylene (XLPE)	90
Natural rubber	60

Once the ambient temperature is also specified, the permissible temperature rise is given by the difference between the two.

The current rating I in ampere could then be computed by the formula:

$$I = \left(\frac{\Delta\theta - W_d(0.5T_1 + n(T_2 + T_3 + T_4))}{RT_1 + nR(1+\lambda)T_2 + nR(1+\lambda_1+\lambda_2)(T_3+T_4)} \right)^{\frac{1}{2}} \tag{3.6}$$

where $\Delta\theta$ is the permissible temperature rise in the conductor, R is the AC resistance per unit length (Ω/m) of the conductor at maximum operating temperature, T_1, T_2, T_3 and T_4 are the thermal resistances per unit length (°Km/W) respectively between one conductor and the sheath, of the bedding between sheath and armours, of the external serving of the cable and between the cable surface and the surrounding medium; n is the number of conductors in the cable (three for a typical three-phase umbilical), W_d the dielectric loss per unit length per phase (W/m), λ_1 and λ_2 are ratio of losses in the metal sheath and armour with respect to total losses in all conductors in the cable.

Thermal resistance and loss ratios are given by formulas that can be found in standard literature [36]. More often their values are directly specified depending on the application and materials.

Ambient temperatures are also specified by standards for a large number of cases and condition of installations. Data for umbilical installations could be referred to existing standards for offshore and ships' electrical installations that specify an ambient temperature of 45°C for calculation of the current rating; however, considering their application to MECs, lower temperatures might be considered (Figure 3.27).

The typical procedure for selection of the conductor cross-section does not need to include calculation of (3.6). Cross-sections values and cable lay-up are normalized in such a way that current ratings are given for a variety of cases in published standards or even in catalogues from suppliers.

Correction factors for different ambient temperatures and different conditions of installations are also provided.

Table 3.3 gives a brief list of current ratings for cable for offshore installations with EPR insulation. Table 3.4 provides correction factors for different ambient temperatures (to be multiplied directly to the defined current rating).

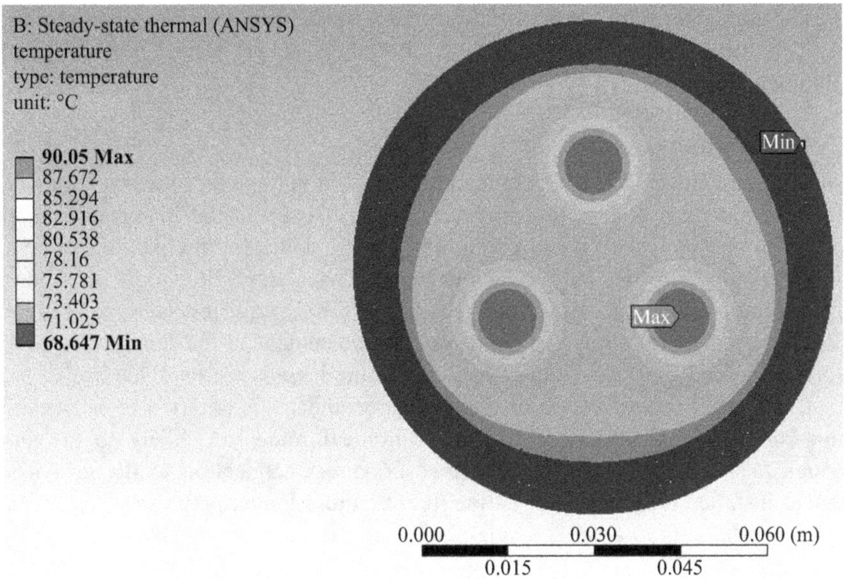

Figure 3.27 Example of cable heat distribution

Table 3.3 Maximum AC continuous current ratings in ships and offshore (based on an ambient temperature of 45°C and a maximum conductor temperature of 90°C)

Conductor size (mm^2)	Single core (A)	Three- or four-core (A)
16	95	**67**
25	125	**89**
35	155	**105**
50	190	**135**
70	240	**170**
95	290	**205**
120	340	**240**
150	385	**270**
185	440	**305**
240	520	**365**
300	590	**415**

Table 3.4 Rating factor for ambient temperature

Temperature (°C)	30	35	40	45	50	55	60	65	70	75	80
Rating factor	**1.15**	**1.11**	**1.05**	**1.00**	**0.94**	**0.88**	**0.82**	**0.75**	**0.67**	**0.58**	**0.47**

Detailed analysis of the thermal behaviour of the cable, particularly in transitory conditions, might be required at a later stage, where finite element models could be necessary (see [41]).

3.3.3.4 Voltage drop

When current flows in a cable conductor, there is a voltage drop between the ends of the conductor which is the product of the current and the impedance. If the voltage drop were excessive, it could result in the voltage at the equipment being supplied being too low for proper operation. The voltage drop is of more consequence at the low end of the voltage range of supply voltages than it is at higher voltages, and generally it is not significant as a percentage of the supply voltage for cables rated above 1000 V unless very long route lengths are involved.

Calculation of the section of the conductor under this criterion is particularly important for long transmission lines. Umbilical routes are likely to be short enough to avoid excessive losses. However, direct connection to the grid from devices installed close to the coastline might require longer paths.

The voltage drop along a single-phase line can be expressed by

$$\Delta U \cong RI \cos \phi + XI \sin \phi \tag{3.7}$$

where I is the current flowing in the line (in A), R the AC resistance (in Ω), X the reactance (in Ω) and ϕ is the phase difference between voltage and current.

If a DC cable was to be considered, the voltage drop would clearly be dependent only on the resistance (the power factor would be equal to 1) and the calculation of the proper section based on this method would be straightforward. In such case the value should be multiplied by 2 for account of the return line.

If we consider three-phase AC transmission the total voltage drop is equal to $\sqrt{3}$ times the value given in (3.7), we obtain:

$$\Delta U \cong (R + X \tan \phi)\frac{P}{U} \tag{3.8}$$

The reactance X is variable with the diameter and the position of the conductors. In absence of precise data on its magnitude, it can be taken equal to a small fraction of the resistance value or to a constant (e.g. 0.1 Ω/km). However, for relatively small conductor cross-section areas (less than 120 mm^2) such as the ones typically assumed for umbilical cables, the contribution of the inductance to the losses can be neglected.

The resistance R is dependent on the temperature and on the material. Its value for AC transmission is slightly larger than the one for DC lines because of the 'skin effect' [25] and the proximity of other conductors. Those factors can be included by assuming a global 2% increment [25]. The resistance is then given by the formula:

$$R = \frac{1.02\rho_{20}(1 + \alpha(\theta - 20))L}{S_c} \tag{3.9}$$

Table 3.5 Resistivity and temperature coefficient for copper and aluminium

Material	ρ_{20}	α
Copper	0.018	**0.00393**
Aluminium	0.029	**0.00403**

where ρ_{20} is the material resistivity at 20°C, α is the coefficient accounting for the variation of the temperature, θ is the temperature of the conductor, L is the length of the cable and S is its cross-sectional area.

Values for ρ_{20} and α are given in Table 3.5 for copper and aluminium.

The cross-section of the conductor can then be computed through

$$S_c = 1.02\rho_{20}(1 + \alpha(\theta - 20))\frac{LP}{\Delta U \cdot U} \tag{3.10}$$

The cross-section of the conductor will be chosen to be equal to the minimum normalized value larger than the one given by (3.10).

The maximum voltage drop acceptable across the line will depend on the specific requirements determined by the electrical installations and is often prescribed by standards and national rules. For long lines it is usual to specify a value per kilometre.

3.3.3.5 Short-circuit verification

The chosen conductor size should also be checked against its ability to carry short-circuit currents. During a short-circuit there is a sudden inrush of current for a few cycles followed by a steadier flow for a short period until the protection operates, normally between 0.2 and 3 s. The temperature in the cable augments abruptly because of the heat generated by the Joule effect and possible damages might occur if the components of the cable are not properly designed.

Since the time involved in short-circuits is assumed to be very short, the insulation can withstand much higher temperatures than the ones allowed for sustained operation. IEC standards specify maximum short-circuit conductor temperatures for transmission and distribution lines (e.g. 250°C for XLPE and EPR insulated cables).

The difference between these and the maximum conductor temperature for sustained operational conditions provides a maximum temperature rise which can be used in calculations to define the current rating.

Although the current experiences during short-circuit a slight decrease due to increase in the conductor resistance, it is assumed steady for conservative reasons in the calculations.

Short-circuit ratings can be calculated using the adiabatic method, which assumes that all of the heat generated remains trapped within the current-carrying component.

A simplified relationship that is commonly adopted [25] is

$$I_{sc}^2 = \frac{K^2 {S_c}^2}{t_c} \tag{3.11}$$

where K is a constant coefficient depending on the material of the conductor and of the insulation (equal to 143 $As^{1/2}/mm^2$ for copper conductors insulated with EPR or XLPE), S is the area of the cross-section of the conductor (in mm^2) and t_c is the duration of the short-circuit (in s).

The value of the short-circuit current to be taken as comparison with (3.11) is generally dependent on the circuit (including the generators and loads characteristics) and on the kind of circuit breakers and protections installed. If a three-phase cable is considered, a preliminary conservative estimation could be given assuming a phase-to-phase circuit. In this case, the short-circuit current would be given by

$$I_{sc} = \frac{U}{2\sqrt{R^2 + X^2}} \tag{3.12}$$

where R and X would be resistance and reactance of a single phase line (as introduced before). A value for the cross-section of the conductor could then be defined according to (3.11):

$$S_c \geq \frac{I_{sc} {t_c}^{1/2}}{K} \tag{3.13}$$

All of the criteria presented provide a procedure for the estimation of this minimum area. For medium-voltage cables, it is likely that the thermal design will be the more restraining and thus the actual driver for the cable design.

Nevertheless, the value obtained by all the three calculations should be assessed and compared particularly if the electric circuit in which the cable is supposed to operate is not conventional (and thus might be characterized by insufficient protection against faults).

3.3.3.6 Insulation thickness definition

The insulation size of solid power cables is generally governed by the occurrence of ionization in voids, particularly for extruded solid-dielectric insulated cables. The electrical breakdown of polymeric insulation as a result of discharge is well understood and much work has been conducted on this subject (see, e.g., [42]).

Failure occurs because of discharge activity, the intensity of which erodes fine channels in the insulation layer. The appropriate thickness could be estimated through standard formulae that allow obtaining the minimum value to avoid partial discharge in service.

However, as it was mentioned before, cable design and manufacture is a highly standardized field and it is generally not necessary to perform any calculation because standards specify the insulation thickness to be applied to the conductor depending on the voltage designated.

Due to the flexibility required by umbilical installations, ethylene-propylene rubber (EPR) might be preferred for connection of floating convertors. If an impermeable design is guaranteed, XLPE could also be an option.

Insulation thicknesses for EPR and XLPE specified by IEC standards are identical at each voltage level above 3.6/6 kV (Table 3.6).

Table 3.6 Insulation thicknesses for different voltage classes according to IEC 60502 [24]

Rated voltage (kV)	XLPE insulation (mm)	EPR insulation (mm)
3.6/6	2.5	**3.0**
6/10	3.4	**3.4**
8.7/15	4.5	**4.5**
12/20	5.5	**5.5**
18/30	8.0	**8.0**

Each insulated core will typically include semiconducting screening layers applied over the conductor and over the insulation. For three-core constructions a copper tape is applied over the outer layer for mechanical protection and earth envelope in case of fault. Nominal radial thicknesses of semiconducting layers can be taken equal to 0.5 mm as typical figure but they may change depending on the manufacturing process and on the required value of electrical stresses. XLPE insulations might require an additional metallic sheath (e.g. lead) to prevent contact with water.

The ensemble composed by the cores and the optical elements, possibly with the inclusion of filler to avoid voids, is generally covered by a polymeric sheath, whose thickness ranges from 3 to 4 mm.

3.3.3.7 Armour and load-carrying components

Armours for dynamic power cables consist in general of one or more layers of galvanized steel wires applied with a fairly long lay.

Other applications might consider steel tape armouring but wires are preferred because they have several advantages such as (a) better corrosion protection and hence longer armour life, (b) greatly increased longitudinal reinforcement of the cable, (c) avoidance of problems due to armour displacement which can occur under difficult laying conditions and (d) better compatibility with extruded thermoplastic oversheathing layers.

In subsea cables, armours are also required as strength members that serve to carry the applied cable loads and to provide mechanical protection to the electrical and optical conductors located within the core.

For this reason, the preliminary definition of the armour characteristics typically depends on the loads the cable will be required to withstand. The majority of the designs, especially for deep-water applications, present two layers of round steel wires contra-helically wrapped around the central core.

Direction of lay

The convention for determining right-hand and left-hand lay is the direction of helices as they progress away from the end of the cable as viewed from either end. A typical arrangement has a right-hand lay inner armour and left-hand lay outer armour. This has become an industry standard having its roots in the logging cables used in the oil industry.

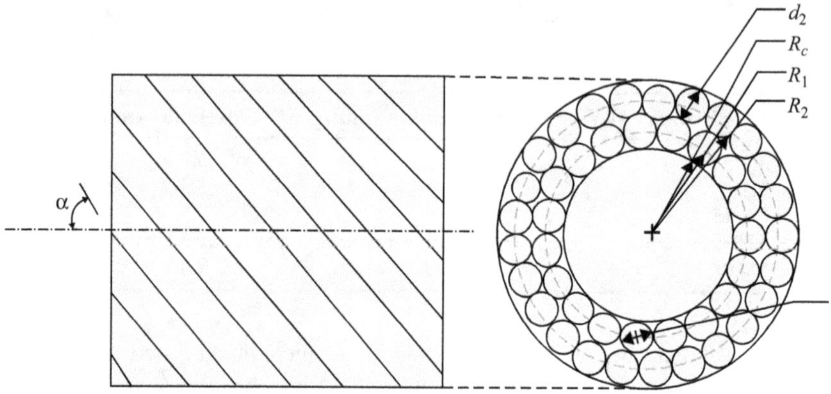

Figure 3.28 Typical arrangement of the armour of an umbilical cable

There is no evidence, however, that a right-hand lay outer armour, with a left-hand lay inner armour, would not provide the same performance characteristics.

Lay angle

This is the angle the armour helix forms with the axis of the cable as illustrated in Figure 3.28; the magnitude of the lay angle is conventionally between 18 and 24 degrees. Different lay angles may be used for different layers, depending on the design characteristics and interrelationship with outer components.

Number of armour wires

The number and diameter of armour wires are selected to cover 95–99% of the surface or as determined by the application. There is a balance between the number and size of wires to obtain this coverage. For the same pitch diameter (the diametrical distance between the centre line of the coiled wires) and metal type the larger diameter armour wires provide greater mechanical stability; this stability relates both to resistance to distortion and abrasion.

For the same pitch diameter and metal type, the smaller armour wires offer a greater flexure fatigue life. The smaller diameter wires will have the smaller outer fibre stress and they will, therefore, have a greater flexure fatigue life.

Helical wire coverage

The wire coverage is a measure of how completely the number of wires n_l fills a layer. For a compact cable cross-section, it is desirable to choose coverage above 90% but less than 100%.

Assuming an elliptical approximation to the shape of a helical wire cut transverse to the cable axis, a formula for the coverage (in percentage) of an armour layer may be expressed as

$$C_l = \frac{n_l d_l}{\pi D_l \cos \alpha_l} \times 100 \tag{3.14}$$

where d_l stands for the wire diameter, D_l for the pitch diameter and α_l the lay angle.

Torque and stress balance

The helically served components that render the cable flexible also induce a torque as each helical wire layer tries to unwind with the application of tension. The net torque produced causes a free-end cable to rotate with some layers tightening and some loosening. This means that some layers will be stressed at higher levels than others, thereby lowering the cable breaking strength. Furthermore, long cables that are restrained from rotating can develop sufficient torque to produce hockling (unstable loop formation) if the cable tension is relaxed. To minimize the cable torque and potential hockling, contra-helical layers can be employed in the cable design.

For structural efficiency, it is important that each armour layer carries a fraction of the total tension proportional to its yield strength. This condition is termed *stress balance.*

It can be seen [43] that torque and stress balance conditions are approximately met if (3.15) is satisfied:

$$\sin \alpha_d = -\sum_{l=1(\neq d)}^{N_L} \frac{n_l A_l D_l S_{yl}}{n_d A_d D_d S_{yd}} \sin \alpha_l \tag{3.15}$$

where n_l, A_l, D_l, S_{yl} and α_l are respectively the number of wires, the area, the pitch diameter, the yield strength and the lay angle of the layer l while n_d, A_d, D_d, S_{yd} and α_d similarly refer to the dependent layer (usually the outermost).

Estimation of the armour layer area and preliminary design

A preliminary estimate of the required armour area, depending on the chosen number of layers, could be based on a linear formulation subject to the following assumptions:

- The ends of the cable are fixed against rotation.
- The core is radially rigid and its axial rigidity is neglected.
- Each layer of armour carries an equal fraction of the applied tensions.
- Wire bending and torsional stresses are neglected.

Within this approach, through purely geometrical relations [44, 45], the cable axial strain can be expressed in function of the applied tension T as

$$\varepsilon_c = \frac{T}{\sum_{l=1}^{n_L} n_l E_l \frac{\pi}{4} d_l^2 \cos^3 \alpha_l} \tag{3.16}$$

where E_l is the elastic modulus of the material of the armour wire layer ‘l’.

The deformation produced by a tensile load on the cable is assumed to be the same across the cross-section. This means that every layer is supposed to experience the same axial strain. Considering the assumption that every armour layer is supposed to carry an equal fraction of the applied tension, the cable axial strain can be expressed as depending on a single armour layer as

$$\varepsilon_c = \frac{T}{N_L n_l E_l \frac{\pi}{4} d_l^2 \cos^3 \alpha_l} \tag{3.17}$$

where N_L is the number of layers.

Assuming a maximum operating loading tension M, a safety factor SF and a yielding strength of the armour wire material σ_{sl} of the layer 'l' and providing again that the total load is distributed evenly on each layer, the relationship in (3.18) can be imposed:

$$\frac{M}{N_L} = n_l SF \sigma_{sl} \frac{\pi}{4} d_l^2 \cos \alpha_l \tag{3.18}$$

Considering the applied tension to be equal to the maximum operating loading tension and combining (3.17) and (3.18), the cable axial strain is

$$\varepsilon_c = \frac{n_l SF \sigma_{sl} \frac{\pi}{4} d_l^2 \cos \alpha_l}{n_l E_l \frac{\pi}{4} d_l^2 \cos^3 \alpha_l} = \frac{SF \cdot \sigma_{sl}}{E_l \cos^2 \alpha_l} \tag{3.19}$$

Every layer might have a different lay angle and a different material; therefore (3.19) should be applied to every layer to find the values of the cable axial strain imposed by the maximum tension on each of them.

Afterwards, the minimum value should be taken as the reference value (as assuming the same deformation the other layers would certainly experience a lower tension than the imposed limit).

$$n_l A_l = n_l \frac{\pi}{4} d_l^2 = \frac{M}{N_L E_l \cos^3 \alpha_l \varepsilon_c} \tag{3.20}$$

Clearly number and diameter of wires should be chosen according to other requirements, as also mentioned in the previous sections.

More sophisticated models and detailed analysis will be needed at subsequent steps with a more precise understanding of the dynamic loads.

Additional requirements

A further, often important, feature of armour design is to provide effective conductance of earth fault currents. If the cable has a metallic sheath, the sheath alone is adequate in most cases; steel tape armour does nothing to extend the sheath current-carrying capacity. In the absence of a metallic sheath it is generally desirable or necessary to use wire armour to deal with fault currents.

Since in umbilical applications, armours are mainly required to sustain mechanical loads, their design is not specifically addressed to match fault current capability, but careful consideration and validation of this event should be performed once the layout of the installation and its electrical characteristics are more detailed.

3.3.3.8 Isolation and protection requirements

An outer sheath is generally needed for protection and isolation of the armour wires from abrasion and crush. It shall be of a polymeric material incorporating protection against UV radiation and ozone. The chosen material is typically continuously and concentrically extruded over the laid-up core to produce a uniform cross-section.

Polyethylene sheaths are frequently used because of their resistance to crushing damages. However, a textile serving is also applied in some cases. A bitumen coated jute or polypropylene string serving could be preferred because of the high friction coefficient. A string serving also has the advantage of being sufficiently elastic to permit an increase in the pitch circle diameter of the armour wires resultant on coiling the cable without adverse effect on the serving. Typical thicknesses for outer sheath layers are comprised between 3 and 5 mm.

3.3.4 Mechanical model and validation of umbilical cables

Having defined a preliminary design of the umbilical power cable and a lay-up of its principal components, subsequent improvements should be addressed by the analysis of the dynamic behaviour during operation, installation and layout.

Detailed design would include simulation of the effect of electromagnetic, thermal and mechanical loads possibly with finite element models and advanced computational packages.

In particular, manufacturers should be provided with precise definition of the installation conditions and the expected operational dynamic loads during the service life.

On the other hand, an appropriate selection of the configuration of the umbilical and its connections is possible only after performing a dynamic analysis based on the mechanical properties and requirements of a particular design. A first analytical estimation of these parameters could be based on the preliminary design defined before, although manufacturers generally provide values based on experience and tested equipment.

3.3.4.1 Definition of mechanical properties

A precise definition of the mechanical properties of an umbilical cable is generally complex due to the general non-linear behaviour of the structure and the presence of several different components.

Observing their typical construction and assuming that the armour layer is generally the most important load-carrying component, umbilical cables could be modelled as a kind of wire rope and theory derived for this case [46] can be applied.

There are several analytical models based on different assumptions that have been proposed for the determination of the structural response of a wire strand (see [47] for an extensive comparison) and some present quite a high level of complexity which would not be suitable for the purpose of a specification for subsequent analysis, particularly considering the degree of approximation that the dynamic model might include itself.

Within this section, simple linear cable equilibrium equations are shown based on the assumptions of pure tension wire, i.e. torsional and bending stiffness of the single wire is neglected in the torsion-tension representation and the whole armour structural response is solely based on the wire axial stresses. Reference to this model is made to Hruska [45] and to Knapp and Cruickshank [44].

Under these assumptions, the constitutive equations of the cable can be written as

$$\begin{Bmatrix} T \\ M \\ B \end{Bmatrix} = \begin{bmatrix} K_T & K_{TM} & 0 \\ K_{MT} & K_M & 0 \\ 0 & 0 & K_B \end{bmatrix} \begin{Bmatrix} \varepsilon \\ \theta/L \\ \phi/L \end{Bmatrix} \tag{3.21}$$

where T, M, B represent tensile load, torsion torque and bending moment, K_T, K_M and K_B are axial, torsion and bending stiffness coefficients while K_{TM} and K_{MT} represent cross-coupled stiffness coefficient between torsion and tension. The cable axial strain is represented by ε, while θ and ϕ stand respectively for the rotation angle and bending curvature angle.

The bending equation is decoupled from the axial-torsional equations as a consequence of assuming a large number of wires symmetrically disposed around the cable axis.

The different stiffness coefficients can be found by considering geometric relationships between the parameters.

For the case of the axial stiffness, a difference should be made as to whether the central core is considered radially rigid (i.e. its diameter does not deform under the action of an axial load) or incompressible (the diameter change as function of the cable axial strain and not depending on the cable twist maintaining its volume constant).

This different option is taken into account with a coefficient Θ, which is considered equal to 0 in case of radially rigid core and equal to 1 for incompressible core:

$$K_T = \sum_{l=1}^{N_L} \left[n_l A_l E_l \left(1 - \frac{\Theta}{2} \frac{D_c}{D_l} \tan^2 \alpha_l \right) \cos^3 \alpha_l \right] + A_c E_c \tag{3.22}$$

Subscript c stands for core. Indeed, for three-phase AC cables, as the electrical cores are typically stranded and helically wrapped, the same approach used for the armour calculation could be used to estimate the axial stiffness of the central core. However, to avoid complication, in these equations the whole contribution of the central core to the loads is accounted in a unique coefficient representative of a simple central cylindrical element.

Depending on the compressibility hypothesis of the core, the stiffness matrix of (3.22) might be non-symmetric. The cross-coupled coefficient that accounts for tension loads due to twist rotation is expressed by

$$K_{TM} = \sum_{l=1}^{N_L} n_l A_l E_l \frac{D_l}{2} \sin\alpha_l \cos^2 \alpha_l \tag{3.23}$$

While the one relating axial strain to torsional load is

$$K_{MT} = \sum_{l=1}^{N_L} n_l A_l E_l \frac{D_l}{2} \left(1 - \frac{\Theta}{2} \frac{D_c}{D_l} \tan^2 \alpha_l \right) \sin\alpha_l \cos^2 \alpha_l \tag{3.24}$$

where we can see that when the core is considered incompressible ($\Theta = 1$) its changing diameter in function of the axial load determines a coupling between torsional and axial tensions which disappears if the core is assumed radially rigid ($\Theta = 0$).

The torsional stiffness of the cable can be estimated as

$$K_M = \sum_{l=1}^{N_L} \left[n_l A_l E_l \frac{D_l^2}{4} \sin^2 \alpha_l \cos \alpha_l \right] + J_c G_c \tag{3.25}$$

where J_c is the torsion constant of the central core (the polar moment of inertia since we are considering a cylinder) and G_c is the shear modulus of the central core material (in the case of three-cores cables the torsional rigidity could be estimated under a first approximation through the sum of the different elements rigidities). It should be noticed how the torsional stiffness of the single wires is not taken into account in this model, since the contribution dependent on the axial strain is expected to be far more relevant.

The bending stiffness coefficient can be given by two different formulations (3.26a) or (3.26b), depending on the fact of considering ideal beam bending (and no wire axial slip in such a way that the bending deformation is homogeneous across the section) or individual wire bending (with axial wire slip in such a way that every component practically bends independently from each other):

$$K_B(\text{beam bending}) = \sum_{l=1}^{N_L} \left[\frac{n_l A_l E_l (d_l^2 + 2D_l^2)}{16} \cos \alpha_l^3 \right] + E_c I_c \tag{3.26a}$$

$$K_B(\text{wire bending}) = \sum_{l=1}^{N_L} n_l E_l I_l + E_c I_c \tag{3.26b}$$

where clearly I_l and I_c represent moments of inertia of the wires and central core cross-section.

In reality some slip is experienced depending on the bend curvature, so a realistic estimation would probably lie somewhere between these two ideal cases.

3.3.4.2 Structural behaviour and critical load analysis

A large number of failure and breakdown modes could possibly occur during the operational life of an umbilical cable since it can be exposed to solicitations of various kinds (electromagnetic, thermal, chemical and mechanical).

Thermal and electromagnetic stresses are typically dependent on the design of the manufacturer, on the correct estimation of the environmental conditions (external temperatures and presence of external fields) and on the electrical installation layout. Insulation breakdown is one of the most usual reasons for failure in electrical cables and apart from being determined by ionization, excessive thermal loads and transient voltage conditions can obviously be related to mechanical damage or ingress of moisture due to partial disruption of protection.

Mechanical typical failure modes can be described as follows:

- Tensile failure occurring if the tension applied to the cable produces yielding or fracture of any of its components.
- Excessive rotation of a free-end cable producing excessive cable elongation and unequal load sharing among the helical armour layers as load transfers from layers that are loosening to those layers that are tightening.
- Excessive torque of a fixed-end cable leading to cable hockling (unstable looping) if the applied tension is insufficient to maintain a straight cable.
- Bending failure frequently occurring near supports, terminations and where the cable bend radius is small enough to produce significant bending stresses.
- Fatigue failure occurring due to the cyclical application of any load, but most frequently as a result of repeated bending, especially near terminations and supports.
- Temperature changes producing damaging internal stresses, especially for plastic materials and delicate optical fibres.
- Fatigue failure and damage due to repeated abrasive friction between the cable and the seabed at the touchdown point.

Detailed dynamic analysis of the cable response to environmental loads and motion of the connected convertor should be performed for check and validation of the design against the aforementioned events.

Over-bending seems to be a quite common cause for failure [48] in the offshore industry applications. Recommended bending diameters are generally provided by cable manufacturer to assure customer expectations for cable life. Typically a reduction on the minimum bending radius prescription has the side-effect of affecting the service life of the cable. Moreover, it is preferable to arrange the layout in such a way that large deflections only occur at the same point since it is better for a cable to have a single deflection than several along its path.

Recommended values for the minimum bending radius are typically between 15 and 20 times the outside diameter of the cross-section.

It should be noticed, however, that typical weak points for bending curvature occur close to the terminations where properly designed stiffeners should be included; therefore a complete dynamic analysis should attempt at addressing also the definition of such components.

3.3.4.3 Verification and optimization based on the dynamic response

For standard subsea transmission cable, the most demanding load condition corresponds to the installation and laying of the cable, which is usually performed by using special vessels and requires the minimum bending radius not to exceed 15 to 20 times the radius of the cable. Additionally, subsea cables have to maintain their position during operation so that they must be capable of withstanding the environmental forces acting on them (bottom stability).

The installation of umbilical cables can be carried out by several means and techniques and it is possibly less demanding because of the reduced size and length of the cable. Nevertheless, the mechanical requirements are very similar in this case to the ones mentioned above. Furthermore, dynamic cables designed for mobile devices are subject to extreme motions and their dynamic response has to be accurately verified. After a dynamic analysis of the umbilical behaviour and a preliminary choice of its layout configuration and installation procedure, the design can be validated and optimized based on the critical difficulties that emerged from the simulations.

A typical key parameter for the dynamic response of an umbilical cable is the diameter-over-weight ratio (DOW) which has an important influence on the dynamic stability. Generally, the lower the value, the more stable the design is and this might be an important issue for shallow-water conditions where seabed stability and clashing with other lines could occur. On the other hand, lower DOWs tend to imply higher tensile loads and higher values might be chosen for applications where environmental conditions are not particularly severe. Examples of possible cable designs for ocean energy applications defined after consultation with manufacturers ranged between 7 and 15 m^2/te.

When the weight needs to be increased to secure a more stable design, there are a number of options that can be followed:

- 'Add layers of armour wires' is the first option since it would provide additional ballast and contemporarily increase the resistance, although the outside diameter would as well increase.
- Since a metal sheath is often needed for further protection of the insulation and additional screening, a relatively simple solution would be to increase the thickness of this sheath (which is often made of high-density material such as lead).
- A free-flooded design might be a solution in case this was not considered before. This would mean to let water entry inside the interstices of the central core of the cable instead of letting voids and applying fillers. This choice would then impose constraints on the selection of the material of the insulation and sheaths.
- Other components could be redesigned and redefined with the purpose of increasing the weight-to-diameter ratio. This could also include the possibility of considering a lower voltage class in case this was not a constraint imposed by the electrical connection (lower voltage classes would impose larger conductor cross-sections and smaller insulation thicknesses).

If the design of the cable needs to be left unchanged because of other practical and economical requirements, possible interventions regard the selection of ancillary equipment such as bend stiffener or bend restrictors. These elements will be briefly presented in later sections. However, it should be said that their design is clearly specific to the device and cable type they apply to and a detailed specification is only possible when precise data on the configuration of the cable is provided.

3.3.5 Connectors for marine energy devices

Electrical connectors for marine renewable energy are components that enable the electrical connection between two cables or one cable and one device. There are two kinds of connectors depending on the working environment:

- *'Dry-mate' connectors*: They work under the water but the connection and disconnection must be performed in dry atmosphere. The resources needed for the connection operation and the refloating are very expensive (e.g. vessels). On the other hand, they are more developed than 'wet-mate' connectors and the cost is lower.

- *'Wet-mate' connectors*: They also work under the water but the connection or disconnection can be performed either in dry atmosphere or under the water. The second option avoids refloating the connector but might require the use of ROVs for maintenance operations unless in shallow water, where divers could work.

Wet-mate connectors are subdivided into three classes:

- ROV-mate.
- Diver-mate (manual).
- Stab-plate multi-connector.

In addition to the connector itself, it should be considered how this will be attached to the cable. There are different methods of achieving this:

- *Elastomeric moulding*: This is the simplest and, normally, the cheapest form of termination. It does however have limitations in that the cable has to be sheathed with a material suitable for bonding (polyurethane or neoprene being the most common). Often, these materials are not best suited for long-term deployment subsea. Polyethylene sheathed cables are better but moulding to these requires specialized tooling and machinery.
- *Field installable and testable terminations*: These devices are mainly used on the umbilical cable where it enters the UTA (umbilical termination assembly). They consist of an oil-filled, pressure-balanced chamber which contains the seal to the cable sheath and boot seals on each individual conductor. Installation is a skilled job, normally carried out by the connector suppliers' technicians at the cable manufacturers' site to eliminate the need for costly cable transportation.
- *Hose terminations*: For comparatively short-length jumper leads (lengths of up to 300 m have been supplied) the easiest form of termination is to use a specially developed hose in which the required cables are run. These hoses are oil-filled which provides a fully pressure-balanced system. Termination of the hose to the connector is either by industry standard couplers or by the connector manufacturers' own proprietary design.

3.3.5.1 Examples of connectors from offshore oil and gas suppliers

To date there are very few commercial connectors available since the marine renewable energy market is an emergent market that has not required them so far. However, oil and gas, military and oceanographic industries have developed solutions for high-voltage connection that might be adapted to wave energy converters. Since some of these products are designed for the military industry, it is

Table 3.7 Electrical connector manufacturers and type of products

Company name	Industrial sector	Connection type	Maximum capacity
MacArtney Under Water Technologies	Oil and gas, oceanographic industry, renewable energies	Wet-mate	**3 kV/250 A**
Hydro Group (Hydro House)	Oil and gas, oceanographic industry, renewable energies, military, aquaculture	Dry-mate	**11 kV/500 kV**
Bennex	Oil and gas, oceanographic industry, renewable energies, aquaculture	Dry-mate	**6.6 kV**
Gisma	Oil and gas, renewable energies	Dry-mate	**6.6 kV/1000 A**
ODI	Oil and gas, oceanographic industry, military	Wet-mate	**3kV/30A**
Vetco	Oil and gas, renewable energies, military, aquaculture, aeroespacial, biomedicine	Dry-mate Wet-mate	**145 kV/700 A** **36 kV/500 A**
Expro Group	Oil and gas	Dry-mate Wet-mate	**15 kV/200 A** **30 kV/1350 A**
J&S Marine	Oil and gas, renewable energies, naval industry	Dry-mate	**33 kV (it is a termination)**
Pfisterer	Oil and gas, energy transmission and distribution, rail industry	Dry-mate	**245 kV/2500 A**
Teledyne DGO'Brien	Oil and gas, renewable energies	Dry-mate	**1 kV AC/1.8 kV DC/4 A to 8A**

not easy to collect information on them. Table 3.7 shows the type of products currently available on the market.

3.3.6 Ancillary components

3.3.6.1 Bend stiffeners

Power umbilical cables are generally connected to a rigid structure (wave energy device, offshore substation, junction box). External loads on the umbilical caused by the sea and the motions of the device determine large stresses in this fixed location. The movement of the umbilical in combination with large axial loads may cause damage to its structure possibly because of over-bending and/or fatigue.

Bending stiffeners are used to avoid this problem. These conically shaped polyurethane mouldings add local stiffness to the cable in order to limit bending stresses and curvature to acceptable levels (Figure 3.29).

The bending stiffener has a conical external profile and a central hollow cylindrical section allowing it to slide over the end of the umbilical. Each bending stiffener is designed individually to protect the umbilical minimum bending radius under a defined tension and angle combinations, meeting the load cases (tension vs. angle) of each application (Figure 3.30).

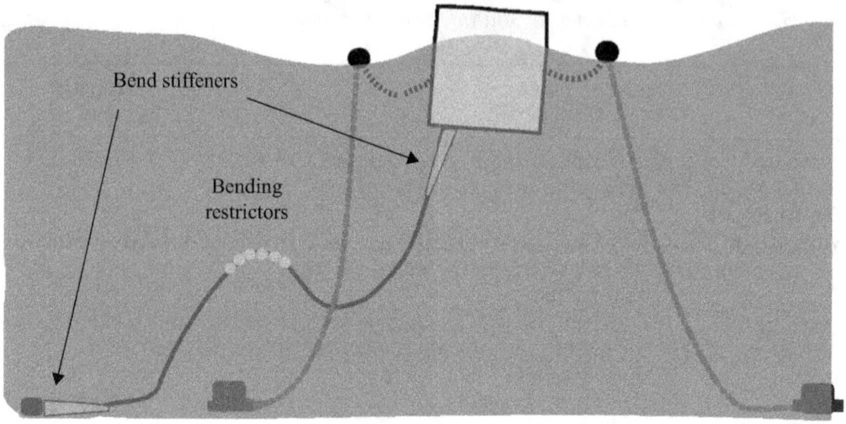

Figure 3.29 Bend stiffeners and bend restrictor examples

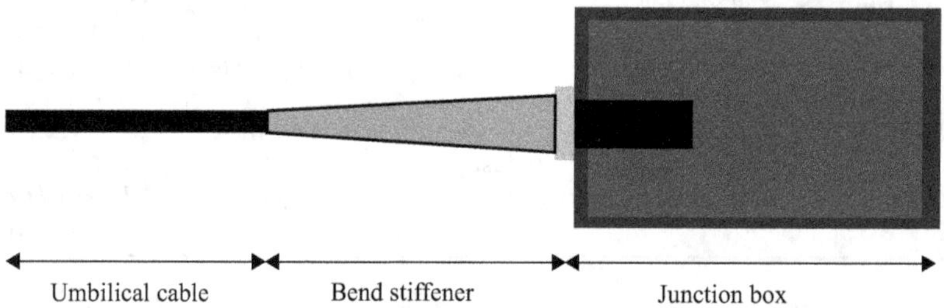

Figure 3.30 Bending stiffener

There are two types of bending stiffeners:

- Dynamic bending stiffeners are designed to protect flexible umbilicals in applications where a long service life is required.
- Static bending stiffeners are used primarily for over-bend protection during installation and over-boarding.

The bending stiffener body is usually manufactured from moulded polyurethane elastomers. The typical choice of polyurethane elastomer is based on its low modulus and high elongation at break. Typical test requirements of the material design used in bend stiffeners are:

- Fatigue resistance.
- Creep resistance.
- Tensile strength.
- Tear resistance.
- Temperature dependency.
- Ageing in air and seawater.
- UV resistance.

The main information required for the design of a bending stiffener include:

- Umbilical diameter.
- Load cases (tensions vs. angle).
- Operational environment (water).
- Interface requirements with load bearing steelwork/end termination.
- Fatigue loads and cycles (for dynamic bend stiffener design).
- Tension and angle combination (for dynamic bend stiffener design).

An alternative to the elastomeric bend stiffener is the Gimbal system [49] which can accommodate large angular deflexion and high axial load by separating the axial load capacity of the assembly from the components.

The Gimbal principle is based on two points:

- The axial load capacity is taken by a mechanical universal joint which is free to move in any direction.
- The axis of the components (e.g. electrical and optical cables/steel tubes) is moved 90 degrees and coiled so the components are at right angles to the plane of angular deflection.

The Gimbal accommodates large angular deflection by changing the orientation of the components in relation to the deflection. Components in the umbilical are normally in line with axis of the umbilical and in the plane of angular deflection. Within the Gimbal, components are coiled around the umbilical axis and are so at right angles to the plane of angular deflection.

A major advantage of the Gimbal system is that connection has minimal stiffness and the bending moment applied to the fixed structure is reduced due to the short length of the connection.

3.3.6.2 Bend restrictors

Within ocean wave energy industry, power umbilical cables provide electrical and optical connections, different rigid structures such as the WEC and subsea devices (transformers substations, electrical hubs etc.). Bending restrictors might be required in order to prevent them from over-bending at the interface between flexible and rigid structures.

A bending restrictor is specifically used where static (or quasi-static) loads act on a cable, rather than dynamic loads when a bending stiffener would be more suited.

The restrictor usually comprises a number of interlocking elements which articulate when subjected to an external load and lock together to form a smooth curved radius known as the locking radius. The locking radius is chosen to be equal to or greater than the minimum bend radius of the pipe (Figure 3.31).

Once the elements have locked together the bending moment present is transferred into the elements and back through a specially designed steel interface structure into the adjacent rigid connection, therefore protecting the cables from these potentially damaging loads.

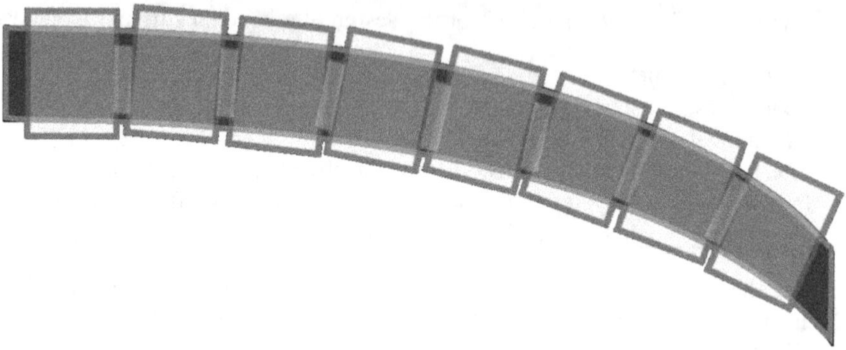

Figure 3.31 Bending restrictors

Bending restrictors need to follow some design criteria such as:

- Split design, allowing installation of the restrictor after umbilical termination.
- Easy installation onshore and offshore.
- Neutral buoyant in water, eliminating self-weight loading on the cable.

The main variables and cable parameters to take into account in order to select optimal bending restrictors for specific utilization are:

- Minimum bend radius.
- Outside diameter.
- Loads (bending moments, shear loads).
- Length of coverage.
- Operating temperature.

The materials used in the manufacture of the rending restrictor components are:

- Elements – structural polyurethane.
- Element fasteners – super duplex stainless steel.
- Interface steelwork – high-strength structured steel.

The structural polyurethane and super duplex stainless steel fasteners are corrosion resistant in seawater. The interface steelwork is the part of the structure that requires corrosion protection. This can be provided by a subsea coating system and either connection to an adjacent cathodic protection system or by attachment of its own dedicated anodes. For polyurethane elements usually yellow or alternatively orange is used because both colours provide excellent subsea visibility.

3.3.7 Dynamic analysis of umbilical connections

3.3.7.1 Configurations and connection layouts

The ensemble of cables connecting a floater at the surface with a connector at the seabed (or alternatively at different positions) might be considered a flexible riser system.

Although dynamic response of a riser system to extreme and operational environmental conditions plays the key role in selection of a particular configuration, other important factors that should also be considered during this phase are:

- Interference with the mooring lines.
- Activity of other WEC and/or support vessels in the vicinity.
- Ease of laying and retrieval and future requirements of maintenance.
- Inspection and workover operations.
- Easy disconnection procedures.

Based on the above the main design parameters at this stage are therefore the choice of the configuration, the length of the umbilical, the system geometry and the sizing of the buoyancy modules, subsurface buoy or arch. The riser should be as short as possible in order to reduce material and installation costs, but it must have sufficient flexibility to allow for large excursions of the floater.

At first stage the following different compliant alternatives are selected. Hybrid alternatives are not considered here because they were found not suitable due to the typical depth range (i.e. S-tether configuration with a single subsurface buoy).

The dynamics of umbilical cables designed for electrical power transmission from floating devices is likely to be strongly influenced by the motion of the convertor; therefore decoupling from this should be sought whenever is possible. This makes 'wave' configurations more attractive for wave energy applications.

Free hanging catenary

The free catenary is the simplest umbilical configuration corresponding to a simple umbilical without any additional structure. It is also the cheapest one to install because it requires minimal subsea infrastructure and ease of installation. Although close to vertical when it leaves the WEC, it is parallel with the seabed at its touchdown point.

The main shortcoming with this configuration is that there is no decoupling between hang-off and touchdown motions. Therefore excessive heave motions at hangoff often cause compression, buckling and minimum bend radius violations in the touchdown region (Figure 3.32).

Lazy S

In the 'S' configurations there is a subsea buoy, either a fixed buoy, which is fixed to a structure at the seabed, or a buoyant buoy, which is positioned by chains. Buoyancy of the arch is provided by one or two large cans under the arch that the umbilical is laid on. The umbilical is laid in a gutter on the arch and clipped in to prevent it falling off (Figure 3.33).

'Lazy' means that the bottom part of the riser lies on the seabed. 'S' refers to the resulting line shape due to the arch structure.

The addition of a buoy removes the problems with the touchdown point because such subsea buoy absorbs the tension variation induced by the floater and the touch down point experiences very small variations of tension. This configuration is suitable where more than one umbilical is required, as several umbilicals can be hung over a single arch and clashing is typically not an issue.

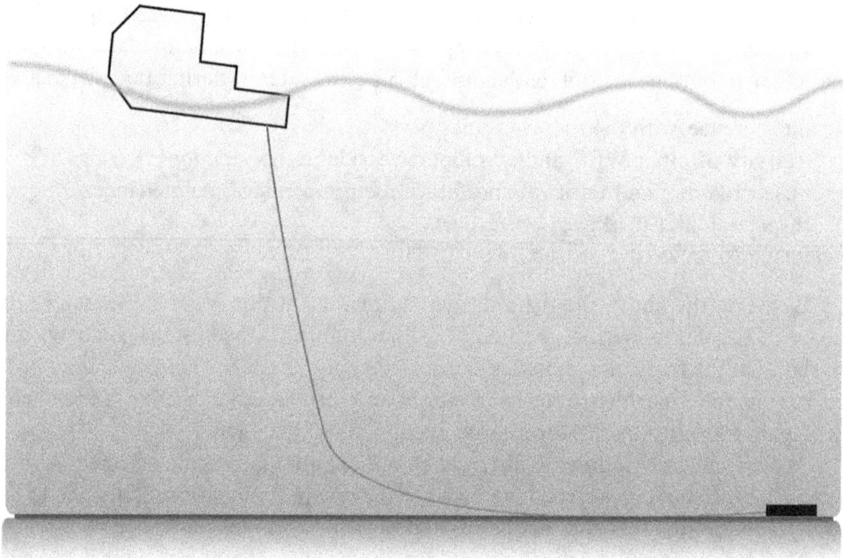

Figure 3.32 Free-hanging catenary layout

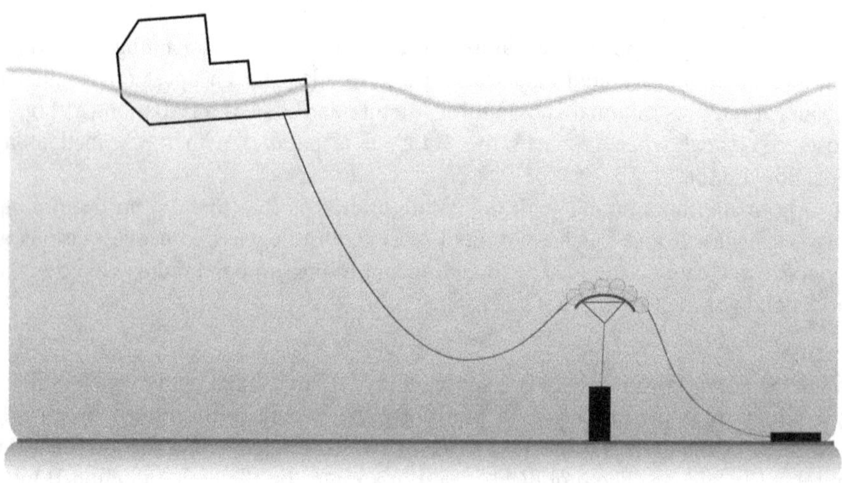

Figure 3.33 Lazy S configuration

Lazy wave
In the wave type, buoyancy and weight are added along a longer length of the riser to decouple the motion of the floater from the touchdown point of the umbilical.

A lazy wave formation is similar to a catenary but has support provided at about mid-depth by distributed buoyancy modules. 'Lazy' means that the umbilical

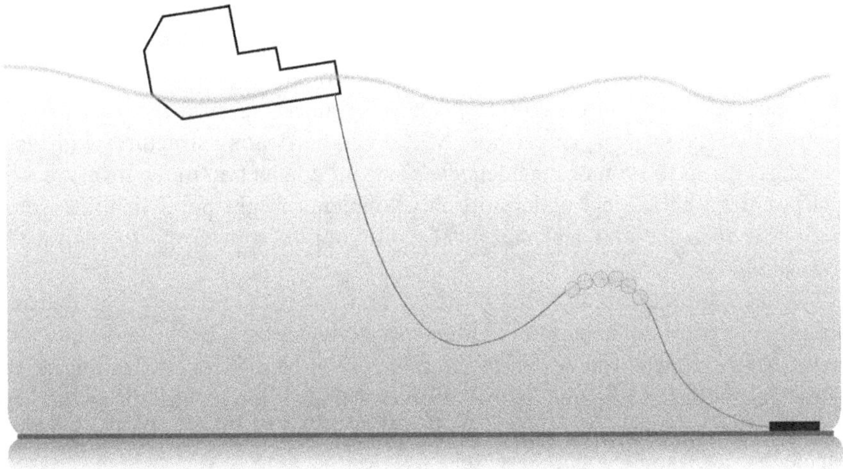

Figure 3.34 Lazy wave configuration

centreline is tangent to the seabed on its end contact while 'wave' describes the line shape due to the buoyancy modules. The necessity for buoyancy modules represents an increase in capital expenditure as well as in installation costs, with each module having to be individually attached to the umbilical during installation (Figure 3.34).

Others

Steep wave: The steep wave solution is very similar to the lazy wave, with buoyancy modules again being used to decouple floater motions and touchdown motions. The umbilical ties directly into the subsea equipment, without touching the seabed first. The steep wave configuration is suitable for congested seabed developments.

Pliant wave: The pliant wave configuration was developed as a hybrid between the lazy wave and steep wave. It retains the advantages of the lazy wave and dynamic behaviour of the steep wave. It can be used in very harsh environments, but the tethered connection to a clump weight can make installation quite complex.

Steep S: The steep S is similar to the lazy S configuration except for the umbilical touchdown arrangement. In the steep S configuration, the umbilical ties directly into the subsea equipment, without touching the seabed.

3.3.8 Recommendations and final remarks

The need for electrical connection of wave energy devices imposes the design and analysis of a specific connecting infrastructure.

The installed power, the distance to shore and the grid connection point drive the choice of the transmission voltage and of the type of configuration. It is likely, e.g., that large-scale arrays will apply two levels of connection: one from the device to an intermediate hub or junction box and the one from the same hub to shore.

The chosen connection configuration will determine the requirements for each component. One of the key issues to address is the selection of the adequate connection point to the device and to the grid infrastructure. Even if some guidelines can be given on how to place the connection on the device (as it is assumed that the majority of offshore MECs will be floating structures with slack moorings) the definition of the adequate point for connection to the infrastructure is still a debated subject and future developments might point to other solutions rather than the classical umbilical configuration going from the surface to the seabed.

As the requirements in terms of power and voltage are defined, the dynamic cable for electrical connection of marine energy devices can be designed based on the procedure shown. The design of the cable might be different depending on the insulation applied. However, global geometrical and mechanical properties are very similar for the two materials considered (EPR and XLPE) except for a smaller outside diameter applicable in case of a wet design.

Finally, the mechanical response of the cable can be improved by applying additional armour layers and assuming larger loads for design input.

Connectors for marine energy technology are still a subject of research but reference can be made to existing commercial products for other sectors. Indeed, some manufacturers are currently looking at the marine energy sector and defining novel concept for the connection.

3.4 References

[1] B. Teillant, R. Costello, J. Weber and J. Ringwood, 'Productivity and economic assessment of wave energy projects through operational simulations', *Renewable Energy,* vol. 48, pp. 220–230, 2012.

[2] P. Ricci, J. López, I. Touzon, O. Duperray and J. L. Villate, 'A methodology for the global evaluation of wave energy array performance', in *Proceedings of the 4th International Conference on Ocean Energy*, Dublin, Ireland, 2012.

[3] M. Folley, T. Whittaker and J. Van't Hoff, 'The design of small seabed-mounted bottom-hinged wave energy converters', in *Proceedings of the 7th European Wave and Tidal Energy Conference*, Porto, Portugal, 2007.

[4] P. Ricci, J. López, J. Plaza, M. Scuotto, J. L. Villate, A. Bahaj, L. Myers, J.-F.... C. Retzler, 'EquiMar Deliverable D5.1: Guidance Protocols on choosing of electrical connection configurations', 2009. [Online]. Available: https://www.wiki.ed.ac.uk/display/EquiMarwiki/EquiMar.

[5] The European Marine Energy Centre, 2013. [Online]. Available at: www.emec.org.uk. Accessed March 2013.

[6] Wave Hub, 2013. [Online]. Available at: www.wavehub.co.uk. Accessed March 2013.

[7] H. Mouslim, A. Babarit, A. Clément and B. Borgarino, 'Development of the French Wave Energy Test Site SEMREV', in *Proceedings of the 8th European Wave and Tidal Energy Conference*, Uppsala, Sweden, 2009.

[8] Sustainable Energy Authority Ireland (SEAI), *Bellmullet wave energy test site*, 2013. [Online]. Available at: http://www.seai.ie/Renewables/Ocean_Energy/Belmullet_Wave_Energy_Test_Site/. Accessed March 2013.

[9] Ente Vasco de la Energía (EVE), *Biscay marine energy platform*, 2013. [Online]. Available at: http://eve.es/web/Energias-Renovables/Energia-marina/bimep.aspx?lang=en-GB. Accessed March 2013.

[10] L. Myers and A. Bahaj, 'An experimental investigation simulating flow effects in first generation marine current energy converter arrays', *Renewable Energy,* vol. 37, no. 1, pp. 28–36, 2011.

[11] K. Budal and J. Falnes, 'Interacting point absorbers with controlled motion', in *Power from Sea Waves*, Academic Press, London, 1980, pp. 381–399.

[12] S. Mavrakos and P. McIver, 'Comparison of methods for computing hydrodynamic characteristics of arrays of wave power devices', *Applied Ocean Research,* vol. 19, pp. 283–291, 1997.

[13] P. Ricci, J.-B. Saulnier and A. F. d. O. Falcão, 'Point-absorber arrays: A configuration study off the Portuguese West-Coast', in *Proceedings of the 7th European Wave and Tidal Energy Conference*, Porto, Portugal, 2007.

[14] B. Borgarino, A. Babarit and P. Ferrant, 'Impact of wave interactions effects on energy absorption in large arrays of wave energy converters', *Ocean Engineering,* vol. 41, pp. 79–88, 2012.

[15] P. Vicente, A. F. d. O. Falcão, L. M. C. Gato and P. A. P. Justino, 'Dynamics of arrays of floating point-absorber wave energy converters with inter-body and bottom slack-mooring connections', *Applied Ocean Research,* vol. 31, no. 4, pp. 267–281, 2009.

[16] D. Ben Haïm, *Progress report WP3 – Wave energy farm grid integration: Tools and methodologies*, Wavetrain2 Initial Training Network, 2010.

[17] T. Ackermann, N. Barberis Negra, J. Todorovic and L. Lazaridis, 'Evaluation of electrical transmission concepts for large offshore wind farms', in *Proceedings of the Copenhagen Offshore Wind Conference and Exhibition*, Copenhagen, 2005.

[18] M. Faria da Silva, *Offshore Wind Parks Electrical Connection*. Master's thesis, Instituto Superior Técnico, 2008.

[19] I. Martínez de Alegria, J. L. Martín, I. Kortabarria, J. Andreu and P. Ibañez, 'Transmission alternatives for offshore electrical power', *Renewable and Sustainable Energy Reviews,* vol. 13, pp. 1027–1038, 2009.

[20] M. Rahm, C. Böstrom, O. Svensson, M. Grabbe, F. Bülo and M. Leijon, 'Laboratory experimental verification of a marine substation', in *Proceedings of the 8th European Wave and Tidal Energy Conference*, Uppsala, Sweden, 2009.

[21] P. Kundur, in N. J. Balu, M. G. Lauby (eds.). *Power System Stability and Control*. Power System Planning and Operations Program, Electrical System division, Electric Power Research Institute, 3412 Hillview Avenue, Palo Alto, CA, McGraw Hill, 1994. ISBN 13: 9780070359581.

[22] ABB, 2013. [Online]. Available at: www.abb.com. Accessed March 2013.

[23] Prysmian, 2013. [Online]. Available at: www.prysmian.es. Accessed March 2013.

[24] International Electrotechnical Commission, 'IEC 60502-2: Power cables with extruded insulation and their accessories for rated voltages from 1 kV (Um=1.2 kV) up to 30 kW (Um=36 kV)', International Electrotechnical Commission, Geneva, Switzerland, 2005.

[25] G. Moore, Ed., Electric Cables Handbook, Blackwell Pub, 1997.

[26] J. Todorovic, *Losses Evaluation of HVAC Connection for Large Offshore Wind Farms*. Master's thesis, Department of Electrical Engineering, Stockholm, Royal Institute of Technology, Sweden, 2004.

[27] N. Barberis Negra, *Evaluation of Losses of HVDC Solutions for Large Offshore Wind Farms*, Master's thesis, Royal Institute of Technology, Stockholm, Sweden, 2005.

[28] J. López, P. Ricci, J. L. Villate, A. Bahaj, L. Myers, C. Retzler and J.-F. Dhédin, 'Preliminary economic assessment and analysis of grid connection schemes for ocean energy arrays', in *Proceedings of the 3rd International Conference on Ocean Energy*, Bilbao, Spain, 2010.

[29] P. Bresesti, W. Kling and R. Vailati, 'Transmission expansion issues for offshore wind farms integration in Europe', in *Transmission and Distribution Conference and Exposition*, Chicago, IL, 2008.

[30] P. Ricci, B. Couñago, A. Rico, P. Ibañez and Y. Torre-Enciso, 'Alternatives for the Design of Grid Connection Infrastructures for Wave Energy', in *Proceedings of the 3rd International Conference on Ocean Energy*, Bilbao, Spain, 2010.

[31] L. Martinelli, A. Lamberti, P. Ruol, P. Ricci, P. Kirrane, C. Fenton and L. Johanning, 'Power umbilical for ocean renewable energy systems – Feasibility and dynamics response analysis', in *Proceedings of the 3rd International Conference on Ocean Energy*, Bilbao, Spain, 2010.

[32] American Petroleum Institute, *API 17E: Specification for Subsea Umbilicals*, ISO 13628-5, American Petroleum Institute, Washington, DC, 2003.

[33] Institute of Electrical and Electronics Engineers, *IEEE Std. 519: Recommended Practices and Requirements for Harmonic Control in Electrical Power Systems,* Institute of Electrical and Electronics Engineers, 1992.

[34] Det Norkse Veritas, *Guidelines on Design and Operation of Wave Energy Converters*, Carbon Trust, London, England, 2005.

[35] International Electrotechnical Commission, *IEC 60092 Electrical Installations on Ships – Part 3: Cables (Construction, Testing and Installation)*, International Electrotechnical Commission, Geneva, Switzerland, 1984.

[36] International Electrotechnical Commission, *IEC 60287 Electric Cables – Calculation of the Current Rating*, International Electrotechnical Commission, Geneva, Switzerland, 2006.

[37] K. Thorburn and M. Leijon, 'Farm size comparison with analytical model of linear generator wave energy converters', *Ocean Engineering,* vol. 34, pp. 908–916, 2007.

[38] Pelamis Wave Power, 'The Pelamis brochure', 2013. [Online]. Available at: http://www.pelamiswave.com/pelamis-technology. Accessed March 2013.

[39] W. Grainger and N. Jenkins, "Offshore Wind Farm Electrical Connection Options" *Proceedings of the 1998 Twentieth BWEA Wind Energy Conference Wind Energy - Switch on to wind*, 1998.

[40] P. Christiansen, K. Jorgensen and A. Sorensen, *Grid Connection and Remote Control for the Horns Rev 150 MW Offshore Wind Farm in Denmark*, [Online]. Available at: www.hornsrev.dk/nyheder/nyh_aug_02/grid_control.pdf.

[41] D. Shackleton, L. Abib and R. Balena, 'Electrical and thermal design of umbilical cables', in *Proceedings of the 7th Jicable Conference*, Versailles, France, 2007.

[42] F. C. Cheng, 'Insulation thickness determination of polymeric power cables', *IEEE Transactions on Dielectrics and Electrical Insulation,* vol. 1, no. 4, pp. 1070–9878, 1994.

[43] R. Knapp, 'Torque balance design for helically armored cables', *ASME Journal of Engineering for Industry,* vol. 83, pp. 61–66, 1981.

[44] R. Knapp and M. Cruickshank, 'Design methodology for undersea umbilical cables', in *OCEANS: Ocean Technologies and Opportunities in the Pacific for the 90s, Proceedings*, Honolulu, HI, 1991.

[45] F. Hruska, 'Calculation of stresses in wire ropes', *Wire and Wire products,* vol. 26, no. 9, pp. 766, 767, 1951.

[46] G. Costello, *Theory of Wire Rope*, New York, NY: Springer, 1997.

[47] S. Ghoreishi, T. Messager, P. Cartraud and P. Davies, 'Validity and limitations of linear analytical models for steel wire strands under axial loading, using a 3D FE model', *International Journal of Mechanical Sciences,* vol. 49, pp. 1255–1261, 2007.

[48] International Umbilicals Manufacturers Federation (UMF), *Experience Report: Control Umbilicals Delivered by UMF Members in Period 1995–2000*, 2008. [Online]. Available at: www.ufd.as.

[49] C. Djumegard and P. Fellows, 'Installation of metallic tube umbilicals in 3000 meters water', in *Proceedings of the 2003 Offshore Technology Conference*, Houston, Texas, 2003.

[50] P. Ricci and J. López, 'Components for ocean renewable energy systems'. FP7 Grant Agreement No: 213633, CORES: Components for Ocean Renewable Energy Systems, 2009.

[51] M. Molinas, O. Skjervheim, B. Sørby, P. Andreasen, S. Lundberg and T. Undeland, 'Power smoothing by aggregation of wave energy converters for minimizing electrical energy storage requirements', in *Proceedings of the 7th European Wave and Tidal Energy Conference*, Porto, Portugal, 2007.

[52] J. Slootweg, S. de Haan, H. Polinder and W. Kling, 'Voltage control methods with grid connected wind turbines: a tutorial review', *Wind Engineering,* vol. 25, no. 6, pp. 353–365, 2001.

[53] I. Martinez de Alegria, J. Andreu, J. Martín, P. Ibañez, J. Villate and H. Camblong, 'Connection requirements for wind farms: A survey on technical requirements and regulation', *Renewable and Sustainable Energy Reviews,* vol. 11, pp. 1858–1872, 2007.

[54] The European Wind Energy Association, *Large Scale Integration of Wind Energy in the European Supply: Analysis, Issues and Recommendations*, 2005. [Online]. Available: www.ewea.org.
[55] R. Dorf, Ed., *The Electrical Engineering Handbook*, CRC Press, 2010.
[56] N. Barberis Negra, J. Todorovic and T. Ackermann, 'Loss evaluation of HVAC and HVDC transmission solutions for large offshore wind farms', *Electric Power Systems Research,* vol. 76, pp. 916–927, 2006.
[57] K. Thorburn, H. Bernhoff and M. Leijon, 'Wave energy transmission system concepts for linear generator arrays', *Ocean Engineering,* vol. 31, no. 11–12, pp. 1339–1349, 2004.
[58] H. Brakelmann and H. Burges, 'Transmission technologies for collective offshore wind farm connections', in *Proceedings of the European Wind Energy Conference*, Milan, Italy, 2007.
[59] D. O'Sullivan and G. Dalton, 'Challenges in the grid connection of wave energy devices', in *Proceedings of the 8th European Wave and Tidal Energy Conference*, Uppsala, Sweden, 2009.
[60] J. Bash, Ed., *Handbook of Oceanographic Winch, Wire and Cable Technology*, National Science Foundation, 2001.

Chapter 4

Grid integration: part I – power system interactions of wave energy generators

4.1 Introduction

This chapter presents an overview of the main issues associated with a wave energy generation system from a power system's standpoint. Issues specifically related to the time profile of power exported from a wave energy power plant are considered, and the impact of this fluctuating power on the power system performance is addressed. Some of these issues are covered in greater depth in future chapters. The need for reactive power compensation equipment, particularly in far offshore farms, is considered and addressed. The off-grid type of operation is also described. Most of the principles are illustrated with simplified models of wave energy generators (WEGs) with sinusoidal type outputs.

4.2 Interaction of wave energy generation with the electrical grid

A. Garces, O. Fosso and *M. Molinas*

Waves are a regular source of energy for a medium-term frame and hence generated power can be accurately predicted and is effectively dispatchable [1]. Forecasting of the available power can be obtained by using the Bretschneider spectrum and other probabilistic techniques as well as artificial intelligence algorithms [2]. A stochastic economical dispatch is required, but wind and solar systems already require this type of approach [3]. The generated power depends on the location [4]. Therefore, economical studies are required according to the location (see Chapter 10).

On the other hand, the type of WEG has an important impact over the power system [5]. On a system level, every generator should comply with the standards related to power quality, reliability, security and stability. Regulations related to wave energy are expected to be similar to those for wind energy. However, wave energy has particular characteristics that should be considered in future studies. One of the main challenges for safe integration of the WEG to the grid is the low-frequency oscillating power [6]. This and other problems are studied in detail in this chapter, namely:

- *Characterization of generated power (for a buoy type WEG)*: The power generated by a buoy type WEG is presented. This model demonstrates the inception of the low-frequency oscillating power.
- *Power smoothing*: Energy storage as well as a strategic placement of the WEGs are used for smoothing the power oscillations. The effect of the power smoothing on the power system is studied by using a small test system.
- *Effect of oscillating power on protection equipment*: Low-frequency oscillations can affect the coordination of the protections on the power system. This issue is briefly discussed considering the conventional schemes of protection.
- *Effect of the oscillating power on the voltage profile and reactive power*: The WEG can be integrated to the grid by using a voltage source converter. This type of converter can generate reactive power. However, its capacity is limited by the maximum current in the semiconductors. Therefore, the reactive power can also be oscillatory, thus impacting the voltage profile in the entire system.
- Effect of oscillating power on the system frequency.
- *Offshore vs. near-shore HVAC transmission*: A comparison between offshore and near-shore WEG system is presented from the transmission technology standpoint.

4.2.1 Properties of the generated power

Wave energy is a developing technology, hence the best way to extract the energy from the ocean is still an open research question. On the other hand, any power system study requires information on the generation source in order to determine its impact on the system. Dynamics and quality of generated power depends on the type of wave energy device. In this case, the model presented in [7–9] will be used as representative of a typical example of a WEG system. The system consists of a buoy type WEG as shown in Figure 4.1. Some of the conclusions obtained in this study could be extrapolated to other WEG concepts.

The dynamics of the mechanical system are given by

$$M\frac{d^2y}{dt^2} + \beta\frac{dy}{dt} + \alpha y = F \tag{4.1}$$

where M is the equivalent mass including the floating mass and the accelerating water mass, β is the equivalent damping coefficient, α is a spring constant, y is the vertical floating position and F is the force applied by the generator.

On the other hand, the mechanical power is given by

$$P_m = \beta_h\left(\frac{dy}{dt}\right)^2 \tag{4.2}$$

where β_h is the hydrodynamic damping coefficient. The optimal operating condition occurs when the system is in resonance. In that case, the damping provided by the generator is equal to the equivalent damping coefficient:

$$M\frac{d^2y}{dt^2} + \alpha y = 0 \tag{4.3}$$

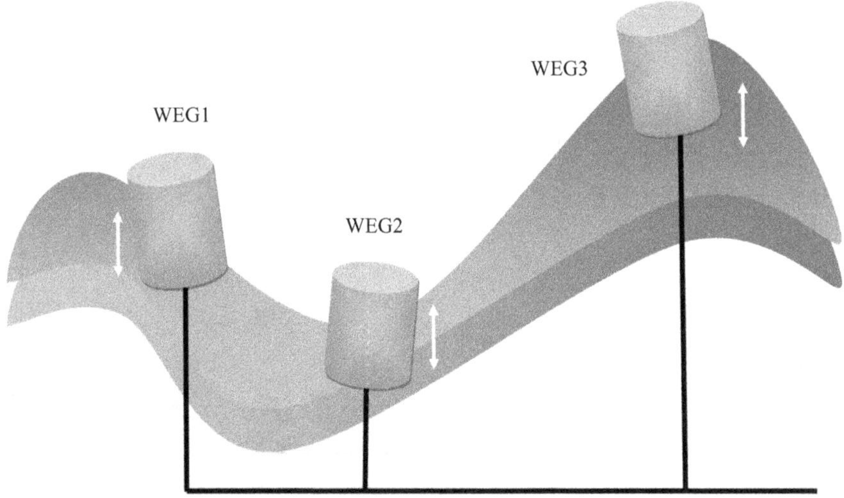

Figure 4.1 Buoy type wave energy generator

Therefore, the generated power for maximum power extraction is given by

$$P = P_{\max} Cos^2(\omega_x t) \tag{4.4}$$

where $P_{\max}$ is the peak power and ω_x is the wave frequency which in turn is given by

$$\omega_x = \sqrt{\frac{\alpha}{M}} \tag{4.5}$$

The equivalent generated power for maximum power extraction is shown in Figure 4.2. This result is valid for ideal conditions with perfect sinusoidal waves. In real applications, waves have a more stochastic behaviour. However, the two main challenges from the power system standpoint are present in this idealistic model: Generated power is oscillatory with a low fundamental frequency and the average power is a half of the peak power.

The peak power as well as the frequency are related to the location. Therefore, customized studies are required according to the wave conditions. Figure 4.3 depicts the generated power in a WEG considering a more realistic model of the waves [2]. Power generation is oscillatory but not periodic and the generated power is limited to 1 pu.

4.2.2 Offshore vs. near-shore and HVAC transmission

The advantages of offshore over near-shore locations for wind energy applications have been widely documented in terms of power capability. As the waves are directly related to the wind, it is expected that similarly advantageous conditions will be obtained for offshore wave energy. However, this is not entirely the case as

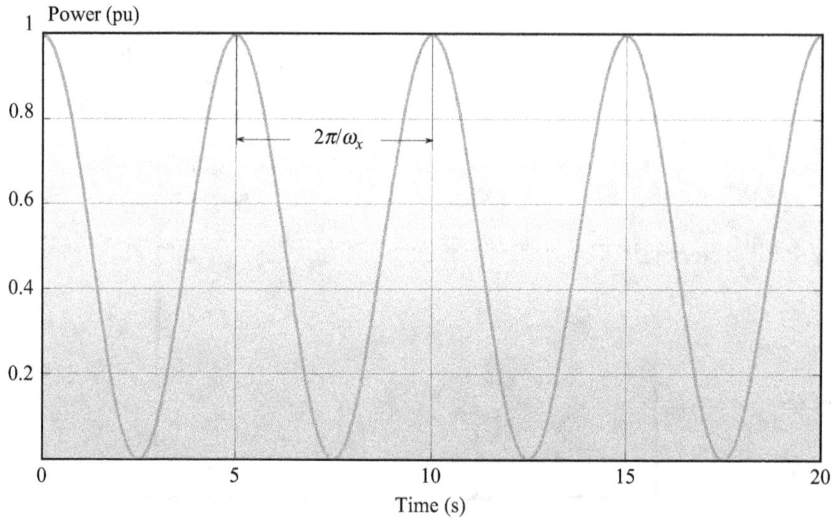

Figure 4.2 Power generated by the WEG in ideal conditions

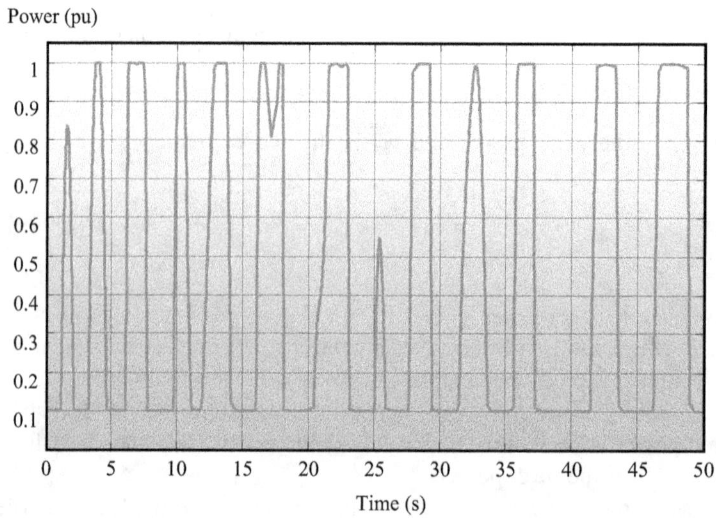

Figure 4.3 Power generated in a WEG

was demonstrated in [10]. The exploitable energy near-shore is almost the same as in offshore locations. Certainly, gross power is higher offshore but the improvement in terms of useful power is not convincing. As a direct consequence of this, HVAC transmission is still a good alternative for wave energy applications. Investment and operation cost are lower in HVAC transmission systems for distances below 40 km as shown in Figure 4.4. This breakpoint is accepted for submarine cables

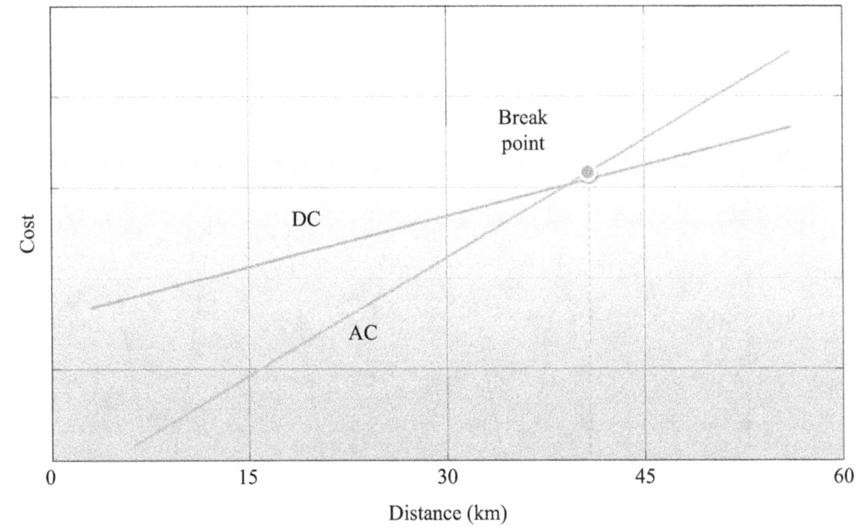

Figure 4.4 Comparison between AC and DC transmission cost

although in overhead lines it is higher. Most of the projects of wave energy are within this limit.

HVAC transmission implies that wave energy presents the same challenges as conventional generation systems in terms of stability, power quality and reactive power compensation. Power electronic equipment is required to control the generated power and to maximize the power extraction from the waves, as discussed in Chapter 2. This equipment as well as the inherent oscillating nature of the generated power influence the stability and power quality of the grid as will be described in next sections.

4.2.3 Power smoothing

The generated power of a single WEG is oscillating as shown in Figure 4.3. However, the total power of a wave park can be smoothed by a coordinated placement of the WEGs. Figure 4.1 shows schematically this concept [11]. When WEG2 is in the crest of the wave, WEG2 is in the valley and therefore the generated power will have a complementary waveform. This effect is similar to the generated active power in a balanced three-phase system where the instantaneous single phase power could be oscillating due to the inductive loads but the resulting three-phase power is constant.

Let us consider a wave farm with N equal WEGs, each of them with a peak generated power $P_{\max}$. For ideal sinusoidal waves, the total generated power in the farm is given by

$$P = \frac{P_{\max}}{2}\left(N + \sum_{k=1}^{N} Cos(2\omega_x t + \phi_k)\right) \tag{4.6}$$

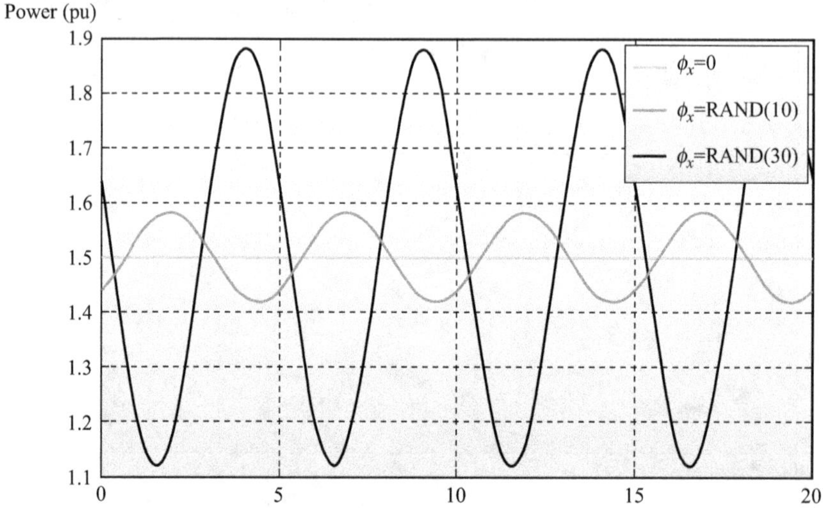

Figure 4.5 Total generated power for a 5 WEG wave park with different errors in the power smoothing strategy

where ϕ_k is the phase added by suitable placement of each WEG as given by

$$\phi_k = \frac{2\pi}{N}k + \phi_x \tag{4.7}$$

ϕ_x is a displacement error due to the random nature of the waves. For $N \geq 3$ and $\phi_x = 0$ the resulting power is constant. However, this condition is not always possible and some random displacements ϕ_x could appear in the park. In that situation, the oscillating power could be reduced but not eliminated. Figure 4.5 shows the total generated power for a wave park with five WEGs and different random displacement. These random displacements were modelled using a uniform distribution of probability.

It is clear that the power smoothing strategy improves the total generated power but oscillating power could not be completely eliminated. Better results can be obtained by correct placement of energy storage devices (ESDs). It is important to notice that the connection of the WEG and the ESDs impact also in the final performance of the power system. For example, only one ESD could be placed for the complete wave farm or conversely distributed ESDs can be placed along the internal grid. In the first case the power injected to the main grid is smooth but the power in each point of the internal grid could be oscillating. It must also be pointed out that such sinusoidal quantities are only approximations of the real sea conditions, and in reality it is significantly more difficult to provide power smoothing due to device placement.

4.2.4 Effect of the energy storage devices

ESDs such as flywheels, superconducting magnetic energy storage (SMES), super capacitors and batteries can be placed in some parts of the electrical system in order to minimize the power oscillations [12]. This is studied in detail in Chapter 7. Similar to the last case, power oscillations can be reduced but not eliminated. Let us consider a 0.20 pu WEG with an ESD. The maximum energy storage capability of the ESS is $E_{max} = 0.15$ pu. Notice in this example that the ESS is almost of the same rating as the WEG itself. This maximum energy is limited by the maximum current in the case of an SMES, the maximum voltage in the case of a super capacitor and the maximum speed in the case of a flywheel. The power injected by the ESS to the grid is calculated in order to compensate the oscillations in the power P_w generated by the WEG. A first-order low-pass filter is required in order to determine the average power.

Figure 4.6 depicts some simulation results. A more realistic model of the waves was used in this case. They are sinusoidal but not constant in magnitude. The energy storage capability limits the compensation of the power oscillation. Despite the ESS is almost the same rating as the WEG the power could not be completely smoothed and the average power is only a half of the rating of WEG. When the accumulated energy is equal to the maximum capability, the ESS is not able to absorb additional energy. In this situation all the oscillating power is injected to the grid. A dual situation occurs when the energy storage is zero and the generated power decreases. In addition, the power compensation is not perfect due to the limitations of the low-pass filter used to calculate the average power. Better filtering techniques are required in order to improve the steady-state performance. However, there are not ideal filters and hence there are not perfect power smoothing even with the most powerful ESS. Consequently, the effect of power oscillations on the power system should be considered in any wave energy project.

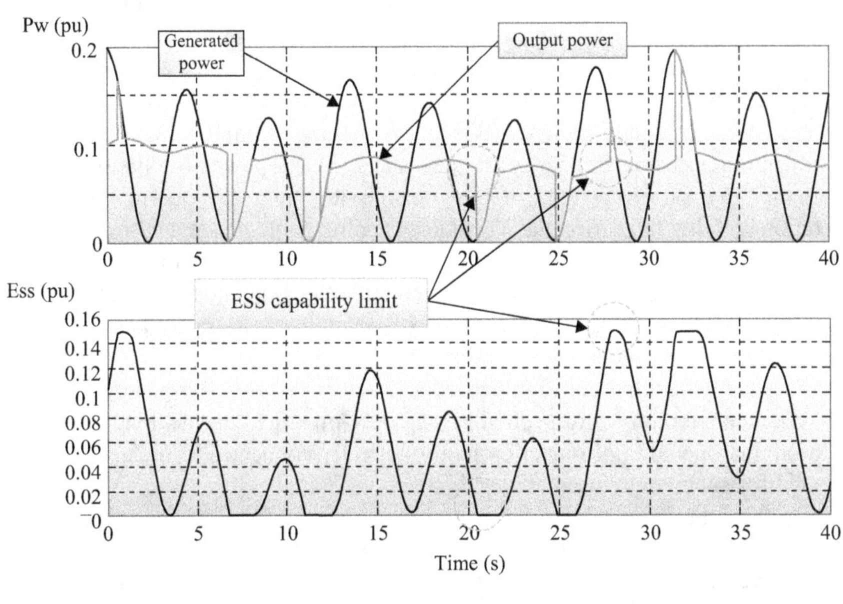

Figure 4.6 Power smoothing with energy storage (ESS)

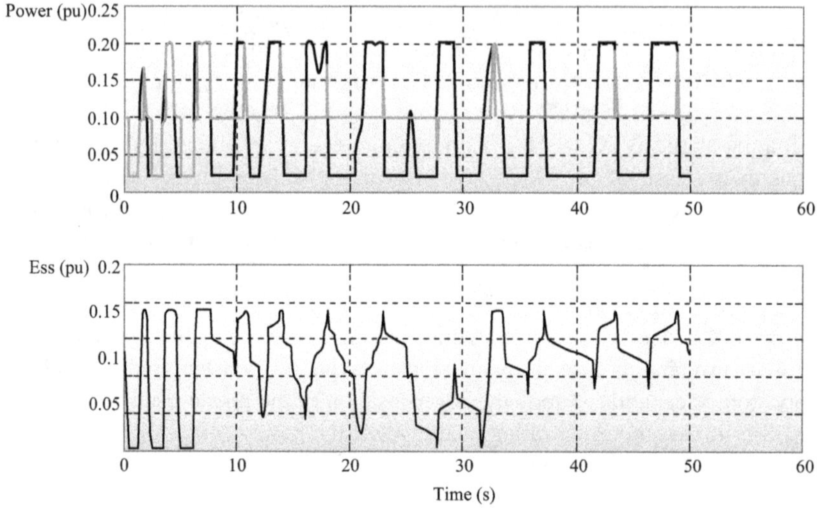

Figure 4.7 Power smoothing with energy storage and a realistic wave profile

Results for the same simulation using the wave profile presented in Figure 4.3 are depicted in Figure 4.7. As expected, the power is smooth but low frequency oscillations cannot be completely eliminated.

4.2.5 Effect of oscillating power on protection equipment

Conventional transmission lines are equipped with oscillating power protection (protection 78). This type of protection is based on the dynamic measurement of the equivalent impedance [13]. Power oscillations in a wave park could generate oscillations in the electrical frequency of the system, especially for large wave parks connected to a weak grid. Electromechanical transients in conventional power systems caused by short-circuits are in the range from 0.1 to 2.5 Hz which is a similar range to the frequency of the waves. These power oscillations can be interpreted by the protections of the transmission lines as abnormal frequency operation creating false tripping. This false tripping could create a cascade tripping effect in the other lines. On the other hand, not only transmission lines are equipped with protections against oscillating power but also generators and transformers connected to the main grid are equipped with volt per Hertz relays to prevent damage due to low frequency conditions.

Thermal generating units are particularly sensitive to low frequency variations that produce vibratory stress on the long low-pressure turbine blades [14]. In addition, the operation of induction motors directly connected to the grid could be affected for such power oscillations.

On the other hand, the power smoothing strategy could reduce the required capacity of the elements such as transmission lines and transformers and reduce the system level impact of power oscillation. Fast communications are required for safe and reliable operation.

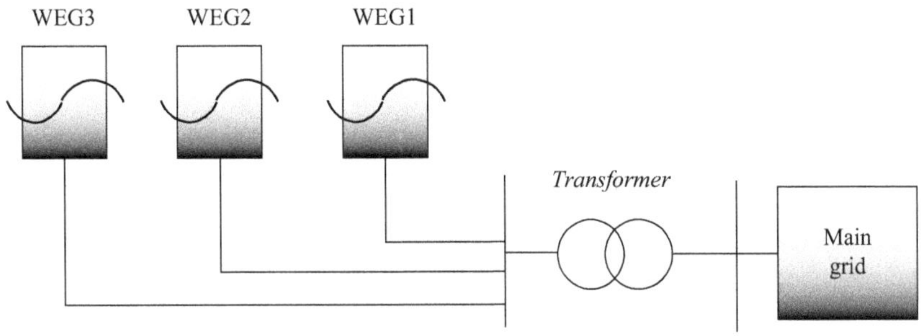

Figure 4.8 Simulated wave park with power smoothing

Let us consider the wave park shown in Figure 4.8. It consists of three WEGs connected to a common step-up transformer. For an ideal power smoothing strategy, the power delivered to the grid is given by

$$P_T = 3\frac{P_{\max}}{2} \tag{4.8}$$

where $P_{\max}$ is the maximum power in each WEG and PT is the total power. The transformer requires to be rated for a nominal power of $P_{\text{nom}} = P_T$. However, if the wave oscillations are not correctly synchronized, the power in the transformer could reach a power of $3\ P_{\max}$. Even the most simple overcurrent protection could interpret this as a short-circuit. Therefore, the transformer must be overrated and the protections dynamically coordinated. Contingency analysis is therefore required taking into account the most critical case ($3P_{\max}$). In addition, contingency analysis should consider island operation of WEG. The wave parks in operation do not consider the possibility of island operation. However, future power system with high integration of wave energy resources must consider this situation at least for a short-term frame after a severe disturbance.

State estimation could be required for wave energy applications in order to monitor the power and voltage oscillations. New technologies such as phasor measurement units (PMUs) will be important in weak grids with high penetration of wave energy. PMUs can be placed in each wave park to measure its power, magnitude of the voltage and angle. This information can be used by the other generators connected to the transmission system which can modify their controls in order to maintain the system within a desired stability margin. In addition centralized ESDs can use this information for a better smoothing strategy of the oscillating power.

4.2.6 Effect of the oscillating power on the voltages

The power oscillations in the WEG also affect the grid voltage. In order to demonstrate this phenomenon, a small wave park is simulated as shown in Figure 4.9. It consists of nine WEGs connected to a strong grid by a feeder with a constant width cable. Impedances of the cable are in per unit over a $P_{\text{BASE}} = 100$ MW. Each WEG generates a peak power of 2 MW and is placed in such a way that the total generated power is constant (perfect power smoothing). Ideal sinusoidal waves are assumed.

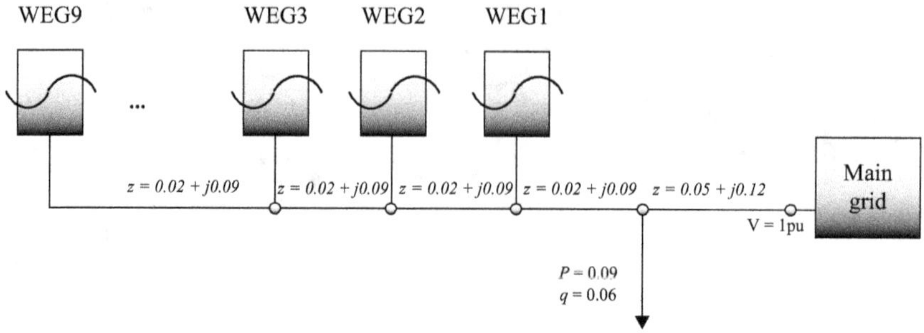

Figure 4.9 Simulated wave park

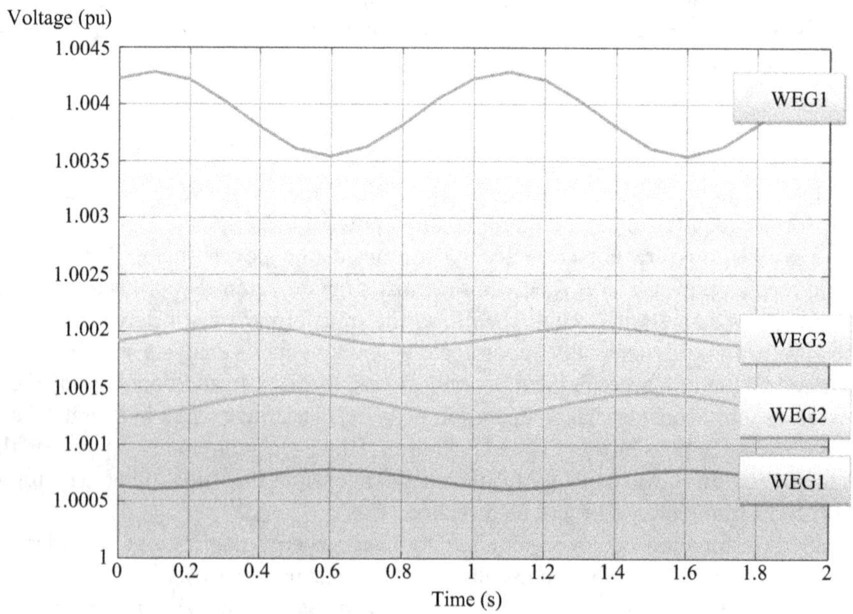

Figure 4.10 Voltages in the wave park

Figure 4.10 shows the voltages in different points of the feeder. Notice that the oscillating voltage increases according to the distance to the main grid.

The effect of oscillating power on the voltage was also tested in a mesh grid. In this case the nine bus Western System Coordinating Council (WSCC) test system was used (see Figure 4.11). Parameters of the impedances and generated power can be found in [15]. A 20 MW WEG was added in node 8 in order to determine its effect on the voltages of the system.

Results are shown in Figure 4.12. Notice that the voltage in other nodes is highly influenced by the oscillating power in node 8 even with a perfect control of the voltage in the other generators. The minimum voltage occurs when the power is in its

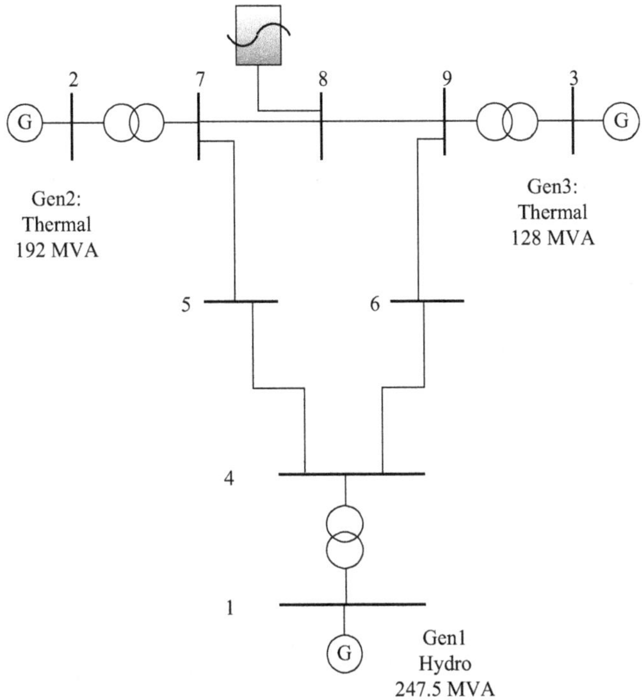

Figure 4.11 One-line diagram for the WSCC nine-bus system [15]

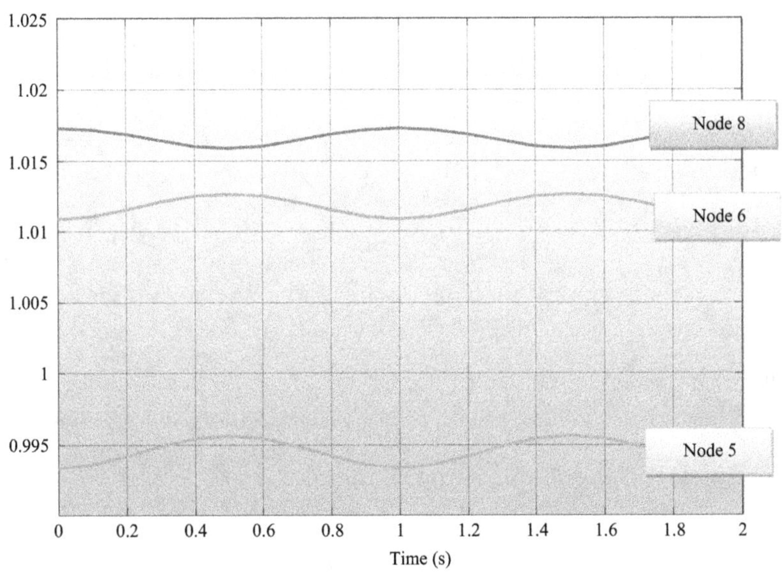

Figure 4.12 Voltages in a small power system with a 20 MW WEG

maximum. This is a disadvantage from the converter point of view. However, the magnitude of the oscillations is quite small since the maximum generated power of the WEG is just a small fraction of the total generation. Systems with high penetration of wave energy will have more variability in the voltage and therefore require customized studies. Sensibility analysis might also be required in order to determine the influence of the nodes where the WEG is connected with respect to the loads.

4.2.7 Control of the reactive power

As mentioned before, reactive power can be controlled by the voltage source converter in a back-to-back configuration. The capacity of the converter depends mainly on the current of the Insolated gate bipolar transistor (IGBTs) since the voltage in the DC link is constant and close to its nominal value. Therefore, the available reactive power can be approximated to

$$Q = Q_{\max} - P \tag{4.9}$$

In the power smoothing process it is possible to add some control for generating additional reactive power. When the active power is at its peak, the reactive power generation is zero, and when the active power is zero, the reactive power is at its maximum, as shown in Figure 4.13. By using this strategy it is possible, in some cases, to decrease the transmission losses.

This strategy was tested in the system shown in Figure 4.11. The total transmission losses for unity power factor and the proposed strategy are shown in Figure 4.14. Transmission losses are reduced since the main grid is relieved due to the internal generation of reactive power. An optimal power flow approach can be used in order to determine the peak reactive power generation in each WEG.

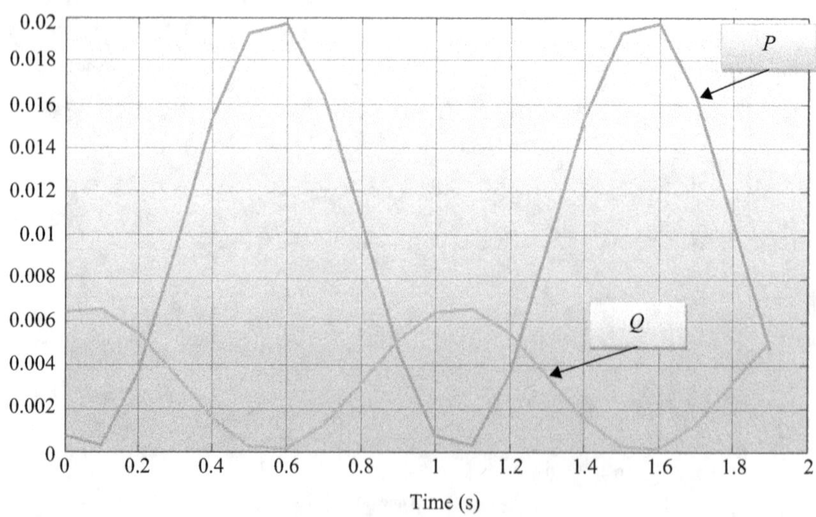

Figure 4.13 Strategy for generation of reactive power in a WEG

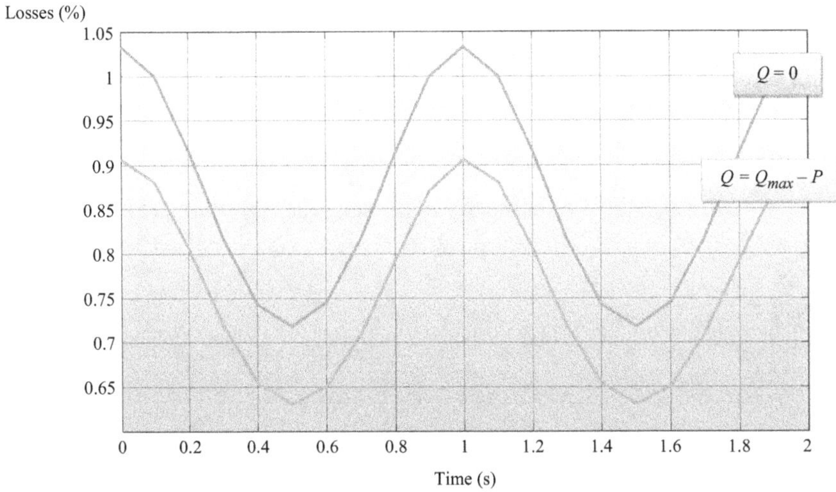

Figure 4.14 Losses for unity power factor and the proposed strategy for reactive power

4.2.8 Effect of the oscillating power on the frequency

The oscillating power generated by the WEG has some negative effects on the frequency of the system. In order to demonstrate this phenomenon, a transient stability simulation was carried out on the nine-bus test system depicted in Figure 4.11. Turbine and governor of the hydro and steam plant are modelled using a generalized representation as shown in Figure 4.15. All parameters of the system are given in [15].

Figure 4.16 shows the change in frequency for a sudden decrease of 10 MW in the load connected to node 8 without any WEG. The simulation considered the model of the turbine, damping and governor in all the generators. The classical quasi-stationary model was used for the simulation. A 5 MW WEG was added in node 8. The transient stability simulation was modified in order to consider the effect of the oscillating power of the WEG. It is assumed to be the low variation on the magnitude of the voltage. Therefore, the DC link in the back-to-back converter is maintained almost constant and generated power is not affected by the dynamics of the power system. This is a strong approximation but it is required due to the lack of a more detailed model of the WEG system for transient stability studies. Only changes on the load are simulated since other contingencies such as short-circuits affect the voltage and the aforementioned approximation would not be valid anymore.

A predictor corrector algorithm was used for the solution of the system of differential equations. The electrical power in the WEG is known and hence an algebraic loop appears in the algorithm. This loop was solved using a modified Gauss–Seidel algorithm. Initial conditions were calculated using a full Newton–Raphson algorithm.

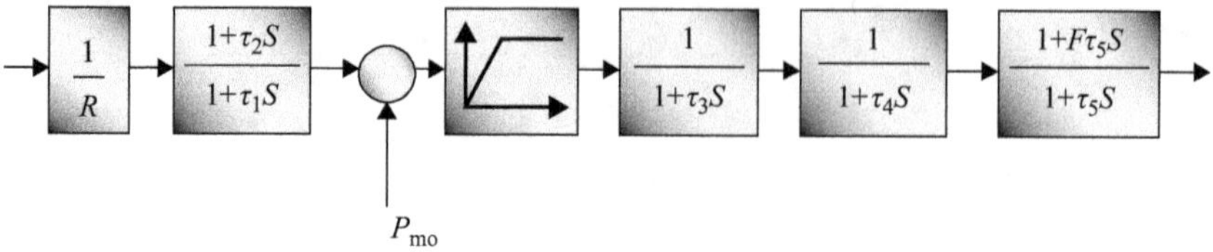

Figure 4.15 General model of the speed governor [15]

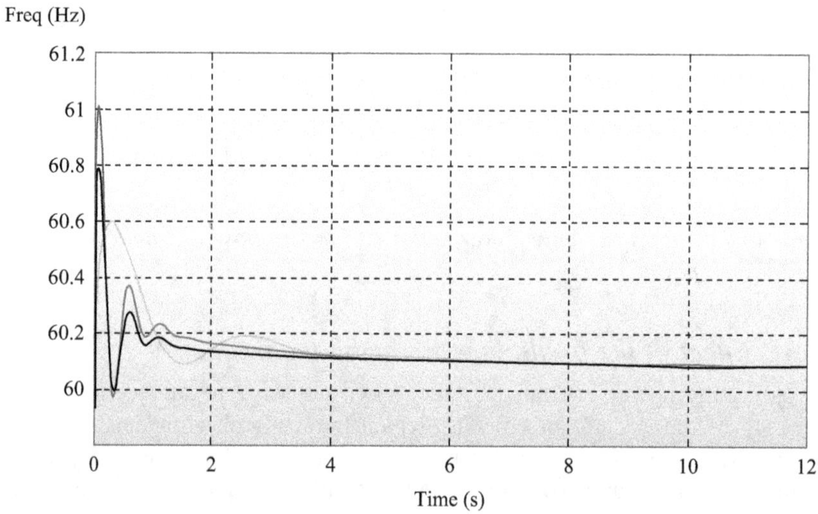

Figure 4.16 Frequency for a sudden decrease of 10 MW in node 8 without WEG

Figure 4.17 shows the change in the frequency for a sudden change of 10 MW in the node 8 considering the WEG. The first swing is higher than in the previous case due to the additional variation of the active power in the WEG. The stationary state frequency is oscillating although the oscillations are not high enough to affect significantly the stability of the system. Notice that in this particular case the oscillating frequency of the WEG is close to the natural frequency of the system. This could produce sub-synchronous resonance in the power system.

Figure 4.18 shows the change in the frequency for the power generation (from irregular waves) depicted in Figure 4.3. As expected, the random behaviour of the waves highly affects the resulting frequency.

Power system oscillations in conventional power systems are damped by the power system stabilizer (PSS). In most cases, the PSS operates mainly during transients. However, for wave energy applications, the PSS could be needed to operate in stationary state. The magnitude of the oscillations will increase as the wave energy penetration increases.

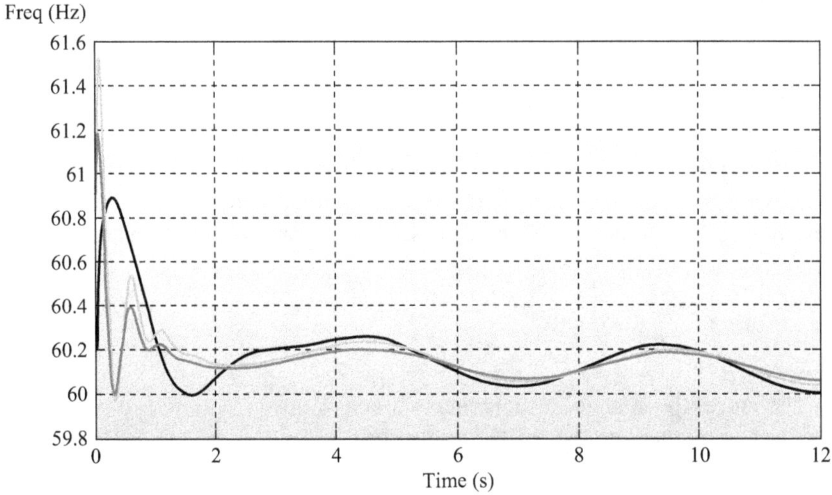

Figure 4.17 Frequency for a sudden decrease of 10 MW in node 8 and with a 5 MW WEG

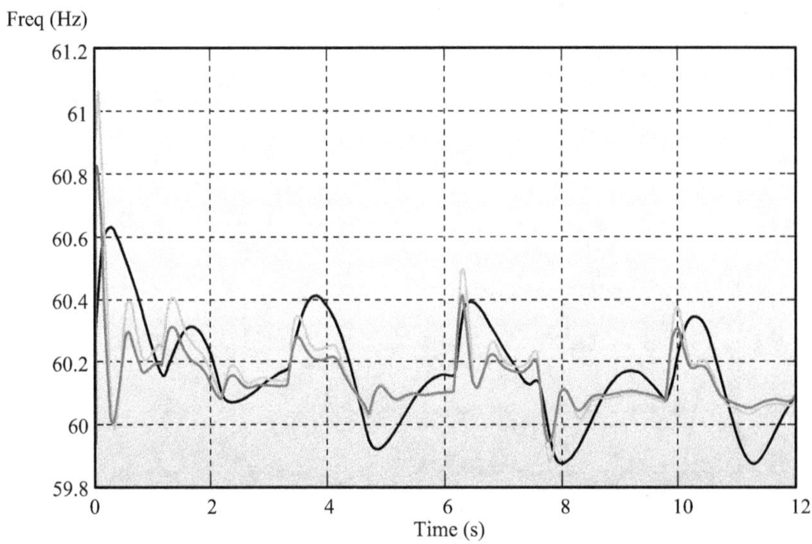

Figure 4.18 Frequency for a sudden decrease of 10 MW in node 8 and with a 5 MW WEG and a more realistic model of the waves

4.2.9 Control of the speed governors

Oscillations in the frequency could be reduced by a proper control of the generators connected to the grid. Let us consider the nine-bus test system depicted in Figure 4.10. By applying the decoupling principle to the Jacobian after the

Kron-reduction we can obtain an approximate relation between the nodal angles and the generated power as [15]:

$$\begin{bmatrix} P_G \\ P_W \end{bmatrix} = [B] \begin{bmatrix} \theta_G \\ \theta_W \end{bmatrix} \tag{4.10}$$

where G represents the generators (thermal and hydro) and W represents the WEGs. By deriving this equation and using the conventional approximations for classic transient stability studies, the changes in frequency can be approximated to:

$$\begin{bmatrix} \Delta w_G \\ \Delta w_W \end{bmatrix} = [H] \begin{bmatrix} \Delta P_G \\ \Delta P_W \end{bmatrix} \tag{4.11}$$

where $H = B^{-1}$.

Therefore, the change in the generated active power required to compensate the power oscillations in the WEG is given by

$$\Delta P_G = -[H_{GG}]^{-1}[H_{GW}]\Delta P_W \tag{4.12}$$

This power is added to the value of P_{mo} in the governor control (see Figure 4.15). Notice that the value of P_W is required in real time in order to determine the compensating power. In addition, H includes the loads connected to the system. This information can be obtained from the state estimator. Figure 4.19 shows the

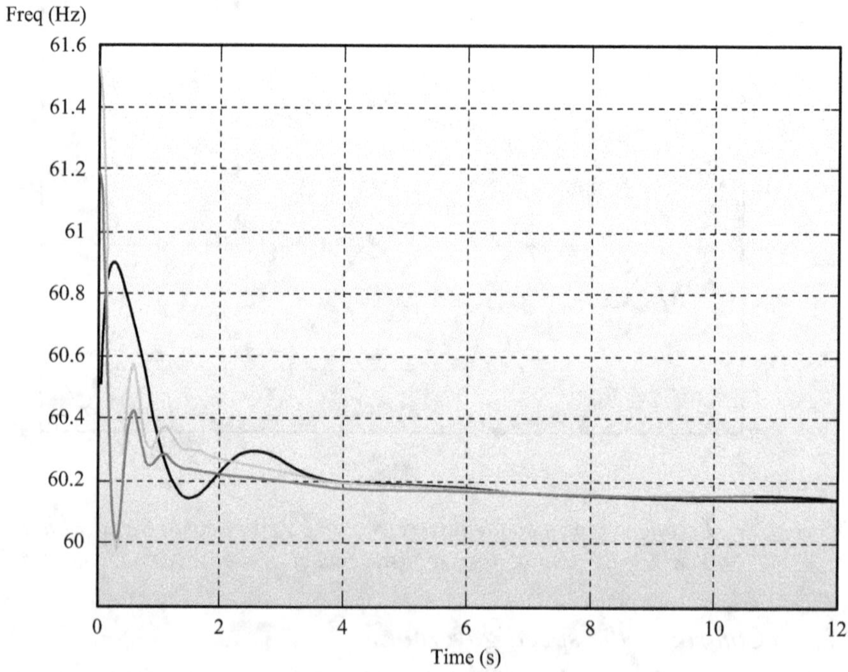

Figure 4.19 Frequency for a sudden decrease of 10 MW in node 8 and with a 5 MW WEG and a coordinated power smoothing strategy

simulation results for sudden decrease of 10 MW in node 8. The proposed control does not eliminate completely the frequency oscillations but it certainly reduces them. A challenge is type of control in the calculation of ΔP_W and the communication required for accurate control. This communication must be fast enough to monitor the changes in generated power in the WEG. On the other hand, the calculation of the spinning reserve must take into account the required power to compensate the power oscillations in the WEG. The PSS are also important elements to compensate these power oscillations in real applications. However, the PSS has a local action on the voltage and could increase the voltage oscillations.

4.3 Off-grid operation of ocean energy systems

P. Kracht and *J. Bard*

There are two scenarios in which it may be reasonable to operate wave and tidal energy converters off-grid, i.e. the device is not connected to a public grid. The first would be the electrification of remote areas and the second the testing of prototype devices. In both cases a so-called island grid is required, allowing the operation of the device. In the last decades island grids powered by renewables have become a quite common alternative for the electrification of remote areas. OECs have also been considered for this application [16] and some literature can be found on implemented projects. In [17] and [18] the integration of OECs in island grids are reported, in which no detailed information about the electrical system is available. In [19] some details of the electrical system for off-grid testing of a WEG are given and in [20] a similar system is described, designed for a planned rural electrification by means of a TEC. In what follows a possible design of an island grid – similar to the examples in [19, 20] – is described and the specific integration conditions of wave and tidal energy converters in the system are discussed.

4.3.1 Systems for electrification in remote areas

Traditionally, island grids are powered by conventional diesel generators, which besides supplying power provide additional crucial functions such as grid-forming and power balancing of the system. Throughout the last decades renewable energy sources have been integrated in such systems, forming so-called hybrid systems. A hybrid system can be classified according to the level of penetration of renewable energy sources. Here the focus is set on high penetration systems, which are defined by the fact that at least temporarily the whole load is supplied by renewable sources and the diesel generator is shut off [21]. For this scenario, additional equipment needs to be incorporated into the island grid, so that the grid-forming and power balancing functions can be provided during the diesel downtimes. One common approach is a configuration as depicted in Figure 4.20.

The core of the island grid is a so-called bidirectional battery inverter (BBI), which forms a three-phase grid and stabilizes it by controlling frequency and

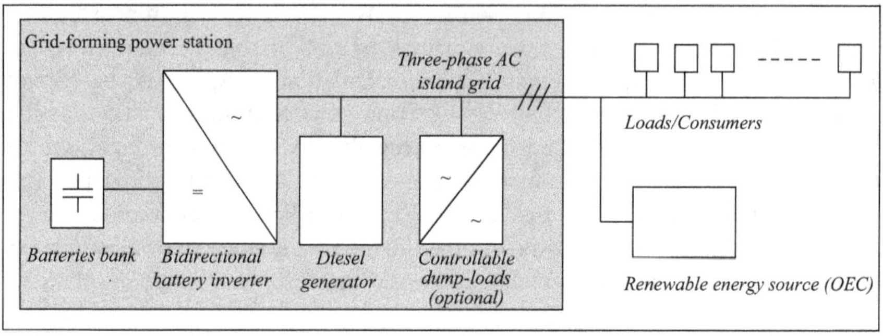

Figure 4.20 Illustrative circuit diagram of a three-phase island grid, based on a bidirectional battery inverter

voltage of the AC bus. The imbalance between power supply and demand at the AC bus is balanced by the BBI via charging and discharging of the batteries. Thus the power stability and quality of the grid are fully controlled by the BBI. From the perspective of the loads and sources, there is no difference compared to a public grid. Appropriate BBIs are available as off-the-shelf components from various manufactures, providing features like supervisory control for the power balancing, an advanced battery management, a high protection degree and a power range up to some hundreds of kilowatts. The diesel generator, usually also governed by the BBI, is used during periods of low availability of the sources or extremely high power demand. A special situation should be taken into account, i.e., when the power supply from the sources exceeds the instantaneous maximum feed-in power of the island grid. This situation can be handled either by additional controllable dump-loads or by reducing the power supplied by the sources. Also in this case the BBI controls the power balancing of the system, by transferring power reference values to the dump-loads and/or the sources respectively.

The system as depicted in Figure 4.20 should be regarded as a basic layout of a possible island grid powered by OECs. Further options to improve the system performance could for instance be the integration of a hydrogen system for long-term storage or a mix of different renewable energy sources. In general it can be said that the aim of the island grid – the delivery of stable power to the consumers at commercially viable costs – highly depends on the choice, design and rating of the components. The system should be designed based on information such as the load demand of the consumers (power profiles) and the projected availability of the resource. When considering the integration of an OEC in an island grid, some particularities of OECs have to be taken into account in the system design. The reliability and predictability of marine current gives the opportunity of designing a highly accurate TEC-powered system. By adjustment of the system components according to power production and demand, a system fully based on the renewable energy source can be designed, completely avoiding the need of a diesel generator. With WEGs the same predictability does not exist. Nonetheless the energy production of a WEG is likely to show a good correlation with the power demand, i.e., a high availability of the energy source during the winter season. One drawback of wave energy is that the converters are likely to show high fluctuations in the power delivered to the grid. This has to be considered when rating the BBI and the batteries (among other components) and might also affect the stability of the island grid. A further aspect not only touching the design of the island grid is the possible requirement of a power adaption of the TEC or WEG respectively. If no controllable dump-loads are integrated in the island grid, the control of the device needs to comprise both the appropriate control schemes for output power reduction and a communication path for a power reference value from the island grid. In [20] the layout of an island grid for rural electrification powered by a TEC can be found, including a description of both the control schemes and a technology used to transfer the reference value to the device by means of the frequency of the island grid.

4.3.2 Systems for off-grid testing of prototype devices

Due to the cost factors associated with offshore grid connections, prototype devices are often tested off-grid. Also in this case some kinds of island grids are required to operate the device. But the requirements on the grid differ significantly from the aforementioned situation. Instead of being designed for constant and stable power supply to the connected consumers, the island grid must be designed to accept any power delivered to the grid by the OEC. In principle a system as depicted in Figure 4.20 also qualifies for this application, but the requirements on the component ratings will differ significantly. First the dump-loads definitely need to be designed for the maximum output power of the device. Second the BBI rating can most likely be reduced, but it must still be capable of stabilising the grid and operating auxiliary devices, such as safety equipment, cooling and communication equipment. Advanced WEG control methods like reactive control (see Chapter 8 'Control Systems – Design and Implementation') depend on a section-wise reversed power flow, i.e. the WEG demands power from the grid. In this case the maximum power demand of the WEG must be considered as one requirement in the design of the BBI. Lastly the batteries and the back-up diesel generator can be downscaled according to the power demand of the auxiliary devices during downtimes of the device. A system appropriately designed according to these requirements allows the testing of the device, including generator system and control, as if it were connected to a public grid [19].

4.4 Conclusions

Wave energy is emerging as a promising technology for clean electric power generation in our future energy scenario. This chapter has however identified a number of technical challenges from the power system standpoint. One of the most critical ones is related to the oscillating nature of the power that could affect the stability and power quality of the entire system. Power smoothing could be achieved by strategic and coordinated placement of the WEGs by using ESDs and by new control algorithms in the conventional generators. None of these approaches eliminates completely the inconveniences of the oscillating power but they can reduce them ostensibly. Probably the final solution will include a mix of these and other approaches.

4.5 References

[1] A. Baha, 'Generating electricity from the oceans', *Renewable and Sustainable Energy Reviews*, vol. 11, no. 7, pp. 3399–3416, September 2011.

[2] C. McLisky, 'Wave-to-wire time domain model for the wave energy converter bolt', Master's thesis. Norwegian University of Science and Technology, 2012.

[3] J. Hetzer, D. C. Yu and K. Bhattarai, 'An economic dispatch model incorporating wind power', *IEEE Transactions on Energy Conversion*, vol. 23, no. 2, pp. 603–611, June 2008.

[4] R. Carballo and G. Iglesias, 'A methodology to determine the power performance of wave energy converters at a particular coastal location', *Energy Conversion and Management*, vol. 61, pp. 8–18, September 2012.

[5] E. Amon, T. Brekken and A. Jouanne, 'A power analysis and data acquisition system for ocean wave energy device testing', *Renewable Energy*, vol. 36, no. 7, pp. 1922–1930, July 2011.

[6] A. Muetze and J. G. Vining, "Ocean Wave Energy Conversion - A Survey," *Industry Applications Conference, 2006*, 41st IAS Annual Meeting. Conference Record of the 2006 IEEE , vol. 3, pp.1410–1417, October 2006.

[7] F. Wu, X. Zhang, P. Ju and M. Sterling, 'Optimal control for AWS-based wave energy conversion system', *IEEE Transactions on Power Systems*, vol. 24, no. 4, pp. 1747–1755, November 2009.

[8] H. Polinder, E. Damen and F. Gardner, 'Linear PM generator system for wave energy conversion in the AWS', *IEEE Transactions on Energy Conversion*, vol. 19, no. 3, September 2004.

[9] P. Sousa, 'Modelling and test results of the Archimedes wave swing', *Proceedings of the Institution of Mechanical Engineers, Part A: Journal of Power and Energy, Institution of mechanical engineers*, 220: 855–868, December 1, 2006.

[10] M. Folley, B. Elsasser and T. Whittaker, 'Analysis of the wave energy resource at the European Marine Energy Center', *Marine Structures and Breakwaters conference. Institution of Civil Engineers Coasts* Edimburg, Scotland, September 2009.

[11] M. Molinas, O. Skjervheim, B. Sørby, P. Andreasen, S. Lundberg and T. Undeland, 'Power smoothing by aggregation of wave energy converters for minimizing electrical energy storage requirements', *7th European Wave and Tidal Energy Conference EWTEC 2007*, September 2007.

[12] W. Du, H. F. Wang, L. Xiao and R. Dunn, 'The capability of energy storage systems to damp power system oscillations', *Proceedings of the Institution of Mechanical Engineers, Part A: Journal of Power and Energy*, vol. 223, no. A7, pp. 759–772.

[13] P. Anderson, *Power System Stability and Control*, Wiley Interscience, United States, 1999.

[14] P. Kundur, *Power System Stability and Control*, EPRI Power System Engineering Series, McGraw-Hill, United States, 1993.

[15] P. M. Anderson, *Power System Protection*, IEEE Press Series on Power Engineering, Publisher: Wiley—Interscience. United states, 1999.

[16] International Energy Agency, *'Comparative Study on Rural Electrification Policies in Emerging Economies – Keys to Successful Policies'*, International Energy Agency, Paris, March 2010.

[17] T. J. Goreau, S. Anderson, A. Gorlov and E. Kurth, 'Tidal energy and low head river power: A strategy to use new, proven technology to capture these vast, non-polluting resources for sustainable development', *Sustainable Energy Conference*, May 31, United Nations Commission on Sustainable Development, 2005.

[18] S. Wang, P. Yuan, D. Li and Y. Jiao, 'An overview of ocean renewable energy in China', *Renewable and Sustainable Energy Reviews*, vol. 15, no. 1, pp. 91–111, January 2011.

[19] F. Thiebaut, D. O'Sullivan, P. Kracht, S. Ceballos, J. Lopez, C. Boake, J. Bard, N. Brinquete, J. Varandas, L.M.C. Gato, R. Alcorn, A.W. Lewis, 'Testing of a floating OWC device with movable guide vane impulse turbine power take-off', *Proceedings of the 9th European Wave and Tidal Energy Conference*, CD-ROM, Paper no. 159, Southampton, UK, 2011.

[20] P. Kracht, J. Giebhardt, M. A. Lutz, M. Vecchio, A. Moroso, J. Bard, "Implementation of a vertical axis marine current turbine for off-grid village electrification in Indonesia," *Proceedings of the 4th International Conference on Ocean Energy*, Dublin, Ireland, 2012.

[21] I. Baring-Gould and M. Dabo, 'Technology, performance, and market report of wind-diesel applications for remote and island communities', *Proceedings of the European Wind Energy Conference*, Marseilles, March 2009.

Chapter 5

Grid integration: part II – power quality issues

A. Blavette and J. MacEnri

5.1 Power quality of waveform

5.1.1 Introduction

Grid codes define the requirements of both power plant owners/operators and the grid operator. The requirements imposed on power plant owners are mainly in place in order to assist in maintaining network stability. These requirements can generally be found in the form of grid codes, as in Ireland and in Great Britain for instance, or in the form of decrees, as in France and Spain. As they were initially designed having conventional thermal power plants in mind, additional and modified requirements had to be defined for renewable power plants, the majority of them being released for wind farms in the first place. These wind energy requirements were designed once a significant level of wind penetration was considered reached and have evolved with the increasing share of wind energy in the energy mix. The similarities between wind and ocean energy in terms of power intermittency/variability and non-dispatchability, as well as in terms of their type of grid connection through partially or fully rated power electronics, make it very likely that requirements similar to those used for the wind industry will be applied to the ocean farms.

Some of these requirements are related to the operation of the power plant itself, both under normal and fault conditions. For instance, they impose limits in terms of voltage, flicker level, power ramp rates etc., which represent either minimum requirements (e.g. minimum voltage level at the point of connection) or maximum limits not to be exceeded (e.g. flicker level, maximum voltage level). Any power generation unit must comply with these requirements. Validating the compliance of actual power plants with power quality requirements is a mandatory step in the process of grid connection. It is usually performed by means of numerical models representing the power plant performances and capabilities. Hence, the impact of both wave and tidal current farms on the power quality of their local network must prove to remain negligible. This impact is measured based on different criteria which are detailed in this chapter.

5.1.2 Voltage

5.1.2.1 Short-duration root-mean-square (rms) variations

There exist three different types of short-duration rms variations, namely voltage sag, swell and interruption, as described in IEEE Standard 1159-2009 [1]. These variations are strictly defined with respect to voltage amplitude and may be qualified as instantaneous, momentary or temporary depending on their duration, as shown in Table 5.1.

Sag

According to IEEE Standard 1159-2009, a voltage sag or dip (these words being interchangeable) is a decrease in rms voltage between 0.1 and 0.9 pu. Voltage sags are traditionally induced by faults, the starting of large motors or the switching of heavy loads. However, power fluctuations due to tidal flow turbulence and wave power fluctuations may also be responsible for momentary to temporary sags.

Swell

Voltage swells consist in a voltage rise above 1.1 pu for a duration ranging from 0.5 cycles to 1 min. In opposite fashion to voltage sags, they are caused by load decrease, either load switching off or load shedding, or by the switching on of a large capacitor bank. However, similarly to sags, they may also be caused by the reflection of tidal flow turbulence or wave power fluctuations on the electrical power output of an ocean farm.

Interruption

Voltage interruption designates a phenomenon for which the remaining voltage is comprised between 0% and 10% of its nominal value and is generally induced by faults, although it may also be triggered by equipment malfunction or loose connection. Voltage interruptions sustained over 1 min, classified as long duration rms variations, are addressed in Section 5.1.4 'Long-duration interruptions'.

It must be noted that grid strength at the connection point plays a major role in the induction of voltage variations from power fluctuations. A strong grid is less prone to be negatively affected by the injection of power fluctuations than a weaker

Table 5.1 Classification of short-duration root-mean-square (rms) voltage variations [1]

		Duration	Amplitude (pu)
Instantaneous	Sag	0.5–30 cycles	0.1–0.9
	Swell		1.1–1.8
Momentary	Interruption	0.5–30 cycles	< 0.1
	Sag	30 cycles—3 s	0.1–0.9
	Swell		1.1–1.4
Temporary	Interruption	> 3 s–1 min	< 0.1
	Sag		0.1–0.9
	Swell		1.1–1.2

grid. Although there is no general definition of grid strength, a grid relatively immune to significant power quality issues is generally characterized by a sufficiently high short-circuit level. The impedance angle Ψ_k. may also have a significant, though less important, influence on the power quality impact and it is expressed as:

$$\Psi_k = arctan(X_k/R_k) \tag{5.1}$$

where X_k and R_k are the Thévenin reactance and resistance of the rest of the network. In practice, rural, low-voltage distribution grids benefiting from a low level of interconnection, such as those supplying the west coast of Ireland or the north of Scotland, are considered as relatively weak compared to general grid strength in western Europe.

5.1.2.2 Flicker

Introduction

Applying a varying voltage to a light bulb can result in significant light intensity variations, thus causing a potential visual disturbance. Although it is normal that voltage may vary during the day, due to for example load switching or motor starting, the impact of these fluctuations must be limited in order not to represent any disturbance to the customers. A statistical index of visual disturbance was developed in order to evaluate the level of annoyance caused by light intensity variations on the average individual. This index, called flicker level, can be evaluated over 10 min (short-term) or 2 h (long-term), and is expressed as P_{st} and P_{lt} respectively.

To ensure that visual disturbance remains negligible to the customer, grid operators limit either (1) the individual contribution in terms of flicker which a power plant is allowed to emit or (2) the total flicker level at the connection point [2–5].

Flicker is a phenomenon which can have very significant consequences on customers' comfort and health. It has been recognized since the 1960s that flicker may potentially represent a risk to individuals prone to photosensitive epilepsy [6]. In addition, flicker can also induce electrical equipment malfunction, as well as non-negligible physical deterioration which may reduce significantly its lifetime. The performances of control systems using electronics drives have been also reported to be adversely affected by excessive and repeated voltage variations [7]. In addition, the influence of these variations on rotating machines' speed/torque control may for instance cause temperature rise and motor overloading issues. As waves and tidal flow turbulence can create rapid electrical power fluctuations, flicker is an issue of particular interest in the context of ocean energy grid integration.

Flicker level is usually evaluated based on the perception of light intensity variations as defined in the IEC Standard 61000-4-15 [8]. The lighting equipment considered in this standard is an incandescent light bulb. This may represent a worst case scenario regarding a number of lighting equipment types such as LEDs or compact fluorescent lamps, given their lower flicker response to low-frequency voltage modulation [9, 10]. However, there is no widely agreed guideline or standard on the flicker response of different types of lighting and electrical equipment as yet. Flicker generation by LED and fluorescent lamps due to interharmonics, non-integer multiples of the fundamental, and produced for instance by power electronics, has not been studied in significant detail in the industry at the time of this book's publication and is, as such, not very well modelled and understood.

IEC Standard 61000-4-15

The computation of flicker level from voltage time series has been strictly defined in the IEC Standard 61000-4-15. This standard describes the design of a flicker evaluation tool, called flickermeter. This tool, which can be either hardware or software, computes flicker severity levels from voltage time series, which may be generated either from field experiments or from numerical simulations.

A flickermeter consists of five functional blocks, as illustrated in Figure 5.1. Blocks 1 to 4 compute the instantaneous flicker level (i.e. flicker perceptibility) from voltage time series, whereas Block 5 computes the statistical index of flicker, referred to as flicker level or flicker severity.

In detail, scaling down the rms voltage amplitude to a per-unit value with respect to the time series mean rms value enables the use of the flickermeter for any voltage level. The luminous intensity produced by an incandescent light bulb for a given voltage amplitude is then obtained by squaring the input voltage in Block 2. Blocks 3 and 4 simulate the physical human perceptibility to light intensity variations by means of filters, whose parameters were defined as an outcome of experiments on groups of individuals. The experiments consisted of exposing a group of people to periodic light intensity variations of different amplitudes and frequencies. An instantaneous flicker level (or flicker perceptibility) equal to unity was considered to be reached when 50% of the group stated it could perceive the light intensity variations. Sensitivity curves, also called perceptibility curves, were produced as an outcome of these experiments. This perceptibility varies as a function of the frequency of a sinusoidal luminous signal, and is maximal for a frequency equal to 8.8 Hz. This is illustrated in Figure 5.2, which shows the normalised gain applied by the weighting filter included in Block 3 to a period signal of fixed amplitude as a function of its frequency.

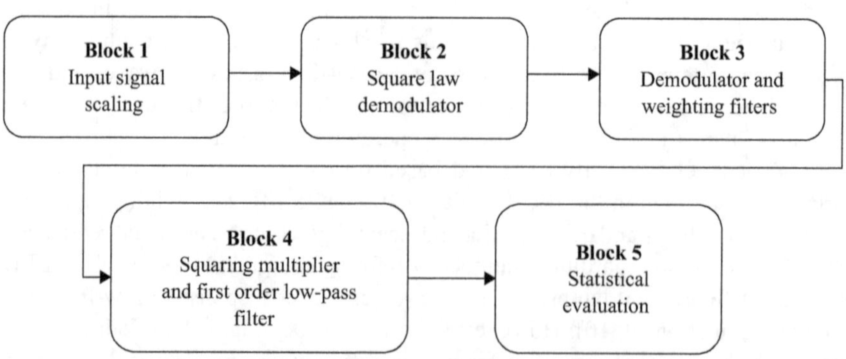

Figure 5.1 Functional representation of a flickermeter

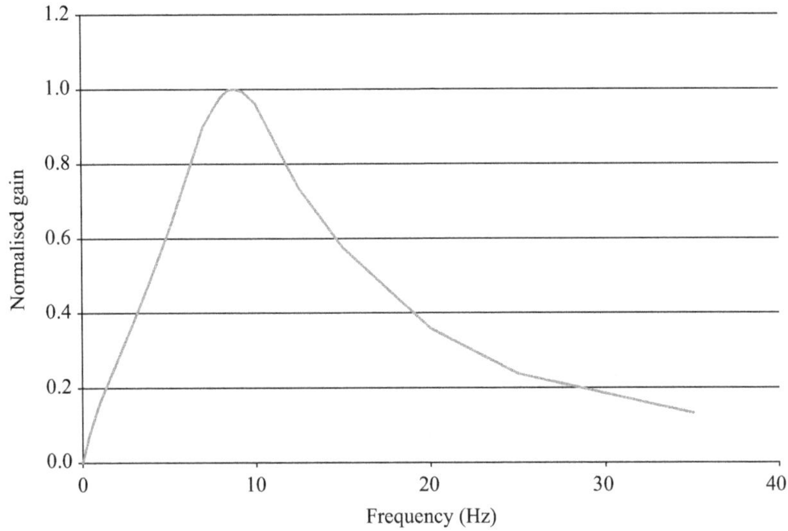

Figure 5.2 Normalised gain applied by the weighting filter included in Block 3

The statistical indices of flicker, which were mentioned earlier as the short-term flicker level P_{st} and the long-term flicker level P_{lt}, are then computed by Block 5 of Figure 5.1 from the instantaneous flicker level. These indices are a measure of the level of disturbance caused by flicker to an average customer. Maximum limits for the flicker emission of a power plant are defined in grid code requirements with respect to P_{st} and P_{lt}.

IEC Standard 61400-21

As mentioned in section 5.1.2.1, weak grids are more affected by flicker issues than stronger grids, as the flicker level increases in an inverse proportion to the short-circuit level of the point of connection. Hence, a flicker coefficient $c(\Psi_k)$ intended for wind turbine power quality evaluation is defined in the IEC Standard 61400-21 as

$$c(\psi_k) = Pst\frac{Ssc}{Sn} \tag{5.2}$$

where *Ssc* is the short-circuit level at the point of connection and *Sn* is the apparent rated power of a wind turbines. As $c(\Psi_k)$ is constant as a function of the short-circuit ratio Ssc/Sn, it constitutes a characteristic of a given wind turbine technology, regardless of the node in the network to which it is connected.

5.1.2.3 Harmonics

Ideally, the voltage waveform in a public electricity supply should contain the fundamental frequency only (50 or 60 Hz). However, due to non-linearity in the magnetic circuits of generators, a certain level of non-fundamental frequency current is introduced onto the system, particularly the third harmonic (150 or 180 Hz).

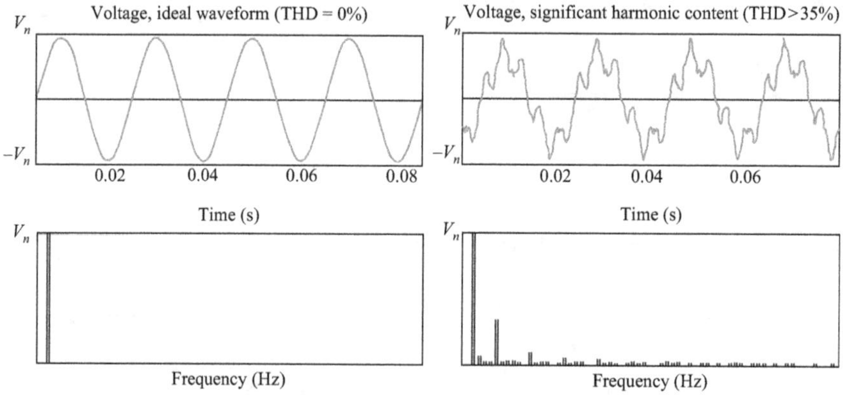

Figure 5.3 Effect of a significant level of total harmonic distortion (THD > 35%) on a 50 Hz fundamental

However in traditional hydro and thermal power plants, good generator design and the use of delta windings on the step-up transformer reduced these harmonics to very low levels. The problem with harmonics is that electricity users can also introduce them onto the supply voltage waveform through the use of non-linear loads. Initially, power systems harmonics were introduced by mercury arc rectifiers used to convert AC to DC current for railway electrification and for DC variable drives in industry. More recently the range of types and the number of units of equipment causing harmonics have risen sharply, and will continue to rise with the proliferation of switch mode power supplies, electronic fluorescent lighting ballasts, UPS units, variable speed drives and AC converters. These harmonics from the non-linear current supplied to their load will appear as voltage harmonics at other supply points on the same branch of the network. Figure 5.3 shows the effect on the ideal 50Hz public supply voltage waveform of 35% total harmonic distortion (THD) imposed on it and typically standards will not permit in excess of 10%THD. Increasingly, inter-harmonics (non-integer multiples of the fundamental frequency) are appearing on our systems introduced by some of these newer technologies listed previously.

The problems created by harmonics on power system themselves can be as follows:

- Neutral overloading (harmonics that are multiples of 3 add in the neutral)
- Transformer overheating
- Nuisance tripping
- Increased skin effect (effective reduction of conductor current carrying ability)
- Overloading of power factor correction capacitors.

In addition to these the following problems can also be caused at customer installations:

- Flickering computer screens
- Overloading and excessive bearing wear of induction motors
- Incorrect functioning of process control equipment
- Computer network lockups.

Utilities specify the harmonic emission levels of a load or supply in their connection agreement. The Standard EN50160 for the quality of public voltage supply specifies the maximum level of each harmonic in the public voltage supply and also a summary metric called voltage total harmonic distortion (V_{thd}) which is the ratio of the amplitude of the harmonics up to the 40th to the amplitude of the fundamental. It is calculated as

$$V_{thd} = \frac{\sqrt{\sum_{2}^{40} Vn^2}}{V_1} \tag{5.3}$$

where V_1 is the fundamental rms amplitude and V_n is the nth harmonic rms amplitude. Virtually all modern approved power quality analysers will report all voltage and current harmonics as well as total harmonic distortion and many include software utilities to analyse the compliance against various standards such as EN50160 or other national or international standards. Most analysers will report all harmonics and inter-harmonics up to at least the 50th. Utilities, for new large loads or generators, will require manufacturer certified harmonic emission levels and/or on-site measurements to ensure compliance. With good design practices all harmonic creating equipment can be made to reduce harmonic output to acceptable levels.

Wave and tidal energy devices need to be designed to withstand the level of harmonics specified under the relevant national standards or EN50160. In terms of their harmonic emissions onto the grid, the level permitted is normally set within the grid code and developers generally will be required to measure and demonstrate (often through ongoing continuous measurements) that the standard is being met.

5.1.2.4 Unbalance

In perfectly balanced conditions, voltage or current magnitude in each phase a, b and c is equal to 100% of the nominal value, each phase being shifted by 120 degrees, as shown in Figure 5.4. Each phase can be represented by its magnitude and angle in a phasor diagram.

However, unbalance occurs when at least one phase is no longer symmetrical to the others. Figure 5.5 shows a three-phase voltage unbalance in which amplitude of two phases is affected whereas the angle between each phase is equal. On the contrary, Figure 5.6 shows phase unbalance between the three phases having the same amplitude. In practice, both amplitude and phase unbalance can occur simultaneously on either one or two phases.

Phase unbalance level is defined as the ratio of the magnitude of the negative sequence component to the magnitude of the positive sequence component,

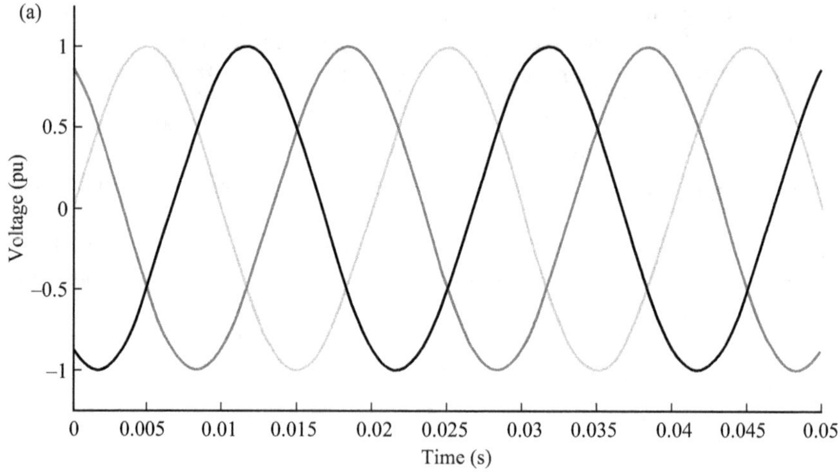

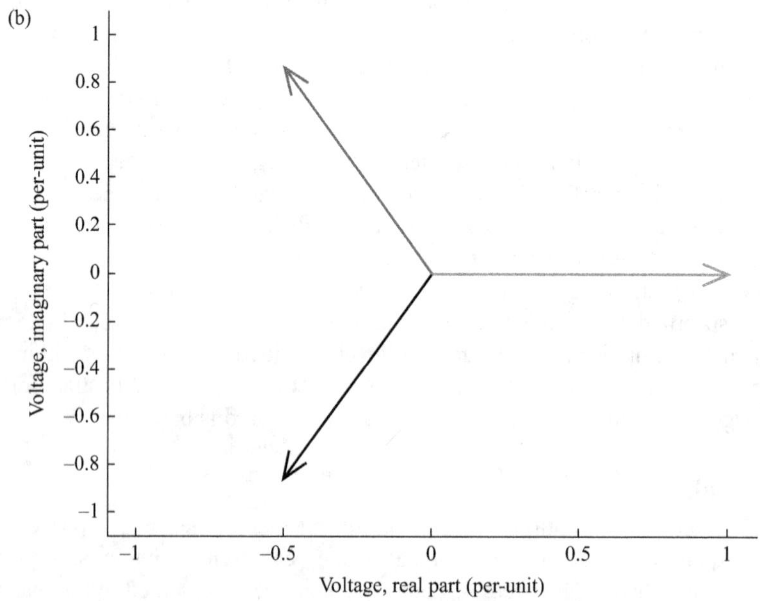

Figure 5.4 Three-phase voltage under balanced conditions: time series (top) and corresponding phasor diagram (bottom)

expressed as a percentage which is typically under 3% for voltage. This ratio can be expressed as

$$\text{Voltage unbalance} = \frac{|V_{neg}|}{|V_{pos}|}100 \tag{5.4}$$

Under normal conditions, unbalance is a phenomenon limited to low-voltage, single-phase networks to which ocean farms are not expected to be connected. However, three-phase networks are not immune to unbalance under asymmetrical

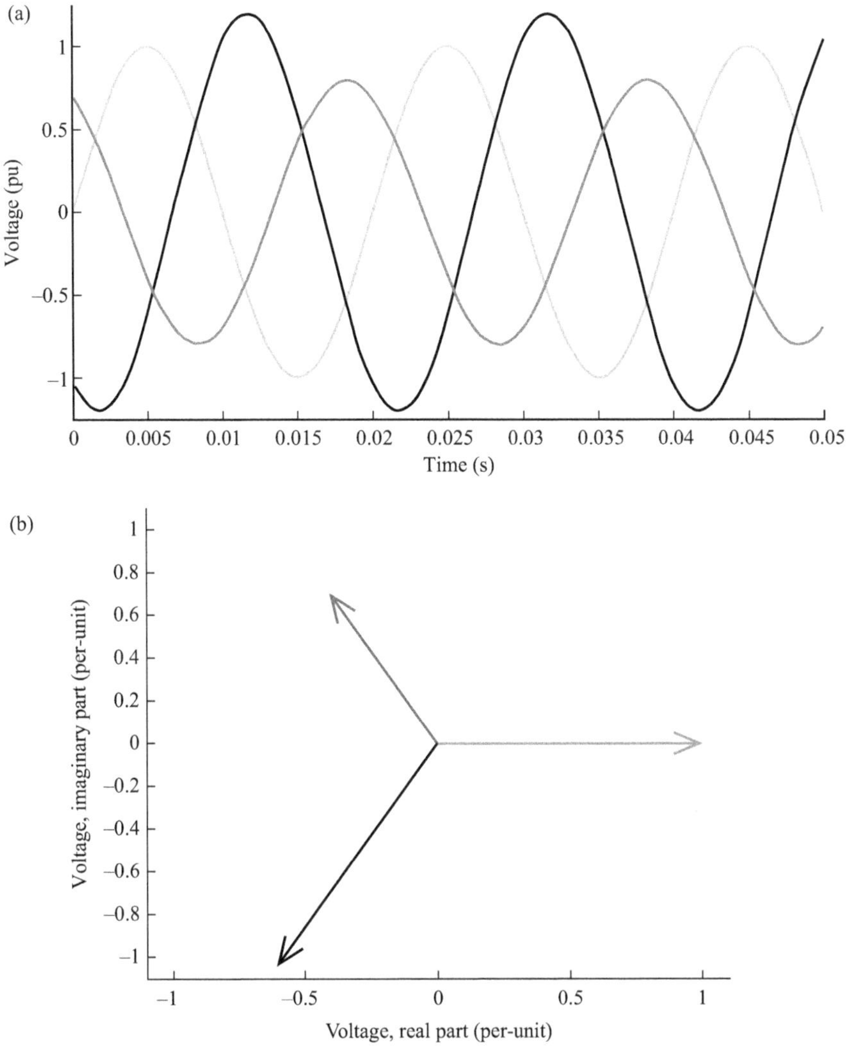

Figure 5.5 Amplitude unbalance between the three phases: time series (top) and corresponding phasor diagram (bottom)

fault conditions. Unbalanced voltage conditions may induce significant overheating and overcurrents, as well as mechanical stress on the device drive train. Torque pulsations are also an issue.

Most ocean energy converter designs, including fully rated power electronics, are relatively protected from the harmful effects of unbalance due to the high level of decoupling from the grid offered by this grid connection configuration. However, other designs not fully decoupled from the grid such as fixed-speed squirrel-cage generators (directly connected) or doubly fed induction generators (connected via partially rated power electronics) may be affected by this issue and forced to

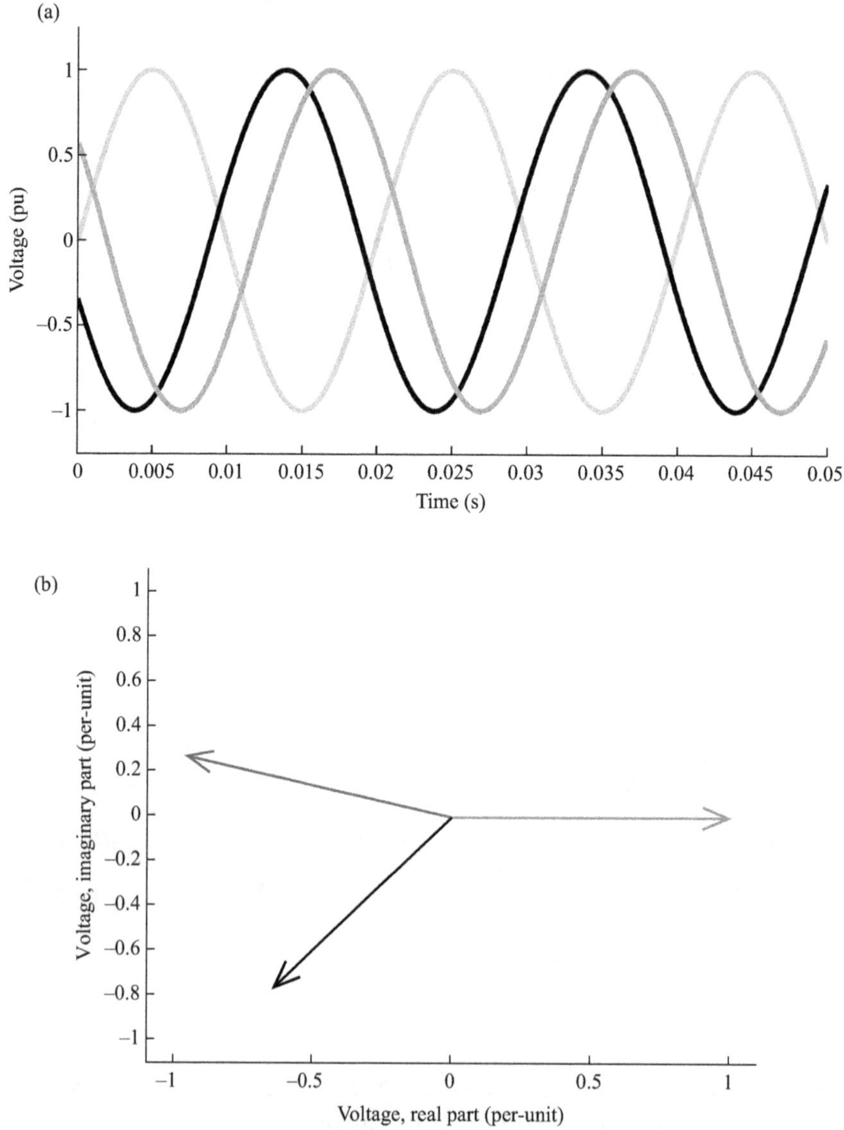

Figure 5.6 Phase unbalance between the three phases: time series (top) and corresponding phasor diagram (bottom)

disconnect in the event of a fault [11]. However, an appropriate control in the specific case of a doubly-fed induction generator (DFIG) [12] or the use of dynamic voltage restorers [13, 14] can facilitate these designs to enhance their fault ride-through and voltage support capabilities to comply with grid code requirements.

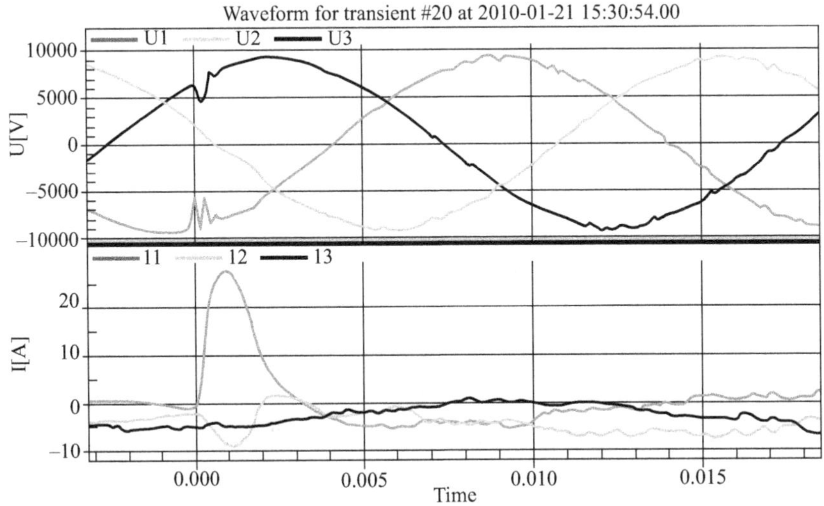

Figure 5.7 Example of a voltage transient recorded on the power system

5.1.2.5 Transient overvoltages

Transients are very short duration waveform events that are commonly called spikes. Their duration is typically from less than 1 μs up to a few milliseconds. They can be positive or negative transients and are only a power quality issue where their magnitude, duration and frequency of occurrence is significant and exceeds the insulation rating of the equipment they are applied to. Typically, a magnitude of greater than 50% of maximum instantaneous value of the fundamental is the cut-off point from where they are viewed as potentially a problem or not. The parameters of interest with regard to a transient are their peak absolute magnitude, their duration (determines their energy content) and the frequency of occurrence. The causes of transients are switching (capacitor switching being most problematic) and lightning. Switching overvoltages have longer duration but lower magnitude than lightning overvoltages. Electrical equipment is protected against such transients through specifying its basic insulation level (BIL) which is the peak voltage withstand capability of the equipment to such switching or lightning transients for defined lightning impulses (using the standard 1.2/50 μs lightning impulse). On power systems, typically, the standard BIL of equipment is around three times the nominal voltage of the equipment itself. Repeated significant transients below the BIL can cause equipment failure due to gradual insulation degradation through the mechanism of partial discharging and hence transients should be mitigated or removed through good design practices. Figure 5.7 shows an example of a transient caused by the switching in of a full converter and by the residual charging of the DC link capacitor to its working export voltage.

Table 5.2 Minimum time periods for which a power plant has to be able to operate without disconnecting [15]

<table>
<tr><th colspan="2" rowspan="2"></th><th colspan="5">Synchronous area</th></tr>
<tr><th>Continental Europe</th><th>Nordic</th><th>Great Britain</th><th>Ireland</th><th>Baltic</th></tr>
<tr><td rowspan="6">Frequency range (Hz)</td><td>47.0–47.5</td><td colspan="2"></td><td>20 s</td><td colspan="2"></td></tr>
<tr><td>47.5–48.5</td><td>Limited > 30min</td><td>30 min</td><td colspan="2">90 min</td><td>Limited > 30 min</td></tr>
<tr><td>48.5–49.0</td><td colspan="2">Limited > 30 min</td><td colspan="2">Limited > 90 min</td><td>Limited > 30 min</td></tr>
<tr><td>49.0–51.0</td><td colspan="5">Unlimited</td></tr>
<tr><td>51.0–51.5</td><td colspan="2">30 min</td><td colspan="2">90 min</td><td>Limited > 30 min</td></tr>
<tr><td>51.5–52</td><td colspan="2"></td><td>15 min</td><td colspan="2"></td></tr>
</table>

5.1.3 Frequency

Although frequency control is not a power quality issue as such, power plant disconnections triggered by sometimes sustained grid frequency drifts below or above the allowed limits may have a significant impact on the local network to which these power plants are connected. Grid codes define very strictly the frequency range and drift durations for which power plants are allowed to disconnect. Table 5.2 shows the numerical values with respect to these parameters for each of the five synchronous areas of Europe, as described in the European 'Network Code on Requirements for Grid Connection Applicable to all Generators' to come into force in 2014 [15].

5.1.4 Long-duration interruptions

Unplanned outages of generating equipment have power quality implications as the voltage sag resulting from their tripping will be experienced by other users on the system assuming they are in a grid-connected system and will cause a blackout where they are the sole generator in an isolated grid. As the unplanned tripping of a generator is effectively a very rapid ramp-down event, it places a large burden on other generators that can have deleterious effects on their condition and performance. They also make management of the system and its voltage profiles throughout very difficult for the system operator.

However, although renewable energy sources, particularly wind and wave, have variable power outputs, they are not expected to impose the same challenge on the system as a large generator that trips frequently due to reliability issues. However, this reliability of performance is another key challenge for marine renewables given the harsh, wet and corrosive environment that the generators are located in.

5.2 Power quality of supply

5.2.1 Earthing/Neutral treatment

The treatment of the neutral (isolated, direct earthing, resistive earthing or reactive earthing) will have some power quality impacts in terms of triplen harmonic current flows, overvoltages and equipment damage under fault conditions. For example, if the neutral is isolated under a single-phase fault the other two phases will rise to line voltage and equipment needs to be rated for that voltage. However, this neutral treatment method has advantages as supply will not be interrupted to customers in the event of a single phase to ground fault. In terms of earthing, it is preferable to have as low a resistance to remote earth as economically possible as this improves protection operation and reduces any touch or step potentials caused by ground loops.

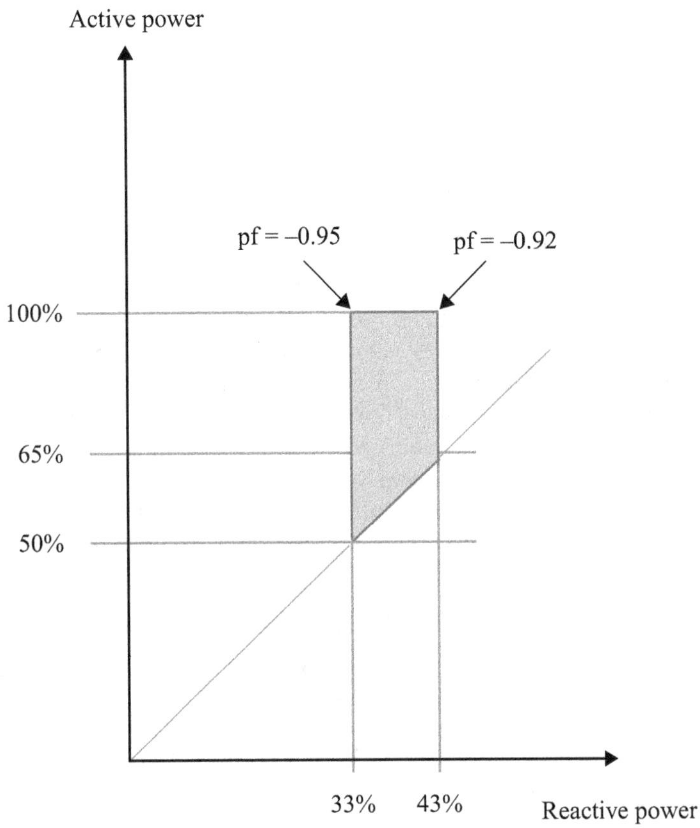

Figure 5.8 Required reactive power capability for wind turbines connected to an existing substation up to 38 kV [2]

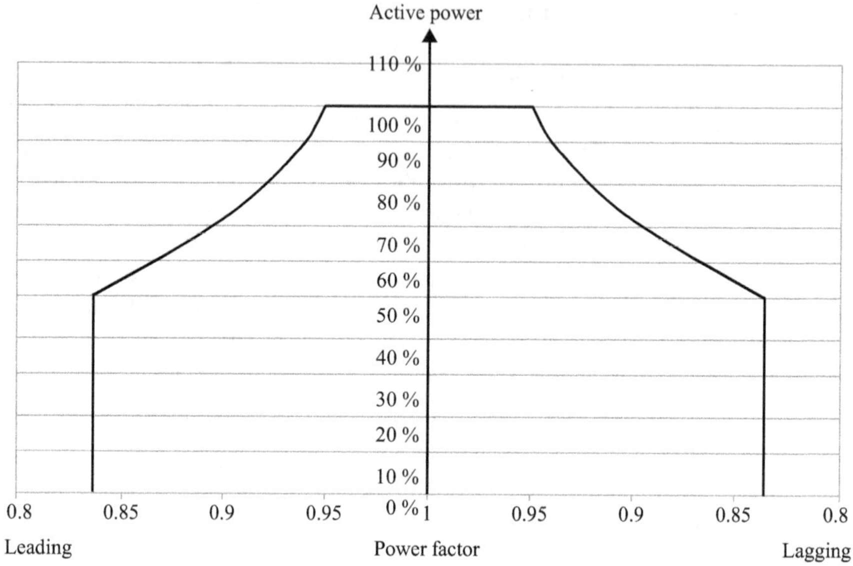

Figure 5.9 Required reactive power capability for wind farms connected at a 38 kV busbar with no existing load customers or at a 110 kV busbar [2]

5.2.2 Voltage control and support

5.2.2.1 Voltage control under normal conditions

Under normal operating conditions, power plants of any generation type are required to take part in local voltage control which is achieved by reactive power control. Power electronic converters, present in a majority of ocean energy device designs, help meeting this ancillary service requirement. Different methods are available for maintaining local voltage within allowed limits specified in grid codes: power factor control, reactive power control, (direct) voltage control or any combination of these. Figures 5.8 and 5.9 show the reactive power capability required from wind farms in the Republic of Ireland depending on their grid connection configuration [2]. Similar requirements are observed worldwide.

5.2.2.2 Voltage support under fault conditions

Power plants must participate in voltage control during fault conditions, also called *voltage support*. In similar fashion to voltage control, reactive power has to be injected in the network to maintain voltage as much as possible during the fault event. This requirement intends to make renewable power plants emulate the beneficial behaviour of conventional, directly connected synchronous generator in this type of situation. Voltage being the unknown variable in this case, the requirements are expressed with respect to reactive current.

The reactive current has to be injected shortly after the fault detection (typically no later than 20 ms [16] to 40 ms [15]) in variable proportions depending on the amplitude of voltage sag or swell. This is achieved by means of power electronic converters. Reactive current injection may also be required to be sustained for the sake of safety for a period of time whose duration is typically of a few hundreds of milliseconds after fault clearance.

5.2.3 Power output controllability

It is important in the performance of a generator that the actual active power output is close to the specified level, or at least does not exceed the maximum level during too long a period of time.

In what concerns tidal turbines, due to the internal dynamics of any generator, a change in input power cannot be immediately regulated in the generator and reflected exactly in the output particularly where the generator is operating at its maximum rated output. Consequently, the turbulence of tidal flow causes an increase in the input mechanical power which is reflected in the power output. This has two effects: first, the device is overloaded and it may cause excessive wear and failure (reliability issue) and, second, it can cause voltage rise issues on the network into which it is feeding. To quantify the output controllability of a tidal turbine generator an accepted method is to measure and record the maximum output active power over different timescales. The shorter the timescale the greater the difference between this value and the specified value will be. At longer timescales they should be nearly identical. Utilities will frequently specify the maximum value these

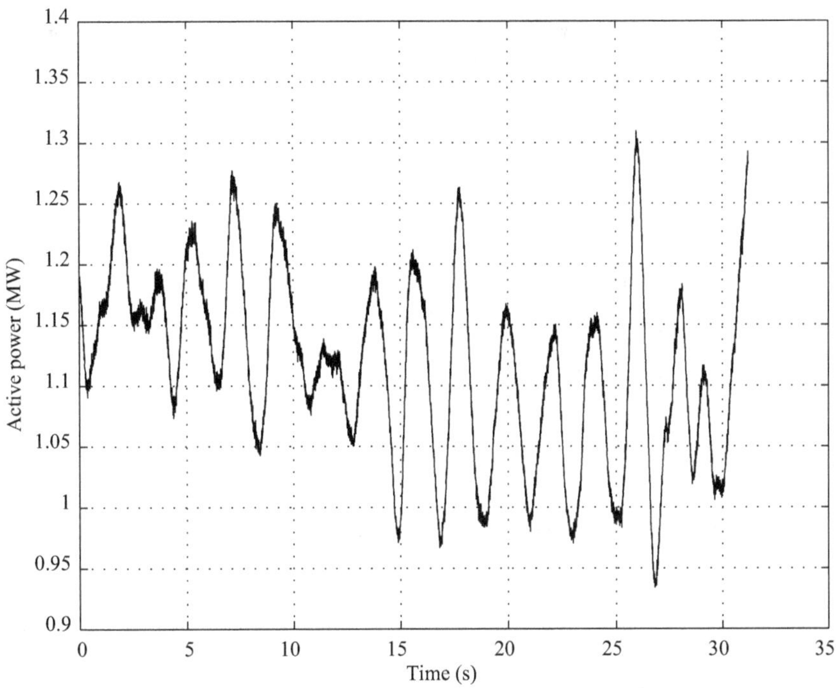

Figure 5.10 Active power output of the SeaGen tidal turbine at its rated power

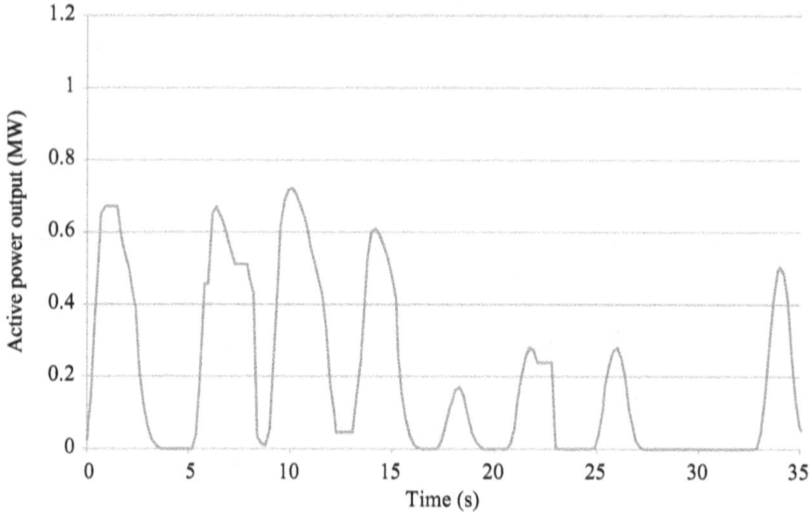

Figure 5.11 Electrical power output of the CORES OWC device

differences can be. The standard three normal timescales used for this purpose are 200 ms, 1 min and 10 min. The graph in Figure 5.10 shows the active power output of the SeaGen MCT turbine (which is described in the next chapter) at its rated output (equal to 1.2 MW). It demonstrates how shorter averaging periods will have values significantly above the rated output.

The issue of output power controllability poses itself in different terms for wave energy devices. Most wave devices present indeed a very fluctuating power output which is very different from typical wind or tidal turbine power profiles. This is illustrated in Figure 5.11 which presents the active power profile of the CORES oscillating water column prototype which is described in more detail in the next chapter. Control strategies, such as the 'peak-shaving' strategy specific to the SEAREV design may be interesting to limit the issues regarding the generation of power peaks of excessive amplitude [17]. However, this strategy does not solve entirely the issue of output power controllability. In conclusion, storage means of significant energy capacity should be envisaged for enabling a sufficient control of a wave device output power.

5.2.4 Frequency reserve response

The electricity generation in a power system must match its consumption in order for the supply frequency to remain within the allowed limits. This is very challenging as the load is subject to constant change which, fortunately, is quite predictable due to the aggregating effects of many independent users and to the development of sophisticated load forecasting tools. The tripping of a large generator may also have a significant impact on the system which, however, cannot be predicted. Where a generator trips out, the load it was providing must be made up by the other generators on the system in the immediate term (milliseconds up to tens of minutes). System operators managed this risk by having some generator's output below their maximum continuous value in order to take up the shortfall if called upon (so-called spinning reserve). In similar fashion, electricity consumption exceeding generation may lead to a significant frequency decrease, forcing synchronous generators to slow down. This automatic reaction provides in return additional energy from the inertial energy contained in the generator's rotation which helps reduce the energy discrepancy. This frequency reserve response is the ability of a generator to increase its active power output when it detects a drop in system frequency due to a mismatch between generation and load (e.g. unit tripping). Heretofore, as renewable plant (wind predominantly) was a small part of the system and due to the serious economic impact of forcing them to always run below the maximum capable output, it has not been a requirement for them to provide this service to the system operator. However, as renewables become a significantly greater part of the system supply (in Ireland currently often up to 50% of system demand [18]), it is not possible to continue as before not requiring renewables to provide some level of frequency reserve response. Some current requirements concerning frequency response prescribe wind farms to operate below full power at normal grid frequency (50 Hz), as shown in Figure 5.12, but it is unlikely that permanently curtailing their output will be economically feasible.

However, the provision of some form of energy storage (ideally additional functionality to the power conditioning equipment – STATCOM or SVC) which

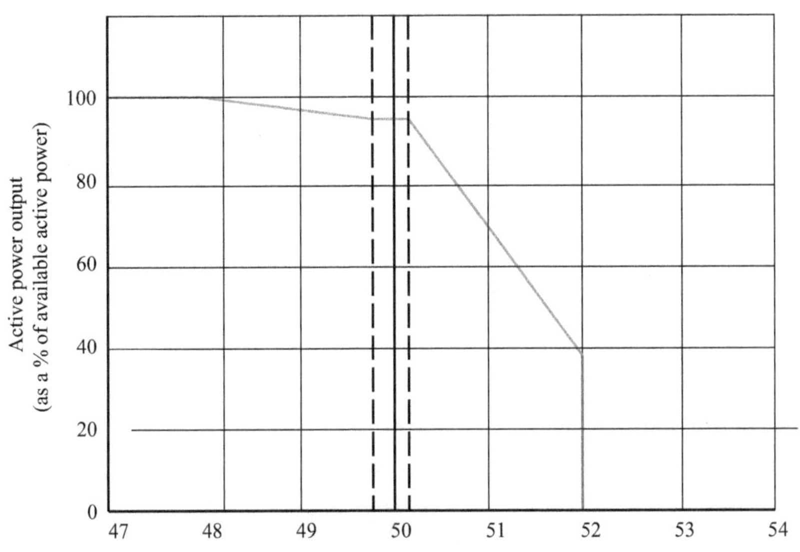

Figure 5.12 Frequency response required from wind farms in the Republic of Ireland [2]

can provide additional active power output at a level and duration specified by the utility is another possible method that may be economic. As tidal is a deterministic source, and its frequency reserve capability would always be known, there is the possibility that it could provide this functionality through a short-term overload capability built into its design. This may be more difficult for wave farms, unless they include controllable storage means of large power capacity, such as an onshore or offshore reservoir. However, unlike tidal turbines, their frequency reserve capability could not be easily predicted, leaving real-time information on that matter as the unique way by which the grid operator may be informed of their potential energy reserve.

5.2.5 Low-voltage fault ride-through

Under normal conditions, voltage must be maintained within an allowed range around its nominal value. The exact limits of this allowed range may differ among grid codes but usually does not exceed approximately ±10% [2, 3, 19, 20]. The grid operator's responsibility to maintain voltage within an allowed range implies in return that the performance of any grid-connected electrical equipment within this voltage range is satisfactory. However, under fault conditions, as in the case of a short-circuit for instance, voltage may drop below the lower limit. In the past, should this event happen, wind turbines were required to disconnect. This decision from grid operators was motivated by the impossibility to control reactive power absorption from older directly connected squirrel-cage generators, which worsen the initial fault conditions. This has been a reasonable strategy as long as the share of wind power in the energy mix was limited to a small capacity. However, nowadays, wind power representing a non-negligible

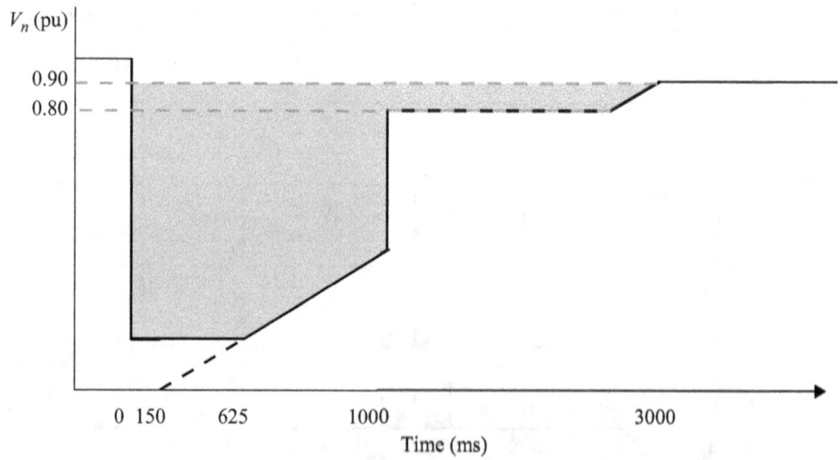

Figure 5.13 Fault ride-through capability as required in the Republic of Ireland [2]

source of electricity generation, it is necessary for the grid stability and reliability that wind turbines remain connected during a fault and that they provide voltage support by reactive power control as well, emulating hence the natural reaction of conventional directly connected synchronous generators. This vital feature, expected to be applied to ocean energy converters in similar fashion to wind turbines, is usually referred to as the 'fault ride-through' requirement. The lapse of time during which wind turbines must remain connected is strictly defined by the voltage amplitude, as shown in Figure 5.13 which presents the required fault ride-through capability as enforced in the Republic of Ireland. The fault ride-through curves as enforced by a large number of grid operators worldwide are similar.

5.2.6 Black start capability

As part of their continuous planning and risk management, system operators have to prepare a detailed and well-rehearsed plan to recover the system in the event that there is a complete system collapse (blackout). With the increasing interconnectedness and complexity of power systems, the risk of such widespread blackout is increasing and there have been some very publicized examples of these in Europe and North America over the last decade or two which have had severe economic and social consequences [21]. Many large system generators are not capable of starting without having a supply from the grid. Obviously, in the event of a complete blackout this will not be available so the system operator has to ensure a sufficient number of generators are capable of starting and maintaining readiness to export power without a grid supply. This is called black start capability and is a part of the services that the generator is contracted to supply to the system. As with frequency reserve response, this traditionally has not been an ancillary service that renewable plant has been required to provide to the system operator. This also may have to change due to the increased presence of renewables on power systems. Depending on the nature of the renewable resource itself, some of which are deterministic, such as tidal stream, offer better prospects in this regard. Some power take-off designs are not self-starting and require a supply to start generating. The black start capability can be provided through the use of a power take-off design that is self-starting (e.g. full converter design) and would also require a local independent supply to meet all internal load requirements until the grid is restored.

5.2.7 Metering/telemetry and telecontrol

The provision of reliable, timely and accurate telemetered data on a renewable plant's actual performance is increasingly important to the system operator as they grow in scale. Traditional distribution connected to renewable power plants provided little, if any, such information.

Due to the proliferation of wind farms connected to distribution system, a lot of grid operators have no direct control over their output. Grid operators today are looking for such telecontrol capability on larger farms and often, where possible, to have it retrofitted to existing wind farms. This greatly assists them in managing the system and consequently the power quality provided to the system users. Typically, such telecontrol will allow them to reduce the output of the wind farm to a certain set point in order to meet transmission constraint issues or system curtailment requirements. Provision of such control will be the norm in the future.

5.2.8 Grid operators' disconnection rights

Uncontrolled power plant disconnection is prohibited by grid operators, in particular during fault event, as they can increase the temporary grid instability. However, grid operators reserve themselves the right to require automatic disconnection from power plants under certain conditions in order to preserve or restore the stable and reliable operation of the grid. Disconnection rights of transmission and distribution system operators, referred to as TSOs and DSOs respectively, are a very sensitive matter since it may affect the revenue of power plant owners.

Although causes implying the automatic disconnection of a power plant may vary greatly from one grid code to another, this paragraph intends to provide general information on the extent of TSO/DSO's rights on that matter. Automatic disconnection on grid operator's order may be justified by abnormal conditions on the network, by the potentially dangerous operation of a power plant and by the need to maintain the instantaneous electrical power generation from non-fully dispatchable renewable power plants under a certain limit. This limit is equal for instance to 30% in France [19]. Regarding abnormal grid conditions, voltage and frequency drifts above or below specified extreme values may force a power plant to disconnect at the grid operator's demand. In addition, in the event of a fault, power plants unable to provide effective voltage support by means of reactive current injection and absorbing reactive power are expected to disconnect [15, 16]. The potentially harmful behaviour of a power plant, such as its loss of stability, should also force a power plant to disconnect, both to preserve its integrity and to prevent any negative impact to the network. Automatic reconnection may be allowed in certain cases under specific conditions, typically based on a resynchronization within few seconds [16].

5.3 Guidelines and standards

5.3.1 Introduction

5.3.1.1 Dynamic modelling

Many grid operators require dynamic models as part of the grid connection process. However, dynamic modelling has been identified by the ocean energy industry and the research community as a major challenge which should be addressed at a very early stage [22]. The experience from the wind energy industry has shown that first generation models were inadequate, led to incorrect results or provoked power system simulator crash [23] (see Chapter 9 'Modelling and Simulation Techniques' for more details). Hence, it appeared clearly that using specific models of each individual wind turbines for preliminary power system studies was a pitfall. An alternative approach, called 'generic modelling', consists of using a common structure for all models which is customisable by device-specific parameters. Its numerous advantages have led the wind industry to converge towards this alternative, which is also of great interest for ocean energy applications, while specific models can be developed for more detailed studies.

5.3.1.2 Inapplicability and irrelevance of some current grid code requirements

Some grid code requirements seem hardly applicable to wave farms, whereas others appear as irrelevant [24]. For instance, any requirements based on the nominal power of a device, equal to maximum power for a wind/tidal turbine, fall in the first category, as no widely agreed notion of nominal power has emerged regarding wave energy converters as yet. The power ramp requirement illustrates the irrelevance of some requirements: the rates of the power ramps a power plant may output are limited, but averaged over a period of time of at least one minute [2] which does not allow this requirement to take the rapid fluctuations of a wave farm into account, although they may have an adverse effect on the network. This inadequacy of some current grid code requirements is believed to be due to the relative lack of a suitable set of metrics enabling to perform a relevant and comprehensive grid impact assessment of wave farms.

5.3.1.3 Grid code evolution

Besides technical obstacles, other issues have been identified regarding the relevance and transparency of grid code requirements themselves. Grid codes, defining the expected behaviour of any power plant on the network have had, and still have, a strong influence on device's design. The decline of the cheap directly connected, squirrel-cage induction generator for wind turbines is a good illustration of this influence. Although cheap and robust, this type of generator is now completely abandoned by manufacturers because of its poor fault ride-through capability and its absence of voltage support means, both of which are required by grid codes. Grid codes weigh thus very heavily on the energy industry.

However, the wind energy industry has claimed, using the European Wind Energy Association (EWEA) as an advocate, that grid code requirements for wind farms were evolving too fast, and with too short notice which was detrimental to the industry [25]. In addition, the application of some requirements was considered as technically unjustified or not economically sound. This is for instance the case for the fault ride-through capability, which was considered unjustified in countries where the penetration level of wind energy was still insignificant according to this organisation. Another concern of the wind industry was the lack of harmonisation of standards which, according to the EWEA 'resulted in gross inefficiencies and additional costs for consumers, manufacturers and wind farm developers.' Finally, they claimed that grid code requirements were 'often not sufficiently clear'.

Hence, the development by the European Union of a harmonised grid code for all types of power plants is fully supported by the EWEA [26]. This alternative, which is intended to increase cross-border electricity trade between European countries, will have the additional advantages to also increase transparency and encourage best practices within the industry. This common European grid code is currently developed by ENTSO-E, the European Network of Transmission System Operators for Electricity [15]. The grid code covers any issues which may have an impact in cross-border trade between European countries. The control of global parameters such as frequency is thus addressed, as well as the control of more local parameters related to power quality which are considered necessary to maintain a safe and reliable operation of the network. However, local issues having an influence considered as negligible on intra-European electricity exchange, such as flicker severity or harmonics levels, are not addressed.

The European grid code presents several advantages, the main of which is the application of grid code requirements to power plants based on their potential impact on the network rather than based on the more ambiguous, and sometimes irrelevant criterion of prime mover type, which is used in most current grid codes. Four levels of potential grid impact are defined in the code with respect to the voltage level at the connection point and to the rated power of the grid-connected installation. A set of grid code requirements is defined for each of these levels, their strictness increasing with the level of potential grid impact. In addition, in order to take into account regional characteristics of the European network, the numerical limits in terms of rated power determining the belonging of a power plant to one level or another depend on the synchronous area considered among the five regions defined by the grid code: Continental Europe, Nordic, Great Britain, Ireland and Baltic.

5.3.2 International standards

The overarching set of standards in Europe and internationally covering the whole area of electromagnetic compatibility (EMC), of which power quality is a part, is the IEC 61000 series which has six parts each with standards and technical reports. These parts are described as follows:

Part 1: General – Safety function and integrity requirements
Part 2: Environment – Description and classification of the environment as well as compatibility levels
Part 3: Limits – Emission and immunity limits
Part 4: Test and measurement techniques
Part 5: Installation and mitigation guidelines
Part 6: Generic standards

The following is a list of some of the most significant standards that are used to evaluate power quality:

Limits on user 'emissions'

IEC 61400-21
Wind turbines – Part 21: Measurement and assessment of power quality characteristics of grid-connected wind turbines

IEC 61000-4-30
Electromagnetic compatibility (EMC) – Part 4-30: Testing and measurement techniques – Power quality measurement methods

IEC 61000-4-7
Electromagnetic compatibility (EMC) – Part 4-7: Testing and measurement techniques – General guide on harmonics and interharmonics measurements and instrumentation, for power supply systems and equipment connected thereto

IEC 61000-3-6
Electromagnetic compatibility (EMC) – Part 3-6: Limits – Assessment of emission limits for the connection of distorting installations to MV, HV and EHV power systems

IEC 61000-3-7
Electromagnetic compatibility (EMC) – Part 3-7: Limits – Assessment of emission limits for the connection of fluctuating installations to MV, HV and EHV power systems

IEC 61000-3-5
Limits on fluctuations and flicker in LV systems

IEEE 1159-2009
Recommended practice for monitoring electric power quality

Quality of supply to users

EN50160
Voltage characteristics of electricity supplied by public distribution systems

This standard is primarily based on IEC 61000 and is defined for LV and MV electricity and is generally the pass/fail level for the quality of supply that utilities operate in order to determine whether remedial measures are required. Utilities will normally require a higher level of performance of power quality 'emissions' from generators in order to ensure that the standard is not breached. Table 5.3 is a summary of this standard and how it relates to the IEC 61000 series.

5.3.3 National standards

UK Engineering Recommendation G5/4
Planning levels for harmonic voltage distortion and the connection of non-linear equipment to transmission systems and distribution networks in the United Kingdom

UK Engineering Recommendation P28
Planning limits for voltage fluctuations caused by industrial, commercial and domestic equipment in the United Kingdom

Table 5.3 Comparison between EN50160 and the IEC 61000 series

No	Parameter	Supply voltage characteristics according to EN50160	Low-voltage characteristics according to EMC standard EN61000	
			EN61000-2-2	Other Parts
1	Power frequency	LV, MV: mean value of fundamental measured over 10 s ±1% (49.5–50.5 Hz) for 99.5% of week −6%/+4% (47–52 Hz) for 100% of week	2%	
2	Voltage magnitude variations	LV, MV ±10% for 95% of week, mean 10 min rms values		±10% applied for 15 min
3	Rapid voltage changes	LV: 5% normal 10% infrequently $P_{lt} \leq 1$ for 95% of week MV: 4% normal 6% infrequently $P_{lt} \leq 1$ for 95% of week	3% normal 8% infrequently $P_{st} < 1.0$ $P_{lt} < 0.8$	3% normal 4% maximum $P_{st} < 1.0$ $P_{lt} < 0.65$ (EN61000-3-3) 3% (IEC 61000-2-12)
4	Supply voltage dips	Majority: duration < 1s, depth < 60% Locally limited dips caused by load switching on: LV: 10–50%, MV: 10–15%	urban: 1–4 months	up to 30% for 10 ms up to 60% for 100 ms (EN61000-6-1,6-2) up to 60% for 1000 ms (EN 61000-6-2)
5	Short interruptions of supply voltage	LV, MV: (up to 3 min) few tens-few hundreds per year Duration 70% of them < 1 s		95% reduction for 5 s (EN61000-6-1,6-2)

6	**Long interruption** of supply overvoltages	LV, MV: (longer than 3 min) < 10–50 kV rms		
7	Temporary, power frequency overvoltages	LV: < 1.5 kV rms MV: 1.7 U_C (solid or impedance earth) 2.0 U_C (unearthed or resonant earth)		
8	Transient overvoltages	LV: generally < 6 kV, occasionally higher; rise time: ms–µs MV: not defined		±2 kV, line-to-earth ±1 kV, line-to-earth 1.2/50(8/20) Tr/Th–µs (EN61000-6-1,6-2)
9	Supply voltage	LV, MV: up to 2% for 95% of week, mean 10 min rms values, up to 3% in some locations	2%	2% (IEC 61000-2-12)
10	Harmonic voltage	LV, MV: see typically between 2–6% for lower order odd harmonics and generally well below 2% for all higher and even harmonics	6%–5th, 5%–7th, 3.5%–11th, 3%–13th, THD < 8%	5%–3rd, 6%–5th, 5%–7th, 1.5%–9th, 3.5%–11th, 3%–13th, 0.3%–15th, 2%–17th (EN61000-3-2)
11	Interharmonic voltage	LV, MV: under consideration		

Engineering Recommendation P29 (1990)
Planning limits for voltage unbalance in the UK for 132 kV and below

These UK standards are largely based on the IEC series as it is interpreted and applied to the regulatory and network topology of the United Kingdom.

5.4 Conclusions

The importance of complying with power quality requirements is essential as failing to meet them is a sufficient reason for any grid operator to deny grid connection applications. However, limits specified by these requirements may differ among different grid connection codes, rendering compliance with power quality requirements somewhat site-specific. This is expected to change due to the increasing level of harmonisation of grid codes.

5.5 References

[1] IEEE. *Recommended Practice for Monitoring Electric Power Quality.* Standard 1159-2009 (Revision of IEEE standard 1159-1995), 26 June 2009.
[2] ESB. Distribution Code, version 2.0, 2009.
[3] National Grid. *The Grid Code*, issue 4, revision 2, 2010.
[4] IEC. Standard 61000-3-7, *Electromagnetic Compatibility (EMC) – part 3-7: Limits – Assessment of Emission Limits for the Connection of Fluctuating Installations to MV, HV and EHV power systems*, 2008-02-22.
[5] NORDEL. *Nordic Grid Code*, 2007.
[6] E. Ossentjuk, C. Elink Sterk and W. Strom van Leeuwen, 'Flicker-induced cardiac arrest in a patient with epilepsy', *Electroencephalography and Clinical Neurophysiology*, vol. 20, no. 3, pp. 257–259, 1966.
[7] M. Bollen and I. Gu. *Signal Processing of Power Quality Disturbances.* John Wiley & Sons, Hoboken, New Jersey, USA, 2006.
[8] IEC. Standard 61000-4-15, TC/SC 77A, *Electromagnetic Compatibility (EMC) – Part 4-15: Testing and Measurement Techniques – Flickermeter – Functional and Design Specifications*, ed2.0 (2010-08).
[9] C. Rong, J. Cobben, J. Myrzik, J. Blom and W. Kling, 'Flickermeter used for different types of lamps', *Proceedings of the 9th International Conference on Electrical Power Quality and Utilisation EPQU*, pp. 1–6, October 2007.
[10] B. Heffernan, L. Frater and N. Watson, 'LED replacement for fluorescent tube lightning', *Australasian Universities Power Engineering Conference, AUPEC*, pp. 1–6, December 2007.
[11] T. Ackermann. *Wind Power in Power Systems.* John Wiley & Sons, Hoboken, New Jersey, USA, 2012, 2012.
[12] A. Hansen and G. Michalke, 'Fault ride-through capability of DFIG wind turbines', *Renewable Energy*, vol. 32, no. 9, pp. 1594–1610, 2007.

[13] C. Wessels, F. Gebhardt and F. Fuchs, 'Fault ride-through of a DFIG wind turbine using a dynamic voltage restorer during symmetrical and asymmetrical grid faults', *IEEE Transactions on Power Electronics*, vol. 26, no. 3, pp. 807–815, 2011.
[14] A. Ibrahim, T.-H. Nguyen, D.-C. Lee and S.-C. Kim, 'A fault ride-through technique of DFIG wind turbine systems using dynamic voltage restorers', *IEEE Transactions on Energy Conversion*, vol. 26, no. 3, pp. 871–882, 2011.
[15] ENTSO-E. 'Network code on requirements for grid connection applicable to all generators (RfG)', 26 June 2012.
[16] E.On Netz. 'Grid connection regulations for high and extra high voltage', 2006.
[17] J. Aubry. *Optimisation du dimensionnement d'une chaîne de conversion électrique directe incluant un système de lissage de production par supercondensateurs: application au houlogénérateur SEAREV*. PhD thesis, Ecole Normale Supérieure de Cachan, France, 2011.
[18] EirGrid. *Facilitation of Renewables*. Technical report, EirGrid, 2010.
[19] Arrêté du 23 avril 2008 relatif aux prescriptions techniques de conception et de fonctionnement pour le raccordement à un réseau public de distribution d'électricité en basse tension ou moyenne tension d'une installation de production d'énergieélectrique, version consolidée du 6 mars 2001.
[20] Standard EN50160, *Voltage Characteristics in Public Distribution Systems*, 2010.
[21] C. Anderson, J. Santos and Y. Haimes, 'Risk-based input-output methodology for measuring the effects of the August 2003 Northeast Blackout'. *Economic Systems Research, Taylor & Francis Online,* Volume 19, Issue 2, p. 183–204, 2007.
[22] D. O'Sullivan, D. Mollaghan, A. Blavette and R. Alcorn. *Dynamic Characteristics of Wave and Tidal Energy Converters, and A Recommended Structure for Development of a Generic Model for Grid Connection*. Technical report. A report prepared by HMRC-UCC for the OES-IA Annex III. Available at: www.iea-oceans.org, 2010.
[23] Y. Coughlan, P. Smith, A. Mullane and M. O'Malley, 'Wind turbine modelling for power system stability analysis: A system operator perspective', *IEEE Transactions on Power Systems*, vol. 22, no. 3, pp. 929–936, August 2007.
[24] A. Blavette, D. O'Sullivan, T. Lewis and M. Egan, 'Grid integration of wave and tidal energy', *Proceedings of the 30th International Conference on Ocean, Offshore and Arctic Engineering OMAE2011*, Rotterdam, the Netherlands, 2011.
[25] European Wind Energy Association (EWEA). *Harmonising Europe's Grid Codes for the Connection of Wind Power Plants to the Electricity Network.* Technical report, 2009.
[26] European Wind Energy Association (EWEA). *EWEA Response on the ERGEG Consultation on the Pilot Framework Guidelines on Electricity Grid Connection*. Technical report, 2009.

Chapter 6
Grid integration: part III – case studies

6.1 Introduction

This chapter details four case studies addressing the grid integration of wave and tidal current energy. The first three studies focus on the power quality issues arising from the connection of a tidal device or of a wave farm on their local network. More specifically, the first study investigates the impact of the SeaGen tidal turbine connected at Strangford Lough, Northern Ireland. The two other studies focus on the grid impact of a medium-size wave farm on its local network. The last study addresses the challenges in terms of grid integration at a power system level and details the capacity value of wave energy in Ireland.

6.2 Case study: tidal energy – SeaGen

J. MacEnri

There was very little published research into the power quality impacts of grid–connected tidal current devices at the time of publication of this research in 2011. From 2011, a number of papers have been published outlining in detail the power performance assessment of the tidal current device, SeaGen, located in Northern Ireland [1, 2]. A summary of the findings is presented in the case study in this section.

Drawing from the conclusions, of the SeaGen research it is unlikely that tidal stream devices will cause significant power quality problems that cannot be ameliorated in some straightforward fashion. However, the results from the assessment have two important caveats: first, wave action is very limited at the SeaGen location (similar to a river) and so ocean waves are likely to have a significant impact on flicker performance as their period is located in the flicker curve frequency range, and second, the aggregated performance of a tidal farm has not yet been measured and analysed as none exist at the time of this book's going to print in late 2013.

6.2.1 Introduction

SeaGen is a 1.2 MW tidal energy converter (TEC) under development by Marine Current Turbines (MCT) Ltd in the United Kingdom. It is located in Strangford Lough, Northern Ireland and has been in regular production since May 2009 and, as of 2013, it is the world's largest grid-connected TEC that has produced over 3000 MWh by mid-2012.

The general structure and outline of SeaGen are shown in Figure 6.1. It has a central column with a crossbeam of 27 m which supports the two horizontal axis turbines each with a swept area of 16 m diameter. The turbines have two blades which can pitch through 270 degrees, with blade pitching used instead of yaw control for flood and ebb production. The turbines are coupled to a planetary gearbox, which increases the shaft speed to a nominal value of 1000 rpm at the two asynchronous two-pole 600 kW generators. All of this equipment are contained in submersed drivetrains at the ends of the crossbeam.

The output from the two generators is fed to two full back-to-back insulated-gate bipolar transistor (IGBT) converters and their grid synchronised outputs are ganged together before the step-up transformer. All of this equipment are located within the central column. Power is exported by means of a submarine cable to the local 11 kV network. The outline of the power take-off (PTO) system is shown in Figure 6.2.

Figure 6.1 SeaGen in Strangford Lough, Northern Ireland

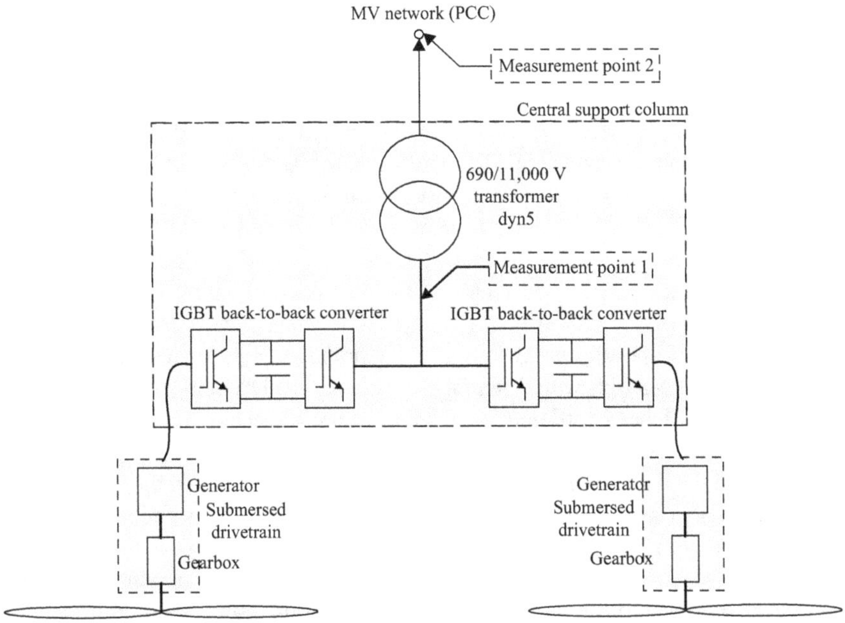

Figure 6.2 Schematic of PTO system

The turbines operate at variable speed below rated tidal current according to the determined optimum $C_p\lambda$ curve (the ratio of blade tip speed to tidal speed that produces the highest blade efficiency). They operate at a nominal fixed speed above the rated tidal current. SeaGen's cut-in and rated tidal current speeds are approximately 0.8 m/s and 2.5 m/s respectively.

The tidal current in the Strangford Lough channel is one of the fastest current around either Ireland or Britain with over 350 million cubic metre of water entering or leaving the lough in each tidal half cycle. The flow peaks at over 4 m/s (7.8 kts) in springs. Tidal flows are semidiurnal with two ebb and flood tides per day.

The water depth at the SeaGen location at Lowest Astronomical Tide (LAT) is approximately 24 m. Mean Low Water Springs are +0.5 m and Mean High Water Springs are +4.1 m.

Previous tidal current surveys indicated that the site chosen for SeaGen would not have unduly turbulent flows due to seabed or shoreline features. As the tidal current device is located nearly 4 km from the open sea, wave action is normally very limited and confined to high-frequency locally generated small waves.

SeaGen is connected by way of a submarine three-phase cable to the local 11 kV grid. The grid connection point is fed by a three-phase 11 kV overhead line from a 33 kV sub-transmission substation. The short-circuit ratio is approximately 9 and the grid impedance angle Ψ_k is 50 degrees. Quality of supply with no SeaGen generation was within EN50160 and is very characteristic for such a remote MV grid point.

6.2.2 Test methodologies

6.2.2.1 Measurement point 1 – SeaGen power quality performance

As no standard exists for the assessment of these power quality aspects of TECs, the closest applicable standard was chosen and modified as required. To this end, the standard used to measure and assess the power quality characteristics of grid-connected wind turbines IEC 61400-21 [15] was chosen as this standard has a comprehensive and widely accepted method to determine power characteristics of wind turbines (WT) independent of the actual electrical characteristics of the grid to which it is connected. These independent (i.e. not grid location specific) power quality characteristics are normally determined for each wind turbine type, so project developers and utilities can determine the actual power quality impact of the proposed wind farm using this turbine type at other grid locations with different short-circuit levels and X/R ratios.

Apart from flicker coefficients and step factors, the active and reactive power, voltage fluctuations, and harmonics power quality characteristic parameters were assessed from the data collected at measurement point 1 following the requirements of [15], as this measurement point was established as the output terminals of the tidal current device. The standard was followed as closely as possible and where deviations were required due to the inherent differences between a wind turbine and a tidal current device these deviations from IEC 61400-21 are specified in detail in the paper published by Thiringer *et al.* on the flicker evaluation of SeaGen [1].

6.2.2.2 Measurement point 2 – impact on Strangford MV grid

At measurement point 2 (the point of common coupling or PCC) the influence of SeaGen on the local Strangford MV grid was assessed against EN50160 [4] as this standard defines the main voltage parameters and their permissible deviation ranges at the customer's PCC in public low voltage (LV) and medium voltage (MV) electricity distribution systems under normal operating conditions.

6.2.2.3 Assessment against EN50160

The results of the evaluation of the impact of SeaGen on the local medium-voltage grid are summarised in Table 6.1. In this test period of nearly 40 days, power was produced for 34 days with 186.1 MWh in total.

Further details on this analysis including graphical representation of the various power quality parameters and analysis of some voltage sag and transient events recorded during the test period are presented in [5].

6.2.2.4 Power quality evaluation of SeaGen

Tidal current speeds vary from around 0.9 m/s up to 3.8 m/s during the measurement periods. For over half the periods both turbines were operating, and only one for the remaining period (both are independent and one or the other was removed from service for various reasons). It allowed for comparison of flicker performance between the two operating modes.

6.2.2.5 Power regulation

Figure 6.3 shows the results of active power for all data files as a 10 min average (P_{600} values). Figure 6.4 shows the reactive power for the same.

Table 6.1 EN50160 (MV) evaluation of SeaGen at Strangford Lough

Parameter	EN50160 (MV) requirement	Measurement (95th percentile)	Remarks	Result
Frequency	$< \pm 1\%$ (99.5% of year)	+0.4%, –0.7%	These values are the maximum and minimum as evaluation period was less than 1 year 95th percentile was +0.12%	Pass
Voltage imbalance	<2% (99.5% of year)	0.57%	Principal contributor to imbalance is the fact that single phase boosters used to reduce voltage rise in the local MV Grid are only on two phases	Pass
Voltage variation	$< \pm 10\%$ (99.5% of year)	+2.92%	Highest voltage rise of 5.63% on phase without booster. Other two phases had voltage variation of less than 2%. Lowest voltage variation was −2%	Pass
Flicker	$P_{lt} < 1$ (99.5% of year)	0.4	Background flicker level with no SeaGen output was typically around 0.2	Pass
Harmonic distortion	$U_{THD} < 8\%$	1.54%	None of the individual harmonics exceeded their limit prescribed in EN50160	Pass

The tidal current device increased its output from around 1.1 MW to the rated output of 1.2 MW, as can be seen in Figure 6.3, due to a faulty speed sensor which was rectified during the series of test measurements. At rated output, the reactive power was around 50 kVAr giving a power factor of 0.99.

IEC 61400-21 requires the powers to be measured and averaged over 600, 60 and 0.2 s. The measured maximum average values are shown in Table 6.2. These values are the maximum values averaged over the different length periods. There is further explanation of their significance when comparing SeaGen power quality performance to wind turbines.

6.2.2.6 Harmonics

Figures 6.5 and 6.6 show the voltage and current total harmonic distortion (THD) for the measurement files. Power quality standards allow for voltage THD of up to 8% on medium-voltage systems and 3% on high-voltage systems and the values measured are well within these standards.

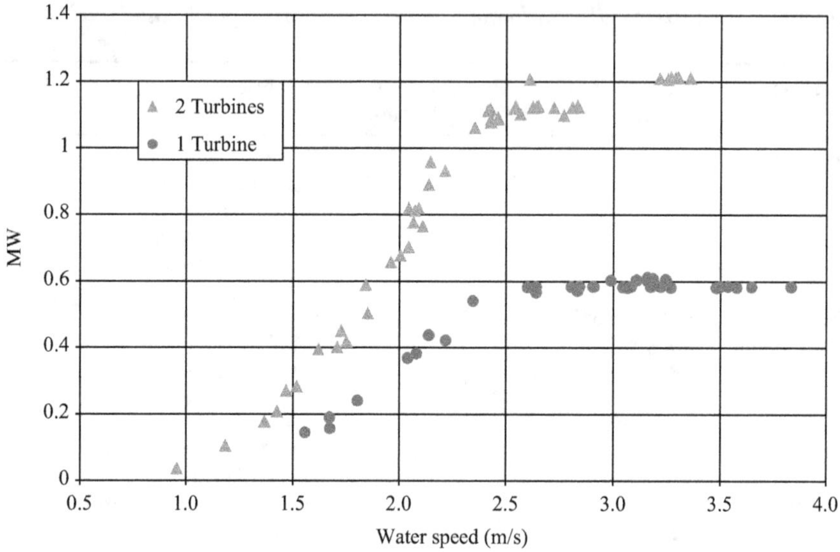

Figure 6.3 Active power

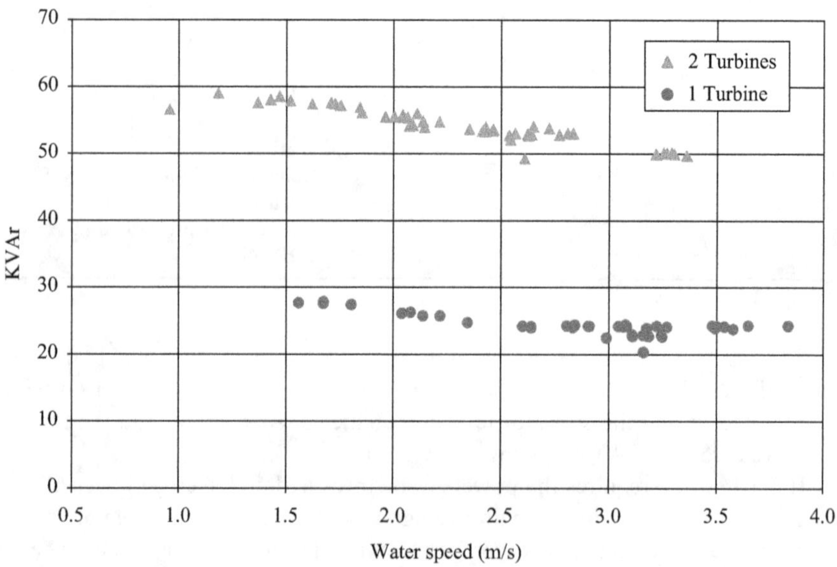

Figure 6.4 Reactive power

Typical individual current harmonic levels up to the 50th harmonics are presented in Table 6.3.

6.2.2.7 Flicker – continuous operations

The flicker coefficients of the tidal current device when operating continuously (or at steady state) were calculated from the voltage waveform using the methodology

Table 6.2 Active powers

Parameter	Maximum value (MW)
P_{600}	1.21
P_{60}	1.30
$P_{0.2}$	1.46

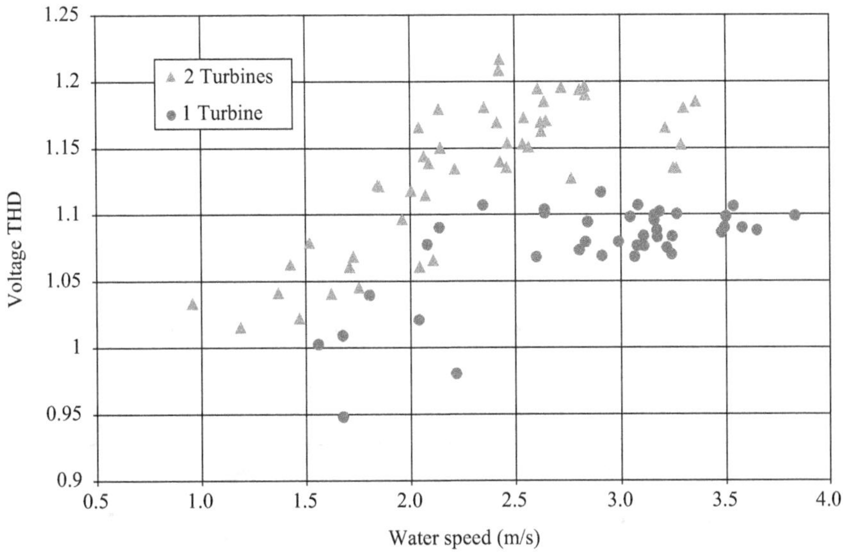

Figure 6.5 Voltage total harmonic distortion (THD)

described in the IEC Standard 61400-21 and are presented as the average of the three phases in Figure 6.7. The results for a grid impedance angle Ψ_k of 50 degrees are shown as it corresponds with the actual angle at the measurement point. Further details on this evaluation method and graphs of the coefficients at different angles are presented in [1]. One interesting outcome of the study of TEC with 1 and 2 turbines operating is that the flicker levels differ between them by a factor of $1/\sqrt{2}$ which corresponds very well with the theory for summing flicker contribution from *N* independent sources as presented in the IEC Standard 61400-21. Further details on this are provided in [1].

IEC Standard 61400-21 requires the 99th percentile flicker coefficients to be reported. Due to the limited amount of data compared with those required under the standard, a conservative approach was adopted and the flicker coefficients reported for SeaGen in Table 6.4 are the maximum values rather than the 99th percentiles.

The power output was analysed in the frequency domain, and a frequency component at just under 0.5 Hz (and its harmonics to the 10th) is clearly visible in the frequency domain in Figure 6.8. This corresponds to the power variations experienced by the turbine as it rotates through the varying tidal speeds at different

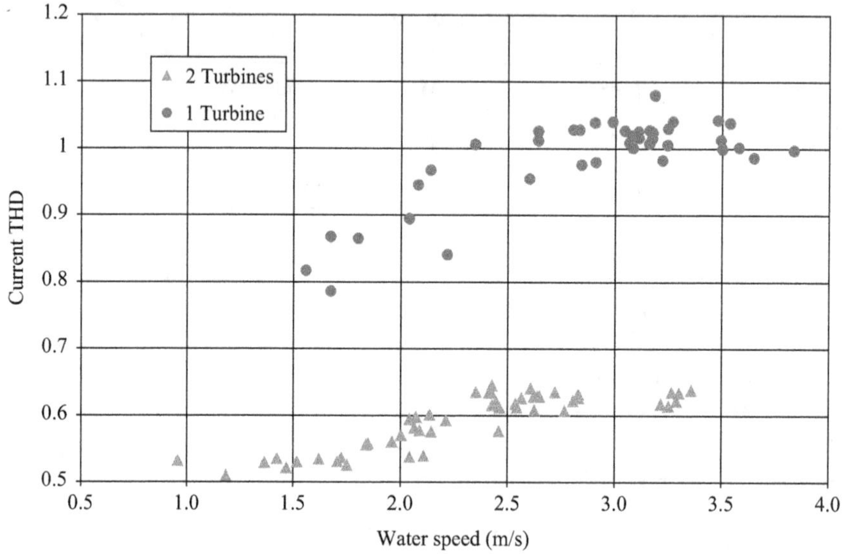

Figure 6.6 Current total harmonic distortion (THD)

Table 6.3 Individual current harmonics at the TEC output terminals

Even harmonics				Odd harmonics			
Order	**% of I_n**	**Order**	**% of I_n**	**Order**	**% of I_n**	**Order**	**% of I_n**
2	0.68	26	0.08	3	0.55	27	0.1
4	0.18	28	0.07	5	1.19	29	0.09
6	0.16	30	0.07	7	0.97	31	0.09
8	0.19	32	0.06	9	0.18	33	0.08
10	0.13	34	0.06	11	0.83	35	0.07
12	0.16	36	0.05	13	0.5	37	0.06
14	0.13	38	0.05	15	0.15	39	0.05
16	0.15	40	0.04	17	0.22	41	0.05
18	0.13	42	0.04	19	0.14	43	0.05
20	0.13	44	0.04	21	0.11	45	0.04
22	0.1	46	0.04	23	0.13	47	0.03
24	0.09	48 & 50	0.03	25	0.1	49	0.03

levels and the shadow effect of the cross-arm. At rated output the power fluctuations on one turbine averaged 28 and 40 kW on two turbines. Further details on this are available in [1].

The relationship between these flicker coefficients and various tidal parameters was studied and it was found that the increasing flicker at higher outputs was due overwhelmingly to the tidal speed rather than any other parameter (see [2]). Further analysis in this reference shows that the higher flicker at higher speeds is predominantly due to the increased turbulence strength at higher speeds rather than power shedding through blade pitching or cross-arm shadow effects.

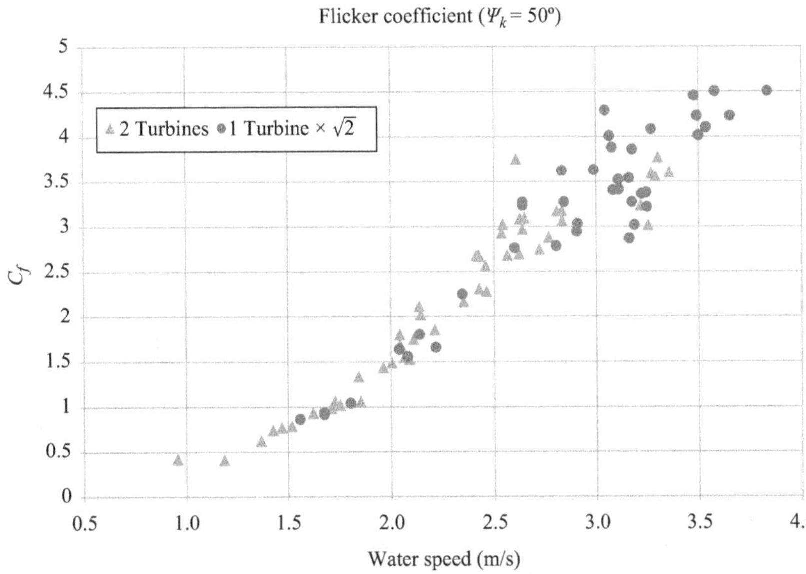

Figure 6.7 Flicker coefficient ($\Psi_k = 50$ degrees)

Table 6.4 Flicker coefficients (continuous operation)

Grid angle (degrees)	Maximum flicker coefficient, C_f (two turbines)	Maximum flicker coefficient, C_f (one turbine)
30	5	8.5
50	3.8	6.5
70	2.1	3.5
85	0.9	1.4

6.2.2.8 Flicker – switching operations

Because wind turbines may be switching in and out at closely spaced frequent intervals due to high or low wind regimes at the envelope of their capability, IEC 61400-21 requires the assessment of a voltage step factor (k_u) and a flicker step factor (k_f) for start-ups at cut-in wind speed and rated wind speed. TECs are fundamentally very different in this regard and would, under normal operations, only start up in a deterministic fashion with no frequent switching. As such, these parameters may not be very relevant to the assessment of tidal current devices in the longer term, but for research purposes this assessment was also undertaken. Fuller details of this assessment are provided in [1] including graphs of waveforms during various switching events. For the purposes of comparison with wind turbines the summary results of this part of the assessment are presented here. Table 6.5 shows the results for a start-up at cut-in tidal speed (~1 m/s) and Table 6.6 shows start-up at a high tidal speed (2.7 m/s).

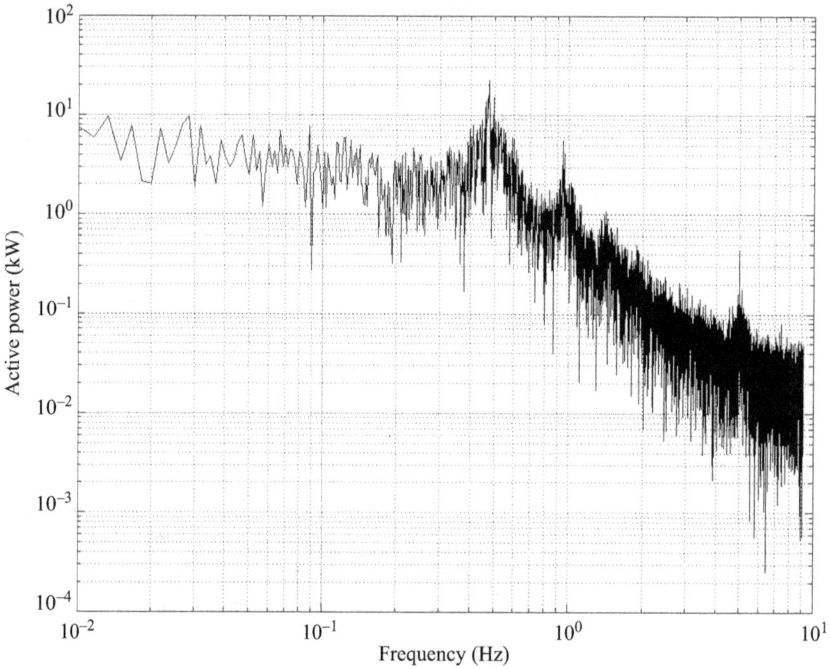

Figure 6.8 Frequency domain analysis of output at high tidal speed

Table 6.5 Start-up cut-in speed

Grid angle (degrees)	K_u	K_f
30	0.52	0.17
50	0.43	0.13
70	0.29	0.08
85	0.16	0.05

Table 6.6 Start-up high speed

Grid angle (degrees)	K_u	K_f
30	0.79	0.18
50	0.63	0.15
70	0.40	0.10
85	0.19	0.06

6.2.3 SeaGen power quality performance compared with similar wind turbines

For comparison, two recent turbine designs were chosen with rated outputs of similar scale to that of SeaGen and with similar PTO and control topologies. The details of the wind turbines are presented in Table 6.7.

Table 6.7 Wind turbines used for comparison

Characteristic	WT1	WT2	SeaGen
Power regulation	Pitch regulated	Pitch regulated	Pitch regulated
Variable/Fixed speed	Variable	Variable	Variable
Power take-off	DFIG	Full converter	Full converter
Rated power	1500 kW	2300kW	1200 kW
Power Factor at rated power	0.99	0.95	0.99
Switching ops 10 min	1	1	1
Switching ops 120 min	10	6	2

Table 6.8 Power regulation and harmonics

Parameter	WT1	WT2	SeaGen
P_{600}	N/A	N/A	+0.8%
P_{60}	+2.11%	+1.9%	+8.3%
$P_{0.2}$	+13.31%	+2.76%	+21.7%
I_{THD}	1.67%	2.42%	0.65%

Table 6.9 Comparison of flicker coefficients for continuous operation

Grid angle (degrees)	WT1 C_f	WT2 C_f	SeaGen-2 turbines		SeaGen-1 turbines	
			C_f	Δ_{avg}	C_f	Δ_{avg}
30	4.41	3.8	5	+21.8%	8.5	+107.1%
50	3.48	3.5	3.8	+8.9%	6.5	+86.2%
70	2.24	3.1	2.1	−21.3%	3.5	+31.1%
85	1.37	2.7	0.9	−55.8%	1.4	−31.2%

Both were tested against IEC 61400-21 and the results used for comparison are from their type test certificates. The comparison of active power control and THD is presented in Table 6.8.

The maximum 10 min power values for the wind turbines are not reported, but given their better power regulation at shorter time intervals it is most likely that they are less than SeaGen's 0.8%. For the 1 min and 200 ms maximum power levels, SeaGen appears to be unable to control the maximum average output power to the same extent as the wind turbines. SeaGen's pitch control system is based on a current successful wind turbine design and so it is likely that the difference is due to inherent differences in the density and compressibility of the fluids they are in. Clearly, the water environment offers more challenges in terms of power output regulation than the air environment and this is apparent from these results of the comparative power regulation. However, even though SeaGen's results are reasonably good in this regard, it is possible this result can be further improved through refinement of the pitching sub-system and analysis of the control strategies and algorithms employed.

The harmonic performance of SeaGen is the best of the three and is probably a reflection of the fact that it is the latest in converter design which can deliver better harmonic performance than designs of just a few years ago. All three are acceptable harmonic performances.

The comparison of flicker coefficients for continuous operations is shown in Table 6.9. WT1 shows a much more significant drop in its C_f values for higher grid angles than WT2 because it operates at a higher power factor very close to unity. SeaGen's C_f values also drop markedly as the grid angle increases because it also operates close to unity power factor and these lower C_f values are very small making any per cent differences between them very large. This is important when inaccuracies in measurements are considered. It is important to realise while comparing the wind turbines and SeaGen that the wind turbine values are based on the 99th percentile while of SeaGen's are based on the maximum values. There can be a significant difference between them when data contains moderate outliers. It is important also to bear in mind that both sets of data are good performances with respect to flicker and would not cause flicker problems when are part of a larger tidal farm due to the cancellation effect across many independent turbines. In trying to get an understanding of the flicker performance of a TEC vis-à-vis a wind turbine, it is possible to draw a number of conclusions from this comparison. First, when looking at the single turbine operation it would appear that a TEC has a flicker coefficient that is nearly double that of a good power quality wind turbine. As in the case of the power regulation, this is probably due to the turbulence of the faster tidal flows and the relatively larger changes in power output that the much denser and incompressible fluid causes on the power output. Second, the cancellation effect is apparent when two independent turbines are in operation, which brings the overall SeaGen unit flicker performance close to that of the wind turbines. Table 6.10 shows the voltage and flicker step factors for start-up at cut-in speed.

The switch-in of the SeaGen converters has a significant current spike for the final charging of the DC link capacitor. There is scope to improve the performance in this respect, albeit with a slower response time for switching in, and this would reduce the voltage step factors at lower grid angles to values closer to the wind turbines. This may become an important consideration in large tidal farms in the future as this type of switching is normal for TECs and would occur four times a day as standard. It could also be managed by staggering the switch-in of TECs in a farm so as to avoid unnecessary voltage dips when the tide starts flowing again

Table 6.10 Step factor comparison for start-up at cut-in speed

Grid angle (degrees)	**Voltage step factors**			**Flicker step factors**		
	WT1 k_u	**WT2** k_u	**SeaGen** k_u	**WT1** k_f	**WT2** k_f	**SeaGen** k_f
30	0.26	0.17	0.52	0.10	0.05	0.17
50	0.20	0.21	0.43	0.11	0.06	0.13
70	0.15	0.27	0.29	0.11	0.07	0.08
85	0.15	0.31	0.16	0.11	0.07	0.05

Table 6.11 Flicker levels caused by start-ups at cut-in speed

Grid angle	WT1 P_{st}	WT2 P_{st}	SeaGen P_{st}	WT1 P_{lt}	WT2 P_{lt}	SeaGen P_{lt}
30	0.18	0.09	0.31	0.16	0.07	0.17
50	0.20	0.11	0.23	0.18	0.08	0.13
70	0.20	0.13	0.14	0.18	0.10	0.08
85	0.20	0.13	0.09	0.18	0.10	0.05

Table 6.12 Step factor comparison for start-up at rated speed

Grid angle (degrees)	Voltage step factors			Flicker step factors		
	WT1 k_u	WT2 k_u	SeaGen k_u	WT1 k_f	WT2 k_f	SeaGen k_f
30	1.01	0.82	0.79	0.11	0.09	0.18
30	0.77	0.58	0.63	0.10	0.08	0.15
30	0.46	0.29	0.40	0.10	0.06	0.10
30	0.20	0.19	0.19	0.10	0.06	0.06

after a slack water period between tides. The flicker step factors are comparable; however, the number of likely switching events is completely different. Using the values outlined in Table 6.7 and the assessment method given in section 8 of IEC 61400-21, the flicker levels caused by switching in at low speeds are compared in Table 6.11 (using a short-circuit ratio of 10).

The switching flicker levels in Table 6.11 are all acceptable and will not cause a power quality problem on a grid to which they are connected.

Step factors for a switch-in at rated speed are presented in Table 6.12. These are similarly analysed for the flicker levels they cause and the results are very similar to those in Table 6.11. With regard to the voltage step factors for the rated speed start-up SeaGen has the best performance, but all would cause problems where a voltage dip of greater than 3.3% is involved. Normally, however, as part of a farm the individual turbine switching should be coordinated to avoid simultaneous start-ups and the standard IEC 61400-21 allows for this. Furthermore, switching on these occasions would be a non-standard event for TECs.

6.2.4 Conclusions

Overall, the conclusion is that SeaGen works well from a power quality perspective and is fully compliant with the relevant standards. It is unlikely to cause power quality problems on transmission systems to which it is connected when making up multiple unit tidal farms. The following more detailed conclusions are also in order:

- SeaGen appears to have inherently higher flicker levels than comparable wind turbines caused by a combination of the turbulence in the tidal stream at higher velocities and the physical layout of the tidal current device itself, namely two-bladed with cross-arm shadow effect.

- SeaGen has a cancellation factor in flicker levels when compared with a 1.2 MW wind turbine as a result of being composed of two independent turbines of 600 kW each. The cancellation factor reduces the flicker level to about 70% of the value it would otherwise be if SeaGen were a single 1.2 MW turbine.
- The long-term flicker level produced by SeaGen is normally 0.2–0.3 at full production.
- SeaGen operates at near unity power factor throughout its output range which greatly mitigates the flicker levels, particularly at higher tidal velocities.
- There is scope to improve the converter's performance at switch-in and this is under consideration.
- The standard for wind turbine power quality analysis provides a good basis for analysing the power quality impacts of TECs.
- The assessment methodology is robust and produces results which compare very well with measured values.
- The harmonic emissions from SeaGen are well within the acceptable limits.

6.3 Case study: wave energy

6.3.1 Introduction

The objective of this section is to establish the importance of grid strength inimpact assessment studies. In a weak grid, the impedance of the grid is high and in such case a small amount of generation can greatly affect the steady-state voltage. That is why a weaker grid will suffer larger voltage variations at the connection point than a strong one. Whenever a wave farm is connected to a grid, especially to a weak one, many grid integration aspects should be analysed to ensure suitable and secure operation both of the wave farm and of the grid.

With the aim of assessing the influence of the grid strength on the grid impact of wave farms, this study two real case studies are considered. The first one refers to the Biscay Marine Energy Platform (bimep), located in Spain, which represents a relatively strong electric grid [6], while the second one models the Atlantic Marine Energy Test Site (AMETS) of the Republic of Ireland (west of Ireland, north of Scotland), which is a representative weak grid with important resources [7].

6.3.1.1 bimep

The bimep is an offshore facility for research, demonstration and operation of real-scale WECs on the open sea [8]. It is located in Northern Spain, south-east of the Bay of Biscay, and it is expected to be in operation in 2013–2014. It is connected to a strong grid with a short-circuit power of 4550 MVA. The facility comprises four offshore berths, rated at 5 MW each, and composed of subsea cables of different lengths between 3 and 6 km. Once onshore, the subsea cables are replaced by four identical onshore land cables up to the substation. The substation consists of two 13.2/132 kV transformers, used for the wave farm connection to the PCC, and a reactive compensation system in order to assure a specific power factor (Figure 6.9a).

The conceptual wave farm consists of four arrays each including 10 generators. Each group of 10 point absorbers is connected to each berth by means of a 0.69/13.2 kV transformer (Figure 6.9b).

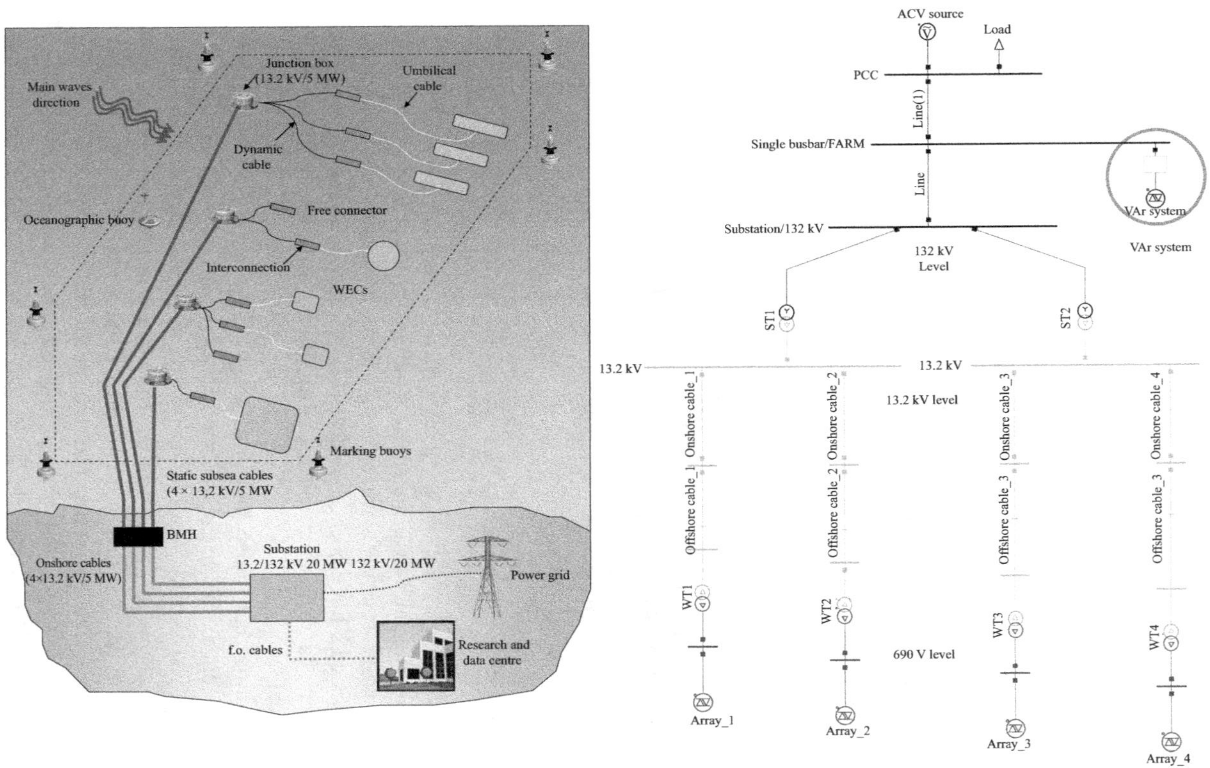

(a) bimep architecture **(b) bimep grid model**

Figure 6.9 bimep

6.3.1.2 AMETS

The AMETS test site is located off the north-west coast of Ireland. It is still under development and it is envisaged that this site will be used by developers for testing final stage prototypes prior to commercial deployment. The grid model used in the current study for the test site is based on cabling design studies performed in conjunction with ESBI in the context of the grid connection application and is shown in Figure 6.10. The conceptual wave farm consists of two clusters each including up to 11 generators. Two clusters are connected to the shore by two AC subsea cables each, one being 6.5 km long, the other being 16 km long. The cluster located at a 6.5 km distance from the shore is referred to as Cluster 1, whereas the other cluster (16 km from the shore) is referred to as Cluster 2. Each cluster consists of two radial feeders (Feeder 1 and Feeder 2) to which wave energy converters are connected. Each feeder cable is connected to an offshore 0.4 kV/10 kV transformer. An onshore substation steps the voltage up to 20 kV. Then, the wave farm is connected to the rest of the national network of Ireland by a 5 km long, 20 kV overhead line.

The rest of the national network is modelled by a 20 kV/38 kV transformer connected to a fixed voltage source in series with a reactor. The impedance of this reactor represents the short-circuit impedance at this node, which was estimated to be equal to 22.8 Ω based on the EirGrid Transmission Forecast Statement [9], that gives a short-circuit power of 63 MVA (Figure 6.9).

6.3.2 Impact of a point absorber farm on the local grid of the bimep and AMETS test sites

M. Santos-Mugica

Among the major challenges of wave energy grid integration are power fluctuations and voltage variations at the PCC, flicker and contingency. In order to analyse the grid impact, of a point absorber's wave energy farm, in various sea conditions, three different sea states have been considered. Low-, medium- and high-energy sea states have been characterised by means of significant wave height H_s and peak period T_p (Table 6.13).

It is worth noting that bimep and AMETS test sites have a quite different natural resource potential. While the average energy density is about 21 kW/m at bimep, at AMETS the density reaches 40–50 kW/m. Considering all the three different sea states allow setting a common framework for the analyses at both sites, once taken into account that low-energy sea states prevail at bimep, while high-energy sea states have higher occurrence at AMETS.

For this study, it has been assumed that the sea states can be modelled by a Bretschneider spectrum [10] at both the locations. Against this background, three different time series have been generated from selected H_s and T_p to represent the required incident wave profiles. From the shape and physical properties of the point absorbers in the array, the excitation forces that the waves exert on the point absorbers have been derived as time domain series and used for the analyses.

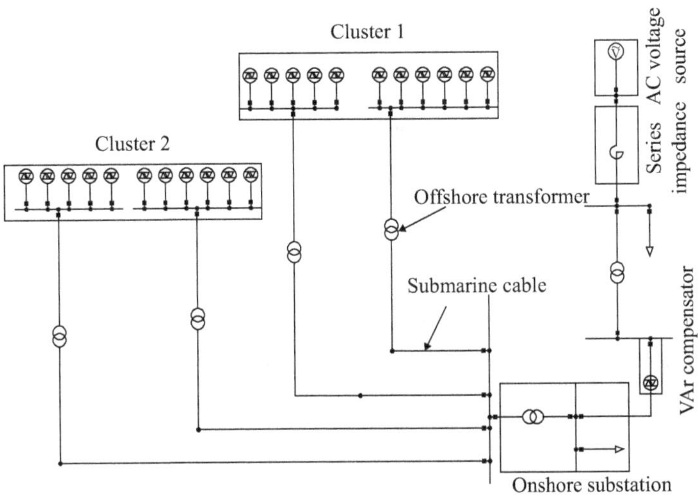

Figure 6.10 AMETS grid model

Table 6.13 Sea states

Sea state	H_s (m)	T_p (s)
Low energy	1.3	**13.8**
Medium energy	2.4	**11.0**
High energy	5.7	**16.5**

To analyse the effect of reducing the oscillations, it has been considered that the wave farm is equipped with a generic energy storage device connected to the PCC. It acts at farm level and it is assumed to have the capability of smoothing the power profile and reducing the variability of the power injected into the electric system without affecting the power capture from the waves.

The implemented generic energy storage device has been modelled as a low-pass filter acting on the instantaneous active power extracted from the WECs, in a similar way to that described in [11] and [12]. Three different cases are analysed, ideally corresponding to different storage capabilities, which correspond to a power smoothing on a time scale of 5, 25 and 50 s respectively. These time constants represent different technology options from turbine inertia to hydraulic reservoirs as indicated in [13].

The basic element of the considered wave farm is a cylindrical point absorber with a hemispherical bottom moving only in heave, which is schematically represented in Figure 6.11. Its main physical parameters are reported in Table 6.14.

In this work the aggregation effect, due to the fact that the waves do not reach all the WECs at the same, has been implemented. For this, the delay between

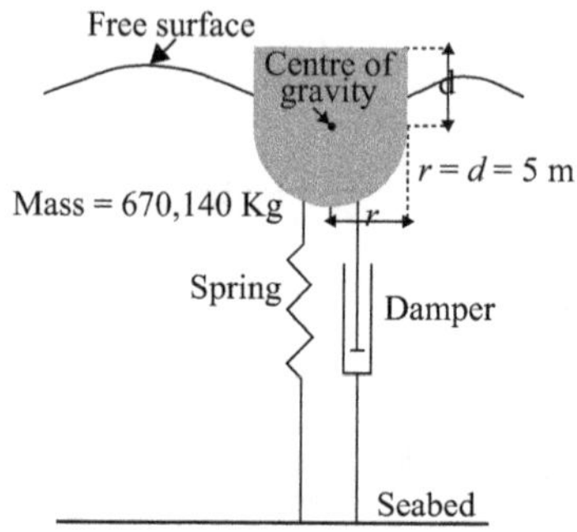

Figure 6.11 Schematic model of the considered point absorber

Table 6.14 Parameters of the point absorber WEC

Quantity	Symbol	Unit	Value
Buoy radius	r	m	**5**
Buoy draught	d	m	**5**
Buoy mass	m	kg	**670,140**
Buoy surface	$\underline{S}$	m^2	**78.6**
Water density	ρ	kg/m^3	**1025**

WECs is calculated on considering the distance among the devices and the peak period of the sea states studied.

6.3.2.1 Power fluctuation and voltage variation

In this subsection power fluctuations and voltage variations due to the variability of the resource and the influence of the storage in both aspects are analysed. Figure 6.12 shows the active power at the PCC for bimep and AMETS for low-, medium- and high-energy sea states when no storage is considered.

The effect of the active power oscillation can be observed in the voltage measured at the PCC (Figure 6.13).

Figure 6.14 shows the active power at the PCC for bimep and AMETS for high-energy sea state and energy storage with three storage time constant: 5, 25 and 50 s. The mean value of the active power with, or without, storage is the same in each test site. However, the ratio between the peak value and the mean value goes from 2.8 to 1.6 in the case of bimep and from 2.7 to 1.6 in AMETS (Figure 6.15). This ratio is lower than expected due to the saturation effect [11]. Furthermore, storage prevents active power output interruptions.

The storage effect on the voltage at the PCC is illustrated in Figure 6.16. In bimep this effect is negligible as the voltage drop without energy storage was nearly zero, while in AMETS, the influence is of 6% in the case of a storage time constant of 50 s.

It can be noted how with the same excitation force profile the voltage variation for AMETS is higher than for bimep. This is due to the weakness of the grid. It is

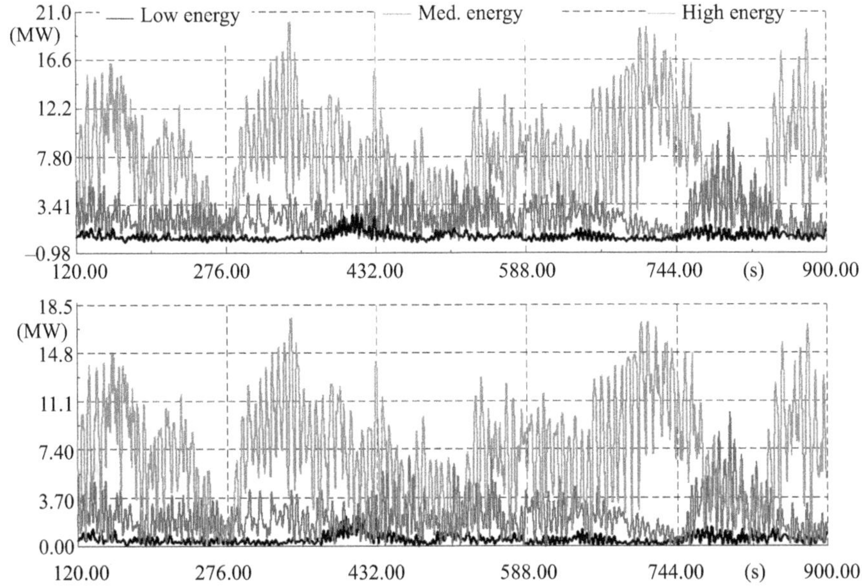

Figure 6.12 Farm active power (MW) for bimep (top) and AMETS (bottom) without storage

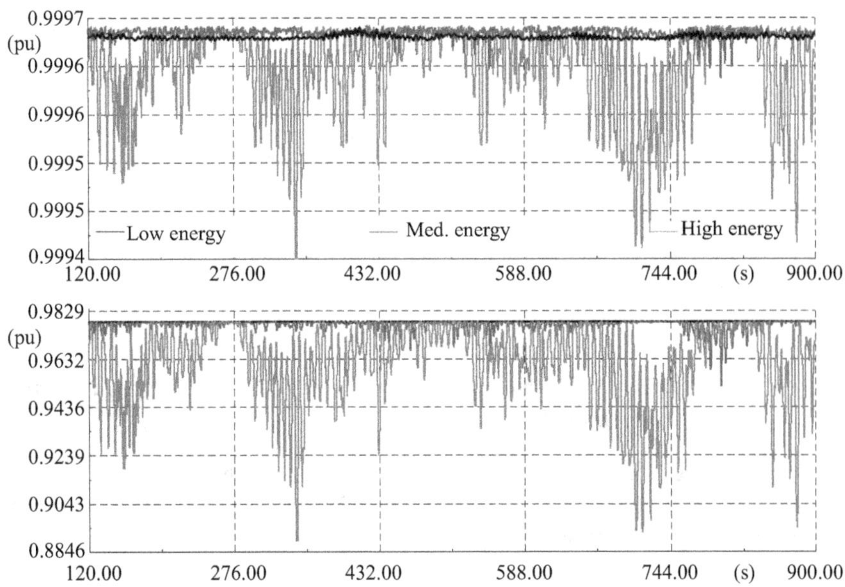

Figure 6.13 Voltage (pu) at the PCC for bimep (top) and AMETS (bottom) without storage

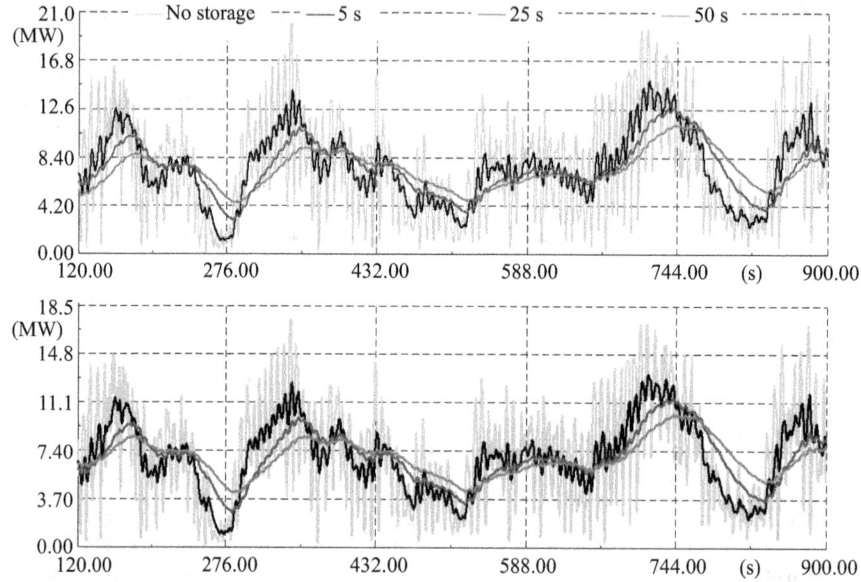

Figure 6.14 Farm active power for bimep (top) and AMETS (bottom), at high energy sea state, for different levels of energy storage

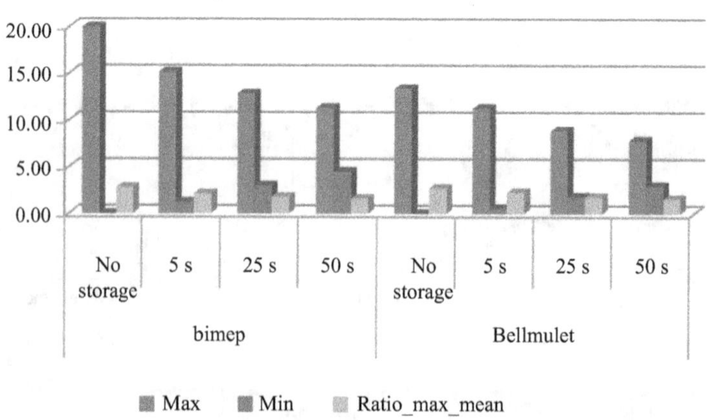

Figure 6.15 Maximum, minimum and peak-to-average ratio for the active power at bimep and AMETS

important to note the effect of the grid on the active power as well, since the generated power is lower in the case of AMETS (Figure 6.17), hence the importance of reducing the oscillations.

6.3.2.2 Farm disconnection

In this case, the effect of a farm shutdown at the PCC is evaluated. Regarding power quality consideration [15] the transient excursion of the PCC voltage when

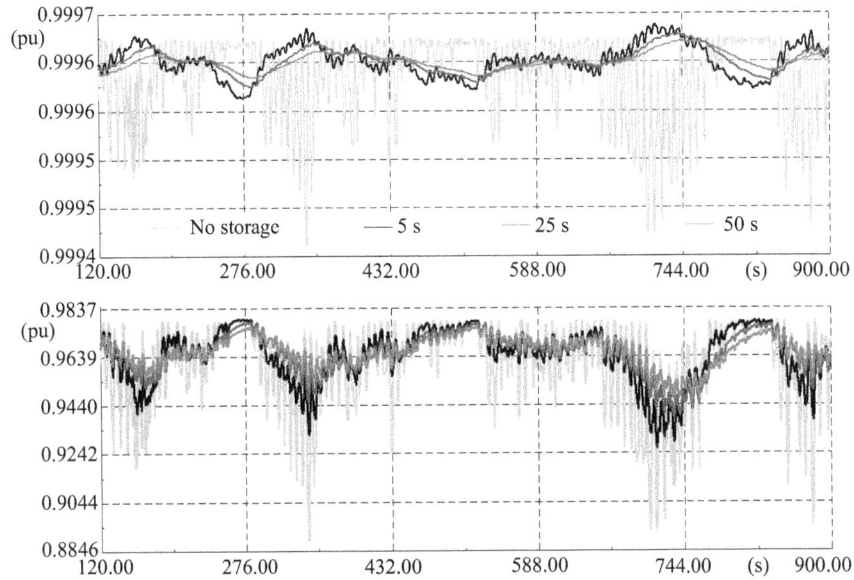

Figure 6.16 Voltage at PCC for bimep (top) and AMETS (bottom), at high energy sea state, for different level of energy storage

connecting and disconnecting a farm is important. Considering the high variability of the active power generated by the farm a shutdown has been carried out when the generated active power is maximum, so the effect is the worst expected for that sea state. In Figure 6.18, the active power profile can be observed when the shutdown occurs both at bimep and AMETS. The obtained results correspond to the high-energy sea state and to the different levels of energy storage.

Due to the shutdown of the farm a voltage step occurs at the PCC. Figure 6.19 illustrates this step both at bimep and AMETS. While at bimep the effect is negligible, the step at AMETS goes from 2.2% up to 5.6% depending on the level of the energy storage.

Obtained results show the importance of the PCC strength to reduce the impact of the oscillations of the produced active power at the voltage. The impact at bimep is quite low even without energy storage, while in the case of AMETS, the energy storage has shown to be mandatory to reduce voltage variations both in steady state and when a shutdown occurs, and the flicker level.

This study has laid the basis for importance of energy storage in a weak grid. Nevertheless, to evaluate the real effect more realistic energy storage system needs to be modelled.

6.3.2.3 Influence of the technology on flicker generation

The main difference in terms of flicker generation between different WEC technologies resides in their variable inherent energy storage capacity which can

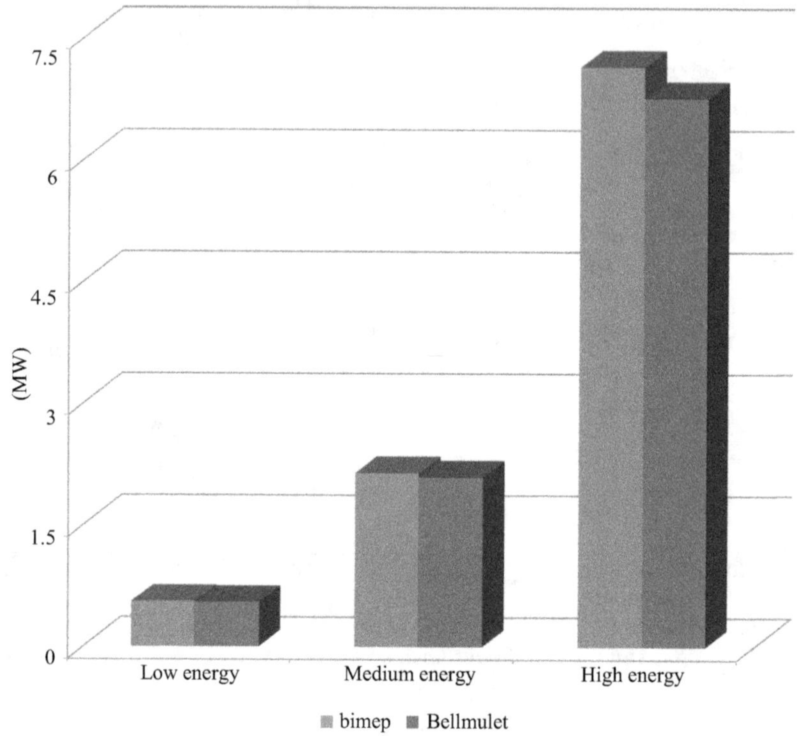

Figure 6.17 Mean value of the active power of the wave farm for bimep and AMETS

take the shape of hydraulics or pneumatic accumulators, hydraulic rams, pneumatic chamber in oscillating water columns (OWCs), flywheel, or a reservoir in overtoppers. This is illustrated in Figures 6.20 and 6.21 in which flicker generation at the AMETS test site appears to be significantly more important in the case of a point absorber's farm than in the OWC's farm. This demonstrates clearly that the OWC air chamber plays a heavy role in smoothing the individual device power output. In both cases, flicker becomes negligible when the farm power output is smoothed by additional storage means whose time constant is equal to few seconds only.

6.3.3 Analysis of the flicker level as a function of a test site's short-circuit characteristics

A. Blavette

6.3.3.1 Introduction

The high-amplitude power fluctuations generated by most wave energy converters without significant amounts of storage or without suitable power output

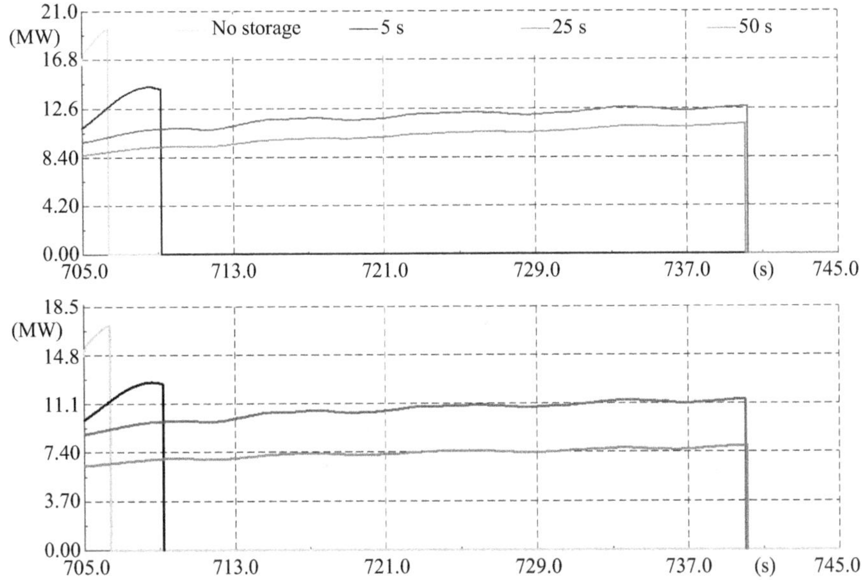

Figure 6.18 Farm active power for bimep (top) and AMETS (bottom) during farm shutdown

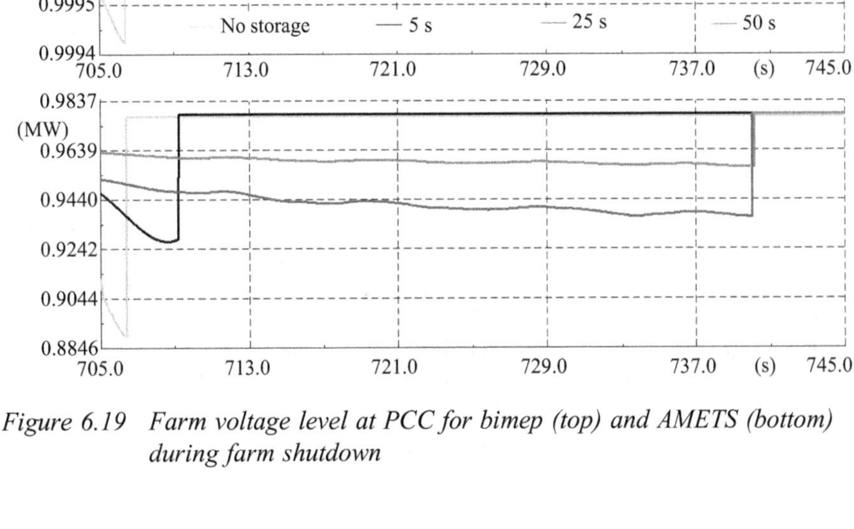

Figure 6.19 Farm voltage level at PCC for bimep (top) and AMETS (bottom) during farm shutdown

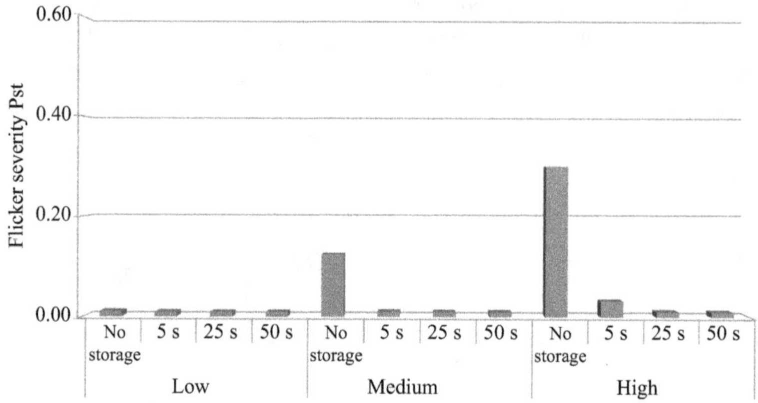

Figure 6.20 Flicker level with an OWC farm for wave energy levels (low, medium, high) and for several storage levels

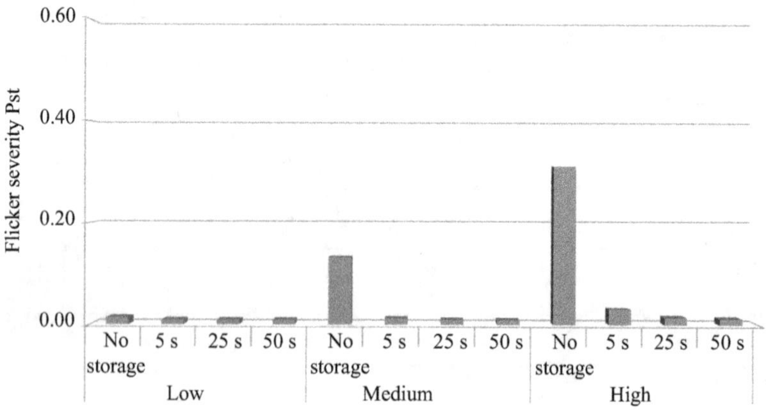

Figure 6.21 Flicker level with a point absorber farm for wave energy levels (low, medium, high) and for several storage levels

control strategies may deteriorate significantly the power quality in the local network to which the wave farms will be connected. Hence, this impact must be carefully assessed before any medium-size wave farm is allowed grid connection. However, the complexity of site-specific grid impact studies usually prevents detailed analyses from being included in the selection process of a suitable site location.

The short-circuit level S_{SC} and the impedance angle Ψ_k are two power quality criteria which help estimate the strength level of a grid, although its absolute strength is always relative to the amplitude of the injected power fluctuations. A weak grid, more prone to be negatively affected by the injection of fluctuating power than a stronger grid, is usually characterised by a low short-circuit level S_{SC} and a low grid impedance angle Ψ_k, which can be expressed as

$$S_{SC} = \sqrt{3}\, V\, I_{SC} \text{ and } \Psi_k = \frac{X_k}{R_k} \tag{6.1}$$

where V is the nominal voltage at a given node, I_{SC} its short-circuit current, and X_k and R_k are respectively the Thevenin reactance and resistance of the rest of the power system. Although several studies have already investigated the potential grid impact generated by a small- to medium-size wave farm, the ratio of the local grid's short-circuit level S_{SC} to the farm's maximum power S_{farm} was either relatively very low or very high [16–21]. At the time of publication, only one study was conducted regardless of the grid strength in [22]. However, only a small-size farm was considered. Hence, it appeared that the knowledge on the grid impact of a medium-size wave farm on connection points of various grid strength levels was relatively fragmented. In addition, the farm's orientation and device layout were not addressed from a power quality compliance perspective. Furthermore, worst case conditions, for instance in the case of a dominant wave direction perpendicular to the wave farm orientation, were not found to be analysed in any of those studies. It is in this context that a comprehensive case study was performed on the grid impact of a medium-size wave farm. The study presented in this chapter is an extract from this work [21].

This case study was intended to give an insight into this potential impact in the case of 20 MW-rated wave farm connected at distribution level. Its scope includes the analysis of the influence of a wave farm orientation and of its device layout on the maximum and minimum voltages it induces at the PCC, as well as on its flicker level. For the sake of illustration, the impact of a wave farm on the power quality of five test sites was also investigated specifically. These sites are either in operation or at a design stage and their grid strength levels are representative of the entire spectrum of strength level. Simulations were performed using DIgSILENT power system simulator 'PowerFactory'.

6.3.3.2 Modelling

Experimental data

The wave farm consists of 22 OWC devices. Input data in the form of generated electrical power output time series was provided as an outcome of the project CORES standing for 'Components for Ocean Renewable Energy Systems'. This FP7 European collaborative research project focused on the development of new concepts and components for power take-off, control, moorings, risers, data acquisition and instrumentation for floating wave devices [23]. The project was based on a floating quarter-scale OWC prototype deployed in Galway Bay, Ireland from March to May 2011. The device was connected to a small on-board islanded grid, shown in Figure 6.22, independent of the national electrical network and maintained by three DC/AC inverters. Generated power was used to charge the on-board battery system or dumped-in resistive load banks. A variable frequency

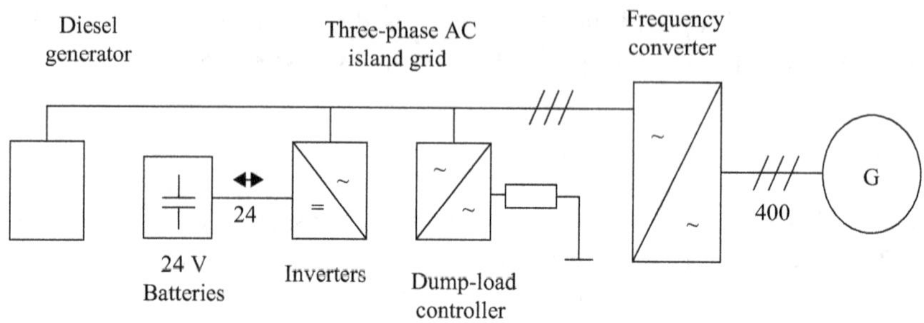

Figure 6.22 On-board island grid

Table 6.15 Characteristics of the different production periods

Production period	Full-scale H_s (m)	Full-scale T_z (s)	Energy level
A	5.7	8.4	High
B	2.4	5.6	Medium
C	1.3	7.0	Low

converter and a diesel generator were also included. The project has allowed the ocean energy research community to gain significant practical experience in the deployment, operation, and maintenance of offshore wave energy converters. It has also generated a considerable amount of time series data on a number of parameters, including electrical parameters at a high resolution of 0.1 s. Contrary to most available data which is averaged over a sea state, a season or even a year, the CORES electrical power time series data can be scaled and used directly for grid impact studies.

Three time series, whose significant wave height H_s and mean zero crossing period T_z are shown in Table 6.15, were selected in order to represent high-, medium- and low-energy sea states representative of different typical wave energy levels around the world. Although the high-energy level time series, named 'production period A' corresponds to relatively moderate sea conditions with respect to the sea climate of the AMETS test site in Ireland [25], it may represent very energetic conditions with respect to other sites benefiting from milder sea conditions. The power corresponding to each sea state is proportional to $H_s^2 T_z$.

During these three production periods, the generator was operated in constant speed control mode in which, unlike in variable speed operation, inertial energy storage by means of speed control is not available. As a result, mechanical power peaks are converted directly into electrical power peaks, which is expected to represent a worst case with respect to power quality. Figures 6.23–6.25 show the equivalent full-scale electrical power time series of the prototype for each production period.

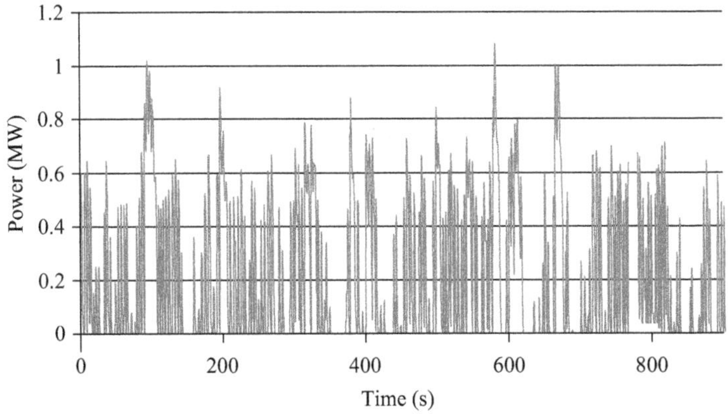

Figure 6.23 Individual power profile for production period A

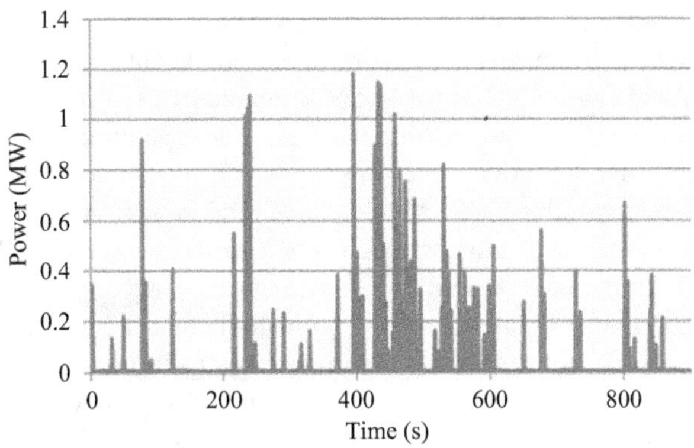

Figure 6.24 Individual power profile for production period B

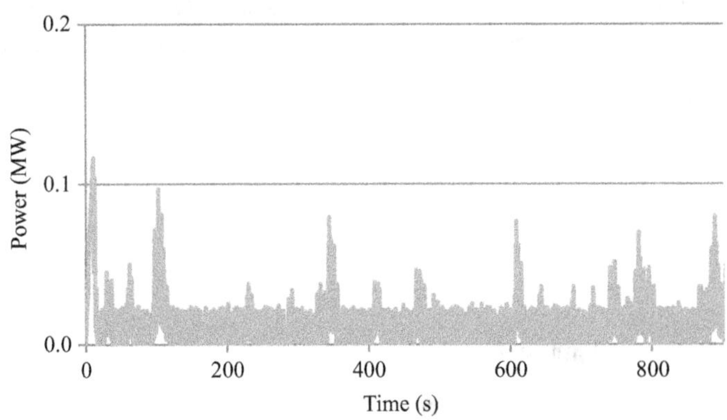

Figure 6.25 Individual power profile for production period C

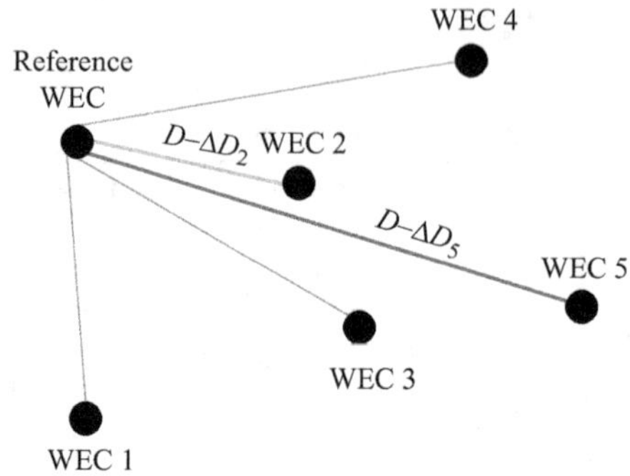

Figure 6.26 Schematic illustration of the time delay method

Device aggregation

Grouping several wave energy converters together is generally thought to be beneficial in terms of power quality by producing a smoother farm power output. This effect was modelled by means of random time delays which were applied individually to each generator power profile but one taken as the reference. Each time delay ΔT_{total} consists of a constant time delay ΔT corresponding to a constant inter-WEC distance D to which a variable time delay ΔT_i is subtracted. This latter time delay is intended to introduce a certain degree of randomness in the spatial layout of the devices within the farm, and corresponds to a variable distance ΔD_i as shown in Figure 6.26. The inter-WEC distance D is usually envisaged to be of the order of magnitude of hundreds of meters, ranging typically between 200 m and 500 m [26]. Each WEC group (i.e. connected to the same subsea cable) consists of a maximum of 5 or 6 WECs which may be laid out like a number of geometrical shapes. Considering the number of devices in each WEC group as well as the typically envisaged inter-WEC distance, it was assumed that most WECs would be located at 1000 m (at most) of the reference WEC, regardless of the geometrical shape adopted for the layout of each wave group. Hence, the fixed distance between the reference WEC and all the other WECs of the group was selected as $D = 1000$ m, to which the variable distance ΔD_i chosen randomly between 0% and 60% of D is subtracted. This leads to an inter-WEC distance $D - \Delta D_i$ ranging between 400 and 1000 m.

Time delays were calculated based on typical value of the wave group speed v_g estimated from the energy period T_e of a given sea state as [27]:

$$\Delta T_{total} = \Delta T - \Delta T_i = \frac{D}{v_g} - \frac{\Delta D_i}{v_g} \tag{6.2}$$

with $v_g = \dfrac{g}{2 \times 2\pi f}$

where v_g is estimated approximately as

$$v_g = \frac{gT_e}{4\pi} \tag{6.3}$$

where T_e is the energy period of the equivalent ideal sinusoidal wave whose power can be expressed as $P = 0.49H_s^2T_e$. The energy period was retained for the study rather than more common parameters such as the peak period T_p or the mean zero-crossing period T_z, as it was considered more representative of the energy propagation speed between two WECs. This estimation method is thought to provide a reasonable order of magnitude for the typical speed of any wave group during a given sea state. The characteristic period T_z for each production period, supplied as part of CORES data, is proportional to the energy period T_e by a factor ranging between 0.71 and 0.83 at the Irish AMETS test site [28]. An average value of 0.77 was hence selected for the studies. Five different time delay sets were used in order to model five different device layout combinations, the maximum flicker level obtained over these five simulations being retained for the study.

Electrical model

The wave farm network model, inspired from the concept design of the national wave test site of Ireland (AMETS) and located off Belmullet town, presents characteristics typical of current and planned test site designs, such as of the WaveHub, EMEC and bimep test sites. The farm is supplied by four subsea cables, two being 6.5 km long and the other two being 16 km long. An onshore substation, connecting the PCC to the rest of the network through a 5 km overhead line, steps the voltage up from 10 to 20 kV. A VAr compensator maintains power factor at the PCC at a fixed value, equal to unity in the base case. Simulations were also performed with different power factor values to study the influence of this control method on flicker level. A series reactor, whose impedance was varied to simulate different short-circuit levels *Ssc* and impedance angles Ψ_k, connected in series with an AC voltage source at 38 kV represents the rest of the national/regional network.

6.3.3.3 Simulation scenarios

The wave farm consists of 22 individual generators. The power profile of each generator is based on the individual electrical power time series presented in Figures 6.23–6.25, to which a random time delay is applied in order to represent the effect of device aggregation on the power output of the farm, as mentioned earlier.

The influence on flicker of the farm orientation with respect to the dominant wave direction was analysed, as well as the influence of the layout of its devices. Two orientations were selected, as shown by Figure 6.27. In the 'lateral' orientation, a greater number of generators output the same power peaks simultaneously (being applied the same time delay) than in the 'frontal' orientation. This former orientation induces a greater impact on power quality in the former case, as the amplitude of the farm's power fluctuations is expected to be higher. This is illustrated in Figure 6.28 which presents the power profile of a wave farm composed of four ideal wave devices outputting a sinusoidal power profile. These power outputs are either in-phase (thus simulating the lateral orientation) or dephased by a random time delay (which corresponds to the frontal orientation). It appears clearly that the amplitude of the power variations is greater in the former case.

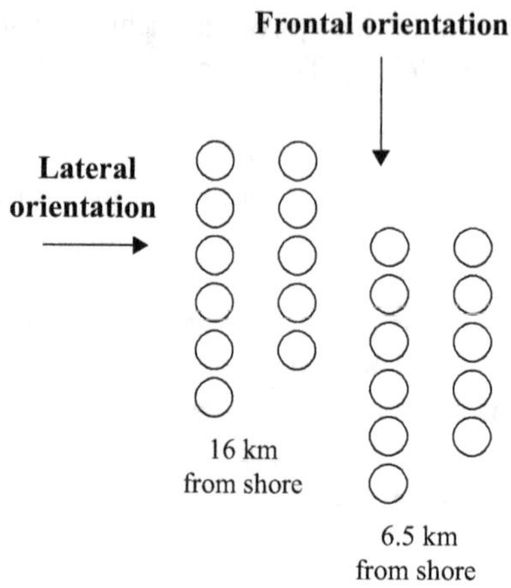

Figure 6.27 Orientations used in the simulations

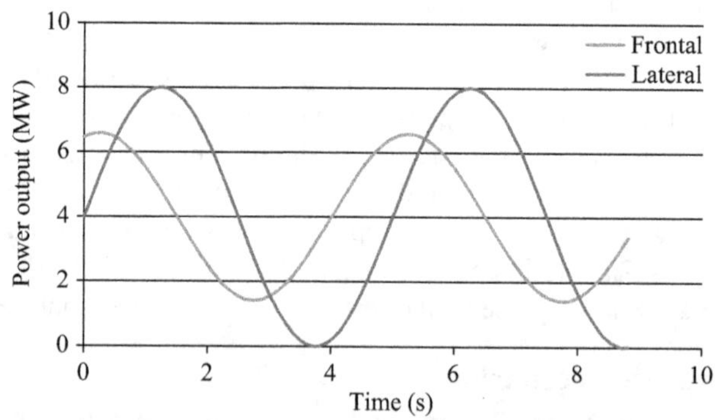

Figure 6.28 Power output of a wave farm consisting of four ideal devices outputting a sinusoidal power profile

Analyses were performed for several values of the series reactor's resistance and reactance, thus simulating several values of both the short-circuit level S_{SC} and of the impedance angle Ψ_k. Short-circuit values ranging between 58 MVA and 10 GVA were considered. The following values for the impedance angle Ψ_k were also used, as recommended by IEC Standard 61400-21 [15]: 30, 50, 70 and 85 degrees. As mentioned previously, compliance tests with respect to flicker requirements were included for five different test sites whose characteristics are presented in Table 6.16.

Estimation of the short-circuit characteristics

With the exception of the bimep test site whose characteristics were provided by Basque Energy Agency, Ente Vasco de la Energia, the short-circuit level and impedance angle at the other sites were estimated. This estimation was based on the short-circuit level at the closest 110 kV [9] or 400 kV [29] connection point for the

Table 6.16 Short-circuit characteristics at the terminals of several test sites

Test site	Country	S_{SC}/S_{farm}	Ψ_k (degrees)
Achill Island	Ireland	2.9	67
Belmullet	Ireland	3.2	69
Killard	Ireland	8.2	81
WaveHub	UK	32.3	78
bimep	Spain	232.0	90

Irish test sites and the WaveHub respectively, and on the impedance of the transformers and overhead lines between this connection point and the closest 38 kV (Ireland) or 33 kV (the United Kingdom) distribution level connection point. Impedances at the WaveHub and bimep terminals were converted into equivalent impedances at 38 kV, the voltage level used in the network model. This method is intended to estimate approximately the short-circuit level at a given test site and does not take into account the loads' consumption between the PCC and the nearest 110 or 400 kV connection point, this data being unavailable usually. However, the corresponding level of error on the short-circuit level and on the impedance angle Ψ_k was estimated as negligible [23]. It must be noted that, due to similarities in terms of power system architecture and typical grid strength between the rural areas of Ireland and Scotland, the results obtained for Belmullet and Achill Island are expected to be relatively similar to those of potential sites located in this latter region. It is interesting to highlight this parallel as these two regions present the highest wave energy potential in Europe.

Description of the test sites

The location of Belmullet is envisaged for two different projects led by the Sustainable Energy Authority of Ireland (SEAI) and by the WestWave consortium [30]. Achill Island and Killard are two other potential locations which are investigated by the latter. Killard, located relatively close to the largest power plant of Ireland (Moneypoint) and to a 400 kV connection point, has a higher short-circuit level and impedance angle than the other Irish sites whose low short-circuit levels are typical of the sparsely populated, rural areas of the north-west of Ireland. The WaveHub is a test site currently in operation in Cornwall, the United Kingdom, and awaiting its first devices. It is also located at a small distance from a 400 kV connection point, which explains its relatively high short-circuit level. The Spanish Biscay Marine Energy Platform (bimep) presents the strongest grid both in terms of short-circuit level and impedance angle and will be located off the Basque country. All of these test sites are currently, or planned to be, designed for a maximum power capacity of 20 MW.

6.3.3.4 Flickermeter design

A flickermeter was built in Matlab for the purpose of this study according to the design specifications of IEC Standards 61000-4-15 [31] and 61400-21. The

Table 6.17 Flicker level computed with several numbers of class

Production period	Number of classes				
	64	3200	6400	12800	25600
A	1.582	1.417	1.416	1.415	1.415
B	0.102	0.093	0.093	0.093	0.093
C	1.132	1.015	1.015	1.015	1.015

flickermeter used in the present study also includes the 'empty class' feature designed by Alcorn [16] and is to be used in the classifier (Block 5) for smoothing the cumulative probability function (CPF) in order to improve its accuracy. It consists of interpolating the ordinate of an empty class (if any) based on the ordinate of surrounding non-empty classes. Additional investigations were also carried out to determine the minimum number of classes necessary to analyse wave-induced flicker. The number of classes is actually recommended to be increased from 64 (as specified in IEC Standard 61000-4-15) to 6400 (by IEC Standard 61400-21) for the analysis of wind farms–induced flicker in order to capture flicker generation from small voltage variations in a sufficiently accurate way. This latter number seems also reasonable for the analysis of wave farms–induced flicker, as shown in Table 6.17 which presents the short-term flicker severity computed from production period A with five different number of classes. This table shows that the flicker level result remains stable from a number of classes greater or equal to 6400.

A high level of accuracy to the different performance tests indicated in IEC Standard 61000-4-15 was demonstrated by the flickermeter. Instantaneous flicker perceptibility equal to unity (within $\pm 1\%$) was obtained by varying the indicative voltage amplitudes by -3% to $+2\%$, hence well within the maximum allowed range of $\pm 5\%$. Similar results were obtained for the tests focusing on Block 5, short-term flicker severity being obtained within -1% to $+3\%$ of unity, hence also well within the maximum allowed range of $\pm 5\%$. Further details on the compliance of this flickermeter design with the IEC Standard 61000-4-15 can be found in [23].

6.3.3.5 Grid code requirements

A review of the requirements in terms of voltage and flicker limits was undertaken. Grid codes usually define the flicker limit in terms of the individual contribution of a power plant to the total flicker level at the PCC, or in terms of the total flicker level at the PCC itself, as shown in Table 6.18.

However, determining the flicker compliance of a wave farm based on only one of these two limits is irrelevant. Using the total flicker level at the PCC as the base for any compliance test necessitates having sufficient information on the flicker level prior to the connection of the wave farm, which is usually unavailable. On the other hand, using the flicker emission limit of $P_{st} = 0.35$ for assessing grid compliance may be unrealistic as, with the exception of the Republic of Ireland, this constitutes the minimum flicker emission limit which a grid operator is recommended to enforce, as defined in the IEC Standard 61000-3-7 [33].

Table 6.18 Flicker level limits

Code, standard or recommendation	Region/Country	Flicker level	Type
ESB's distribution code	Ireland	0.35	Individual
National Grid's code	Great Britain	1.0	Total
IEC Standard 61000-3-7	N/A	0.35 (*min*)	Individual
		0.8	Total
Nordic Grid code [32]	Norway, Sweden, Finland, East Denmark	1.0	Total

The standard defines also a method to determine the individual flicker limit to be assigned to a power plant. This limit represents fraction of the total allowed flicker emission at the connection point, which is estimated from the percentage of the plant rated power compared to the total, already installed power capacity connected at this point. However, this method may lead to unrealistically low flicker emission limits for installations having a small rated power, which is the reason why a minimum flicker emission limit equal to 0.35 is defined.

The most stringent as well as the most permissive flicker limits equal to 0.35 and 1.0 respectively were selected for the compliance test. A total flicker level at the PCC smaller than 0.35 indicates that flicker is not an issue and that the farm complies with the flicker level requirements in any case. A flicker level ranging between 0.35 and 1.0 indicates on the contrary that flicker may be an issue. In this case, detailed information on the background voltage variations at the connection point considered are needed to draw more definitive conclusions. Exceeding the 1.0 flicker level limit implies that the farm fails definitively to comply with the flicker level requirements even with a minimum pre-connection flicker level.

6.3.3.6 Compliance with flicker requirements

Flicker level at the PCC can reach very high values under worst case conditions, namely with production period A and with the lateral orientation, as shown in Figure 6.29. The minimum short-circuit ratio recommended for preventing flicker from exceeding unity is approximately equal to 5 for 30–50 degrees connection points, while points with a greater impedance angle are likely not to be affected by this issue, regarding their typically high short-circuit ratio.

As expected, the flicker level decreases as a function of the impedance angle Ψ_k which is characteristic of the operation at unity power factor. This can be easily explained by the greater influence active power fluctuations have on the voltage of resistive networks. This phenomenon is in agreement with the voltage variation ΔV which may be observed in a simple, two-bus system and which can be expressed by:

$$\Delta V = \frac{PR + QX}{V} \tag{6.4}$$

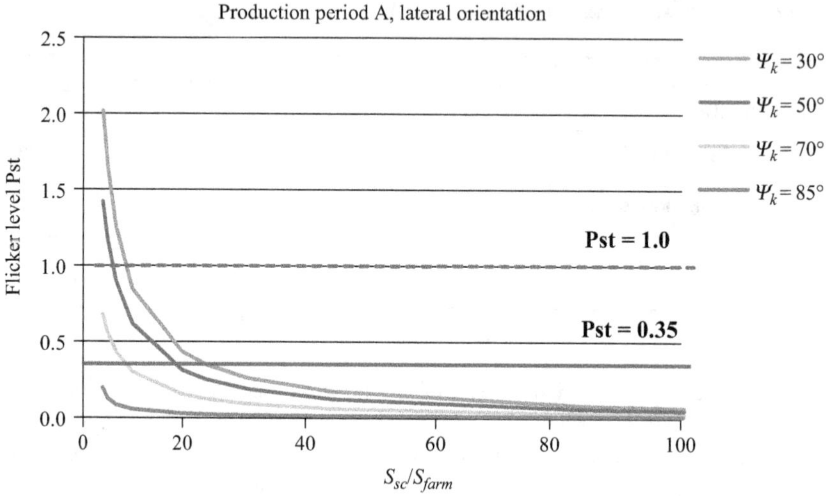

Figure 6.29 Flicker level P_{st} as a function of the short-circuit ratio (production period A, lateral orientation)

where P and Q are the active and reactive powers flowing through the impedance $\boldsymbol{Z} = R + jX$ connected between these two busses. The results seem reasonable compared to the typical minimum short-circuit ratio recommended for wind farm connections and equal to 25 for preventing flicker level from exceeding unity [34]. As expected, the recommended short-circuit limits are smaller in the case of wave farms. This can be explained by the lower perceptibility of wave-induced voltage fluctuations, due to their typically lower frequencies (and thus lower level of perceptibility) compared to those generated by wind farms, as shown in Figure 6.30 and suggested in [16].

However, with the exception of 85 degrees points, all connection points can be affected by flicker exceeding the most stringent limit equal to 0.35. The minimum recommended short-circuit ratio is approximately equal to 20 for 50 and 30 degrees respectively, falling down to 6 for 70 degrees points.

Flicker at the PCC can be partially mitigated by power factor control, as expected from the experience gained from the wind energy industry [35]. Figure 6.31 presents the flicker levels obtained at a connection point presenting the lowest short-circuit ratio investigated in this study and equal to 3. The following power factor values were investigated: unity, 0.95 and 0.92 lagging (referring here as absorbing reactive power). Results show that flicker level at 50 degrees points can be reduced down to 0.36, which is extremely close to the most stringent limit ($P_{st} = 0.35$). However, as power factor control can also induce a significant flicker level rise in highly reactive networks (i.e. with a high impedance angle Ψ_k), it is not recommended in this case. In addition, it must be stressed that complementary flicker mitigation means, such as storage and/or suitable power output control strategies, should be used when a farm is connected to highly resistive networks

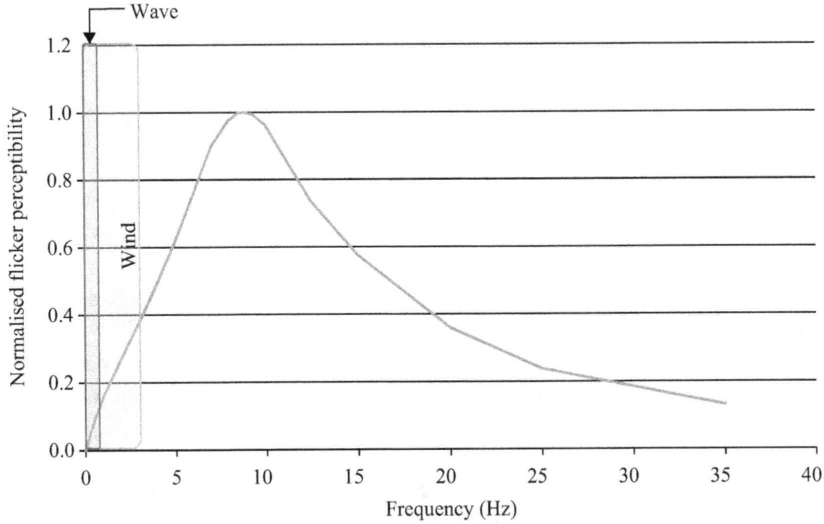

Figure 6.30 Flicker perceptibility curve as defined in IEC Standard 61000-4-15 [31]

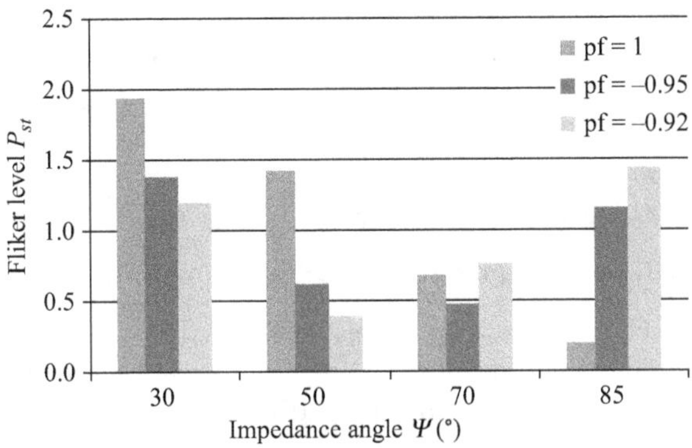

Figure 6.31 Flicker level P_{st} as a function of the impedance angle Ψ_k for different values of the power factor (pf) applied at the PCC

(i.e. low impedance angle Ψ_k) as power factor control is not sufficiently efficient in this case. Further analyses on this topic can be found in [23].

Regarding the different test sites, both the Belmullet and the Achill Island sites can be considered as failing to comply with the most stringent limit $P_{st} = 0.35$ which is equal to the limit enforced in Ireland. Hence, power factor control is insufficient in this case. However, wave farms connected to similarly weak connection points in countries or regions where a more permissive limit is enforced may be compliant provided that the pre-connection flicker level remains sufficiently low. However, evaluating the total flicker level taking into account both the background flicker and the wave farm's contribution may be relatively difficult. Flicker being a non-linear phenomenon, the total flicker level is not equal to the algebraic addition of the different flicker contributions by which it is generated. Hence, precise data on the background voltage fluctuations at potential connection points would be necessary to evaluate the total flicker level, focusing in particular on the fluctuations induced by load switching which represents the major source of flicker in distribution networks. This should be provided for instance in the form of voltage time series recorded prior to the wave farm connection, or in the form of a voltage spectrum. However, this type of data is usually not available. Hence, further investigations are needed when the flicker level lies between 0.35 and unity.

In conclusion, this study shows that wave farms without further mitigation means than power factor control at the PCC are not recommended for connecting to points whose impedance angle is less than or equal to 30 degrees. This recommendation is based on power quality considerations, but seems doubly justified by the expected power system congestion issues due to the typically low power transfer capacity available on low-voltage distribution networks to which these points usually belong. Connection points with an impedance angle of 50 degrees are also affected by this issue, which can however be mitigated to a certain extent by applying a lagging power factor. This technique may in many render cases enable the wave farm compliant with the flicker requirements, at least the most permissive. However, if the most stringent flicker limit is enforced, additional mitigation means such as storage or suitable power output control strategies may be required, as it may also be the case for farms connected to 70 degrees connection points. Connection points whose impedance angle is equal to 85 degrees are not affected by this issue.

More generally speaking, the results imply the following:

- Medium power capacity wave farms can be safely connected to relatively weak grids (down to 50 degrees) provided the power transfer capacity is sufficient and that suitable power factor control is applied. The possible utilisation of 50 degrees connection points is very interesting for regions or countries having a relatively weak power system, as it is the case in developing and some emerging countries. This potential utilisation is also interesting for providing power to partially damaged electrical power systems, for instance in the case of natural disasters or of man-made events [36].
- In addition, it appears that wave farm owners may not necessarily need to connect their plants to a very strong connection point which may be located very far inland, as it is the case in the rural areas of Ireland and Scotland. Hence, they may avoid the costly installation of a long overhead line between the test site's onshore substation and the inland connection point. From a financial point of view, this means that the expensive power systems' reinforcement necessary for facilitating the large-scale integration of wave farms can be postponed until the wave energy industry reaches a certain degree of commercial maturity. This represents a major asset.

Influence of the wave farm layout on flicker level

This section investigates the influence on flicker of the farm layout design in terms of orientation with respect to the dominant wave direction and in terms of different device layouts. This study was performed based on the observation that, although the wave direction is relatively constant throughout the year, it may vary widely over shorter periods of time that are yet long enough to give rise to power quality issues [37].

As expected, the results show that the impact of the farm layout decreases as a function of the short-circuit ratio and of the impedance angle at the connection point. They also demonstrate that the farm orientation and its device layouts may have a very significant influence on the flicker level, the effect of the farm orientation being predominant. Both may be determining factors for the farm's compliance to flicker requirements regarding a number of short-circuit ratio and impedance angle combinations. As shown in Figures 6.31 and 6.32, although flicker level exceeds the most permissive limit for a relatively large number of short-circuit ratio/impedance angle combinations with the lateral orientation (Figure 6.32), none of them exceeds this limit when the frontal orientation is used (Figure 6.33).

Results shown in Figure 6.33 demonstrate that the influence of the wave farm orientation is always very significant if it is connected to 30–50 degrees points, considering their typically low short-circuit level. The difference observed between the frontal and the lateral farm orientations is still greater or equal to 0.1 (i.e. 10% of the most permissive limit $P_{st} = 1.0$) up to a short-circuit ratio equal to 13 for 70 degrees connection points, thus including both the Belmullet and Achill Island test sites. The influence of the farm orientation on connection points whose impedance angle is as high as 85 degrees can be considered as always negligible. The influence on flicker of the device layout within the farm is far more limited, as shown in Figure 6.34, although it may be significant for 30–50 degrees connection points.

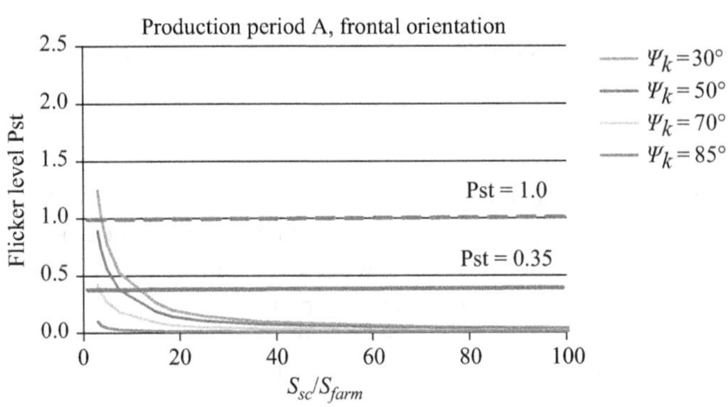

Figure 6.32 Flicker level P_{st} as a function of the short-circuit ratio (frontal orientation)

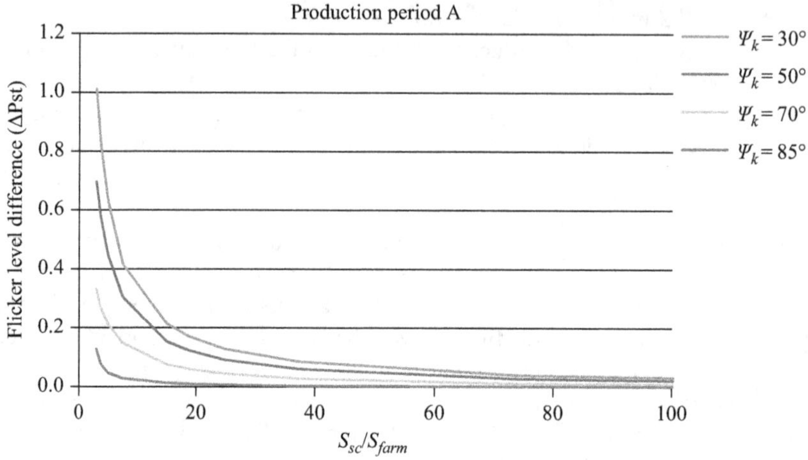

Figure 6.33 Maximum flicker level difference between the lateral and the frontal farm orientations

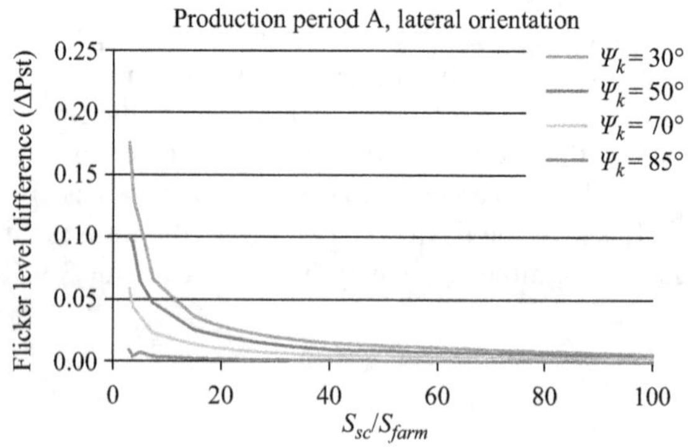

Figure 6.34 Maximum flicker level difference between five different device layouts

In conclusion, the orientation of the wave farm has a strong influence on the flicker level at the PCC when the farm is connected to points whose impedance angle is less than or equal to 70 degrees, whereas the influence of the device layout within each WEC group is rather limited to connections with an impedance angle connection points up to 50 degrees. Wave farm orientation, and to a lesser extent the device layout, can be determining factors in the wave farm success or failure in complying with flicker requirements. Hence, power quality is a constraint that should be taken into account when designing a wave farm, in addition to other aspects such as energy capture maximisation or cable length minimisation.

6.3.3.7 Conclusions

This case study detailed the grid impact in terms of flicker level of a medium-size wave farm connected to points presenting a wide range of grid strength levels, demonstrating that flicker may constitute an issue under certain conditions. More specifically, this study shows that this type of wave farm can be safely connected, from a flicker perspective to points whose impedance angle may be as low as 50 degrees, provided that suitable power factor control is applied at the PCC. However, this technique being not always sufficient, additional mitigation means, such as storage and/or suitable power output control strategies, may be required in some cases.

In addition, this article addressed the influence of the wave farm orientation and of its device layout on flicker level, the influence of the former factor proving to be highly predominant. Both the orientation and the device layout were also demonstrated to be potentially key factors in the farm meeting the flicker level requirements.

Major questions remain, however, concerning the maximum power capacity which can be connected safely to a single point of the distribution network from a power quality perspective. Beyond this limit, wave farms will be required to connect to the transmission system and the very important investments necessary to realise this kind of connection will be accessible to fully mature technologies only. In that respect, this limit may constitute an important milestone in the large-scale development of the ocean energy industry: its estimation would thus contribute in refining the long-term vision of the ocean industry and research community. Another concern regards the criteria used in the assessment of power quality impact. Due to the very fluctuating nature of wave farm electricity, it becomes clear that some current power quality criteria, for instance power ramp rates averaged over several minutes, are not relevant nor applicable in the context of wave farm grid connection. In addition, they may fail in capturing the farm impact in a sufficiently comprehensive way. Hence, new assessment methodologies must be developed, as well as novel power quality criteria, both of which are currently investigated as part of the IEC technical committee 114, PT 62600-30 [38].

6.4 Capacity value of wave energy

D. Kavanagh, *A. Keane* and *D. Flynn*

6.4.1 Capacity factor and capacity value: measuring generation system adequacy

Power systems are developed to achieve a balance between the demand requirements of consumers and the available power generation capacity on the system. System demand is not constant, it can vary on an hourly, daily and seasonal basis. A power system needs to be designed so that sufficient generation capacity is available at times of peak demand. The generation portfolio on a power system is normally composed of a large number of generation units, often with dissimilar technical and economic characteristics. Reserve requirements are drawn up so that loss of the largest single power infeed can be covered by other generators. Unit commitment seeks to achieve the optimal dispatch of generating units for any given system conditions.

The exploitation of renewable energy resources provides ever greater challenges for power system operators. These resources vary, not just with the technology concerned, but also with the location in which they are to be utilised. Power system planners are primarily concerned with the maintenance of electrical supply with a high level of system reliability. This presents many challenges with regard to the integration of variable energy resources onto power systems, but are more evident on relatively small isolated power systems. One such system is taken as an example here to demonstrate some of the capacity factor and capacity value issues associated with large-scale deployment of renewable energy resources.

The maintenance of electrical supply above a predefined level of system reliability is a fundamental concern of any power system operator. Power systems are designed such that they can handle most contingency events while still being able to economically supply the electrical demand placed on them. The generation adequacy of a system measured by a metric is called the loss of load expectation (LOLE). This metric is an important standard when assessing whether there is sufficient generation capacity in place to meet demand and to cover for unit outages. The generation adequacy standard for the case study system in this chapter is 8 h LOLE per year [39].

A useful metric employed to assess the value of a generation unit over time is the capacity factor of that unit, defined as the ratio of actual generation to the maximum theoretically possible generation during a particular study period, i.e. the average power output. While it cannot describe the variation in output that may be experienced during shorter timeframes, it is nevertheless a useful indicator which can be used for comparing different units. The capacity factor of a wave power generation unit is found by averaging the normalised power output across an entire time series. However, a more insightful metric for establishing the value to a power system of a generation unit is the capacity value of that unit.

A generation unit's capacity value (also known as its capacity credit) is generally determined by its effective load carrying capability (ELCC) [40–42], and is broadly defined as the additional load on a power system which can be served by that generation unit while maintaining the existing level of generation adequacy, dependent on the unit's reliability. A high-capacity value would increase the system's ability to meet its peak demand, aiding in system operation, making wave power an important component of the power system. However, a low-capacity value may lead to it being viewed as an energy resource with little to provide in terms of capacity on the power system. Fundamentally, if wave power devices are to have a high-capacity value, they must be able to generate consistently during times of elevated demand. If this is not the case then additional plant capacity or interconnection to other systems may also be required. This will lead to various challenges with regard to their system integration in different power systems based on the underlying shape of their demand curves. A typical daily demand curve for a Northern European country in Winter is shown in Figure 6.35. From both an

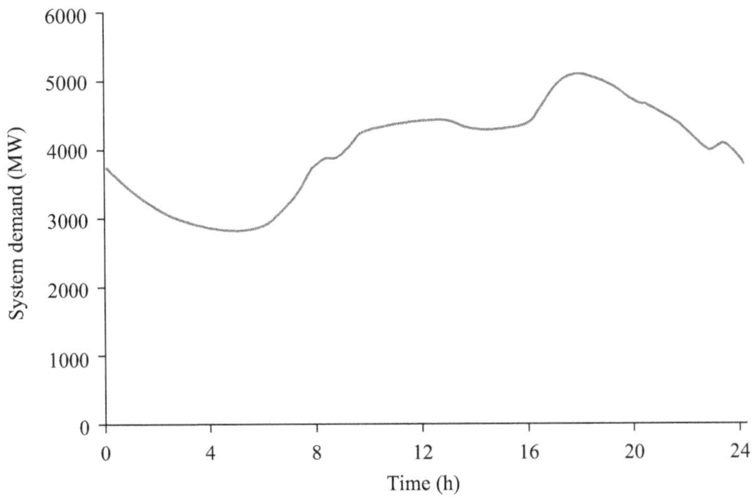

Figure 6.35 System demand curve on a day of peak demand

operational and an economic perspective it would be preferable for wave power to mirror trends in the system demand curve, both on a daily and seasonal basis, hence reducing the impact on conventional generation sources.

Co-siting of wind and wave farms has been suggested in order to reduce local variations in power output from wave and wind farms [43, 44]; however, the wider effects that will be experienced across the power system are of interest. Factors that affect the capacity value of a wave energy resource in the longer term include any spatial and temporal diversity which may occur. Additionally, global changes in climate may reduce this resource [45, 46], so inter-year variability must be accounted for when verifying the results of any capacity value calculations. Potential coincidence of wave power generation and system peak demand would be a crucial factor in this regard. Wave power generation may complement the system-wide wind power generation on any power system, or indeed may exacerbate any variability problems experienced with regard to power system integration, depending on the underlying correlation between them, and with the system demand.

6.4.2 Capacity value calculation

The capacity value of a generation unit is taken as being the ELCC, as described by Garver [40]. The ELCC of a generation unit is described in (6.5), where L_{old} represents the load which may be served with the original generation portfolio and L_{new} represents the load which may be served with the additional generation included in the generation portfolio, while maintaining the same LOLE.

$$\text{ELCC} = L_{\text{new}} - L_{\text{old}} \tag{6.5}$$

The capacity value is a probabilistic measure of system adequacy which indicates whether sufficient generation is in place to meet future demand requirements for a predefined LOLE level. It effectively indicates the generation adequacy of a power system, predicting whether sufficient generation capacity will be available during peak demand periods [47]. A conventional generation unit's ELCC is dependent on the forced outage rate (FOR) of that unit. Calculation of the ELCC of a system with variable generation resources can be carried out in different ways, with both chronological and probabilistic approaches being available [48]. The chronological approach creates a probabilistic capacity value that is derived from system observation in the time domain, whereas the probabilistic approach determines the capacity value by convolving the variable resource's probability density function with conventional power plant probabilities.

Common data requirements between both approaches include a correct load time series (usually hourly) for the period of investigation, a target reliability level, and an inventory of all conventional generation units on the system along with their respective FORs. In addition to this, the chronological approach requires an unbiased time series of power output from the variable generation resources for the same period as the load time series, with data recorded at the same time resolution. The probabilistic approach however requires a probability density function for the variable generation resource that can accurately represent the same period as the load time series. It is usually preferable to use the chronological approach as the probabilistic approach will not sufficiently account for the variability of the resource and is less accurate as a result. Use of the probabilistic approach normally occurs in situations where there is a lack of data required for the chronological approach.

The method employed here is the chronological approach described by Hasche *et al.* [41]. In an iterative process, a CPF is established given the probability of each possible generation level being available on the system. The system LOLE is determined in (6.6) given this CPF and the time series of system demand, where F_C is the cumulative probability distribution function of conventional generation capacity, while l_t is the load level at time step t, for all time steps T. The loss of load probability (LOLP) is determined for every hour of the year, with the sum of the LOLPs amounting to the LOLE for the year. To obtain the system's ELCC, the system demand is incrementally increased/decreased, as appropriate, until the target LOLE level is attained. The demand surplus/deficit necessary to achieve this gives the ELCC for the system.

$$\mathrm{LOLE} = \sum_{t=1}^{T} F_C(l_t) \tag{6.6}$$

Due to the uncontrollable nature of variable energy resources such as wind and wave power, and also due to uncertain correlations with demand patterns, the ELCC of these resources is not easy to calculate. As such, variable energy resources such as wind and wave power are treated as instantaneous negative load on the system, reducing the overall net load served by the system's conventional generation units at any given time. In order to determine the capacity value of wave power, the ELCC of the whole system is established both with and without wave energy devices installed, with the difference between the two figures being the capacity value of wave power. This method automatically captures the underlying relationship between system demand and wave power.

The ELCC of wave power generation may be increased through spatial diversity. It is likely that wave farms will be deployed as geographically dispersed clusters. Hence, it is of interest to analyse what impact, if any, this may have on the capacity value of wave power. While it may seem advantageous to minimise the electrical infrastructure required for installation, an increase in the capacity value resulting from such a spatial variation may pose greater benefits.

As a measure of conventional system reliability, capacity value is dependent on predetermined factors affecting the FORs of conventional generation units. It is not time dependent. Variable generation resources such as wind and wave power are uncontrollable. Hence, any wave power capacity value calculations are valid only for the period in which the data is analysed. The capacity value of wave power will be system specific, dependent on factors such as the correlation between system demand and wave power generation, and the variation in wave climates experienced coincidentally at different sites offshore.

6.4.3 Case study: Ireland

Ireland is taken as an example case study due to its position as a relatively small isolated power system with the potential for large wave power installations. Existing interconnection consists of a 500 MW HVDC submarine link from Northern Ireland to Scotland, and another 500 MW HVDC submarine link from the east coast of Ireland to the north coast of Wales. Conventional generation plant consists of a portfolio of mainly thermal plant, with a small amount of hydropower. Various wave generation device portfolios are analysed to determine whether a significant capacity value benefit occurs with increased spatial variation. The portfolios are composed of data from four offshore wave data measurement buoys (L1–L4), part of a network of buoys maintained by the Irish Marine Institute. The location of each buoy is highlighted in Figure 6.36.

Global climatic processes are the driving force behind the wave resources experienced off the coast of Ireland. Therefore, global phenomena can have a large impact on any potential ELCC results. The capacity value of wave power in Ireland is compared over a number of years in order to analyse any variation over that period. A constant LOLE of 8 h per year is used throughout. The power matrix of the Pelamis P-750 wave energy device is publicly available [49], and hence is employed here as a typical device's power matrix for the generation of a time series of wave power output from 500 MW of wave generation capacity. The capacity factor of wave power farms off the Irish coast would display a similar seasonal variation, in shape (high in winter, low in summer), rather than magnitude, to wind generation on the Irish system, as illustrated in Figure 6.37, with power output shown in normalised per unit values.

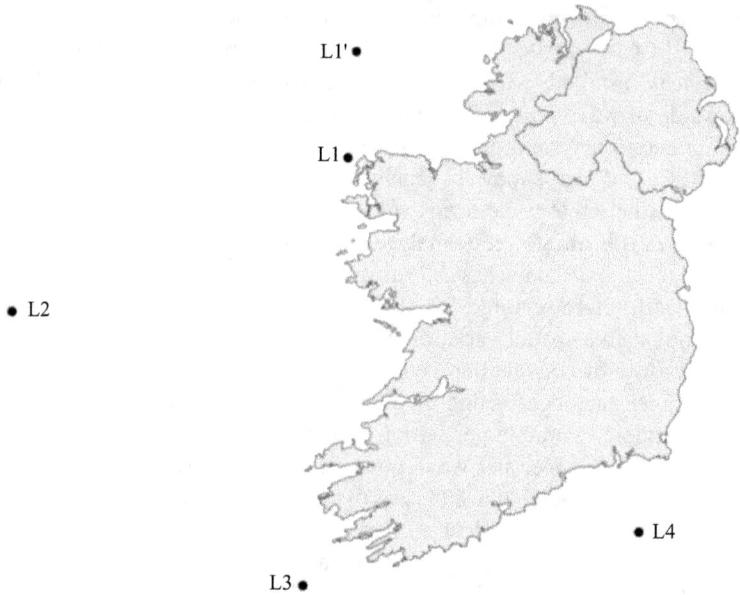

Figure 6.36 Location of offshore measurement buoys

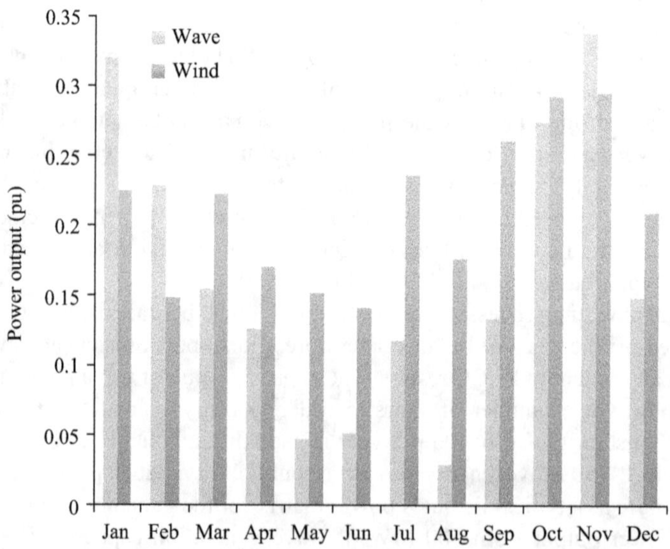

Figure 6.37 Seasonal (2010) variation of capacity factor for wind/wave power at site L3

Wave power has a large seasonal variation which mirrors variation in the system demand, and makes a higher capacity value more likely since high wave power would tend to coincide with periods of high demand. The hourly variation can be comparable to other variable energy resources such as wind power, with variations of up to 5% being experienced in less than an hour [50]. The seasonal variation in wave power output is greater than that for wind power. While the capacity factor for a wave farm at a site off Ireland's west coast (site L3) over 1 year would have been 0.16 per unit (pu) for the year shown, this can mask the fact that average hourly capacity factors in August and November 2010 could have varied between 0.03 pu and 0.34 pu respectively.

Significant seasonal variation occurs while the hourly variation is similar to that experienced by wind power over the same period. It is important to note that the variability seen for a limited number of wave generation sites is comparable to results obtained for system-wide wind generation, which benefits from aggregation of wind farms experiencing different weather conditions across the country. The hourly variation is out of phase with wind power however, raising the possibility of beneficial aggregation of wave and wind power output. Another variation experienced by wave farms is a spatial one, in relation to whether individual devices are positioned in seas with large swell waves from the Atlantic Ocean or in seas where the majority of waves are locally generated wind waves.

Some diversity in output occurs from different locations off the west coast of Ireland, as illustrated by Figure 6.38, indicating that the wave conditions closer to shore are not coincident with wave conditions experienced further offshore. Through spatial diversity it may be possible to site wave farms such that the hourly variation in their aggregated output is reduced from the perspective of the system operator, also aiding in the forecasting of their combined power output. Finally, different devices may react differently to similar sea states. As such, it is possible that certain devices can extract more wave energy in some locations due to developers employing different electrical downrating strategies [52].

Overall, the wave energy resource can be characterised as having a large seasonal variation, with a low average hourly variation. Careful selection of wave farm sites may act to minimise variations due to the benefits of geographical aggregation. Aside from the obvious economic implications for wave farm owners, such variations will pose many challenges with regard to power system integration. This is particularly the case in relation to Ireland, as it is a relatively small system with limited interconnection. However, the wave resource tends to peak during months in which the system demand peaks, with the reduced wave resource also coinciding with months of lower demand.

With the target LOLE being 8 h per year, the capacity value of wave power in 2010 at sites off the west coast of Ireland is found to be comparable to system-wide wind generation over the same period of time. Figure 6.39 illustrates the capacity value for various wave farm locations, with wave power considered to consist of 500 MW of installed capacity, and FORs between 5% and 30% being applied.

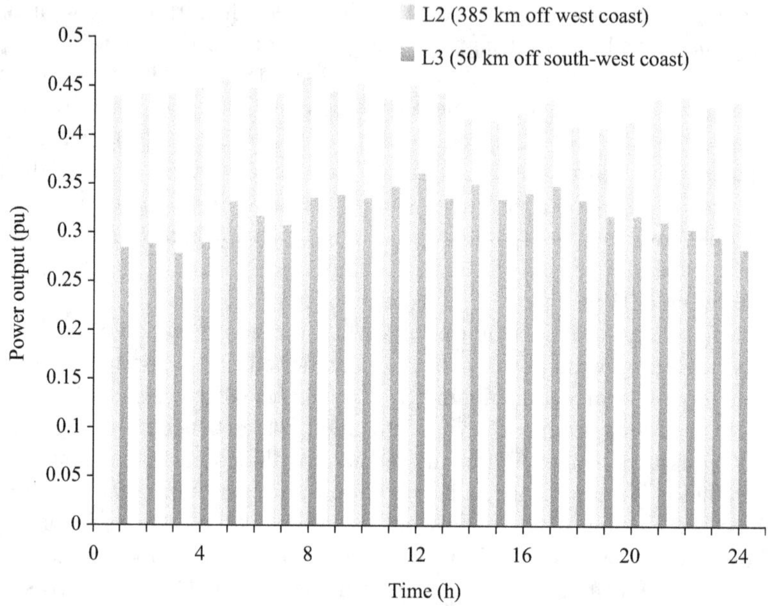

Figure 6.38 Average hourly variation of wave farms at different sites off Ireland's west coast during January 2010

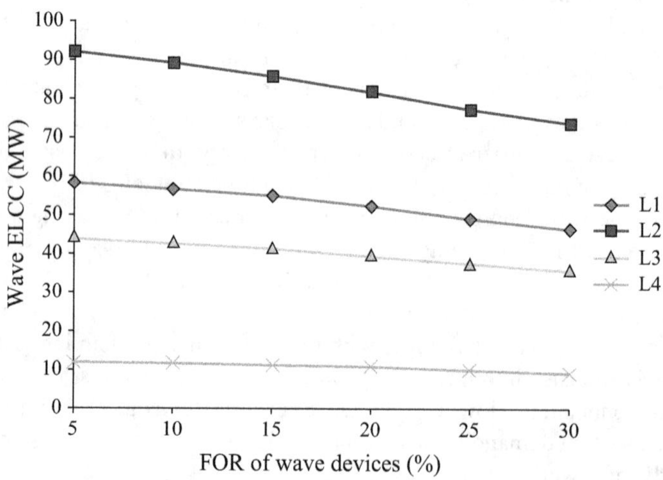

Figure 6.39 Variation of capacity value with increased FOR (single sites, installed capacity of 500 MW)

Installed dispatchable capacity reached 6829 MW in 2010, while system load varied between 1579 MW and 5090 MW. Taking L3 as an example, 43 MW (8.6%) of wave generation capacity could be relied upon during peak demand periods in 2010. This compares to a capacity factor of 16% during the same period,

Table 6.19 Capacity value with 500 MW wave farms in 2010 (10% FOR)

Location of wave farm	Capacity value of wind power	Capacity value of wave power (wind on system)	Capacity value of wave power (no wind on system)
L1	180 MW	57 MW	64 MW
L2	180 MW	89 MW	112 MW
L3	180 MW	43 MW	59 MW
L4	180 MW	12 MW	21 MW

highlighting the necessity to understand the capacity value of a potential energy resource, not relying solely on its capacity factor as a measure of its value to the system. Electricity prices will also tend to be higher during peak demand periods, affecting the economic potential of wave energy devices, while in some countries, such as Ireland, capacity payments, in addition to energy payments, are also made which further values the timing of energy production [51].

Table 6.19 demonstrates the spatial diversity across the various sites. Of particular interest is the capacity value obtained for the L1 site, as this is the location of a full-scale wave energy test site currently under development. A 500 MW farm deployed here would have experienced a capacity value of 57 MW (11.4% of 500 MW). Meanwhile, had 500 MW farms been deployed at sites L2, L3 and L4 they would have witnessed capacity values of 89 (17.8%), 43 (8.6%) and 12 MW (2.4%) respectively. Comparison can be drawn with the capacity value of system-wide wind power generation during the same period of 180 MW (12.8% of installed capacity). It must be noted, however, that 2010 was a very poor year for both wind and potential wave generation across the system, with lower production being observed across the year than in previous years. More typical values would be in the region of 20–25% for wave and wind power.

Table 6.19 and Figure 6.39 highlight the diminishing marginal capacity value of increased levels of variable energy resources on a power system. Had no wind farms been installed, the capacity value of 500 MW of wave power devices would have been significantly higher at all sites (up to 75% greater off the south-east coast compared to values obtained with wind power on the system). One of the largest increases observed with system wind power excluded is the capacity value of wave power at L3, off the south-west coast, which is indicative of a high correlation with wind power across the system. The highest observed values in all scenarios occurred at L2; however, given L2's remote location it is unlikely that significant levels of wave power capacity will be deployed here. The lowest observed increase is at L1, off the north-west coast, which suggests that L1 is correlated less with wind power, an advantage which can be exploited further in the deployment of wave energy devices. While the capacity value of wave power may be higher without wind farms on the system, large amounts of wind generation capacity have nevertheless been installed over the past decade, necessitating its inclusion in any further analysis of the capacity value of wave power.

Wave power could provide a capacity value for the power system comparable to that provided by wind power. Certain sites can provide quite high-capacity values for wave power. While these values appear favourable when compared to wind power, it is important to note that wind power data is aggregated across the entire power system and is the net result of a mixture of both the best and other sites for wind power generation. Some wind power sites may very well have capacity values which exceed those seen for wave power.

Given the spatial diversity highlighted earlier in Figure 6.38, the potential impact of deploying the 500 MW of installed capacity across multiple sites can be expected to result in a less variable overall power output due to the impact of aggregation across the various sites. To illustrate this, five potential wave power capacity portfolios are analysed, listed as Portfolios 1–5 in Table 6.20. A strong weighting is given towards a distribution of wave devices just off the north-west and south-west coasts, with these locations experiencing a high wave energy resource while also benefitting from being relatively close to shore (minimising offshore electrical network requirements).

The north-west coast is allocated a slightly greater emphasis, given its location as a wave energy test site, and the lower correlation with system wind power suggested above. Hence, the presence of facilities such as an offshore grid connection, along with other benefits such as the relatively large wave energy resource, may encourage further device deployment in the surrounding areas. L3 also has a strong resource, but would require additional electrical network facilities, and hence this area is assigned a capacity less than that assigned to L1. L4 is off Ireland's south-east coast, and hence has the lowest potential resource of all the sites, but is assigned a larger capacity than L2 in some portfolios as it is not as far offshore, minimising installation and maintenance issues. The results of the capacity value analysis for portfolios 1–5 are presented in Table 6.21, and are illustrated in Figure 6.40.

Table 6.20 Wave power capacity portfolios

Portfolio	**1**	**2**	**3**	**4**	**5**
L1 wave farm (MW)	300	300	200	300	250
L2 wave farm (MW)	0	0	50	50	75
L3 wave farm (MW)	200	150	150	150	175
L4 wave farm (MW)	0	50	100	0	0

Table 6.21 Capacity value varying by wave farm capacity allocation

Portfolio	**1**	**2**	**3**	**4**	**5**
Wind power (MW)	180	180	180	180	180
Wave power (10% FOR) (MW)	53	50	50	58	59
Wave power (20% FOR) (MW)	48	46	46	53	54
Wave power (30% FOR) (MW)	43	41	40	47	48

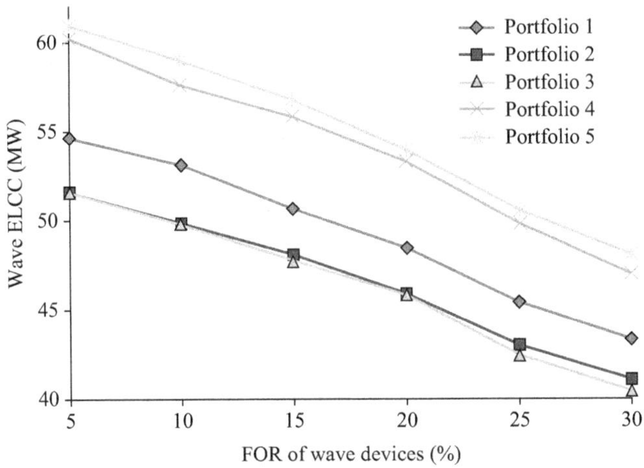

Figure 6.40 Variation of capacity value with increased FOR (multiple sites, installed capacity of 500 MW)

An approximately linear reduction in capacity value is observed for increases in the device FORs in all cases. The results suggest little difference between the highest and lowest capacity values of the portfolios analysed, which is indicative of the limited aggregation benefits to be found across various sites. Higher capacity values would be experienced if a greater weighting was given to L2, far offshore; however, such a benefit would be significantly offset by the logistical and technical difficulties of deployment in the area. These difficulties lead to a greater emphasis being placed on sites closer to shore, with less variation between portfolios being experienced as a result.

There is a large, reliable wave energy resource to be found further offshore. However, the difficulties involved with deploying devices so far from the coast act to negate these benefits. Nevertheless a significant capacity value is also experienced closer to shore at sites which can be considered feasible for the deployment of large numbers of wave energy devices. The effect of aggregation across multiple sites yields marginal benefits in terms of an increased capacity value overall, compared to the capacity value of wave power at single sites. Given the likely deployment of a small number of clusters of large wave farms, there shouldn't be a large loss of capacity value from a scenario involving a wide geographic distribution of wave energy devices.

The results presented up to now are for the year 2010. Four years of data is taken for analysis: 2007, 2008, 2009 and 2010. The buoy at L1 was installed in late 2009; therefore, no data is available for analysis prior to that date. Given its relative proximity to L1, another measurement buoy (L1′ in Figure 6.36) located north of L1 is deemed to be a suitable substitute for the years 2007, 2008 and 2009. L4 was offline for an extended period in 2008. As such two of the five portfolios (portfolios 2 and 3) used thus far are only suitable for analysis in the years 2007, 2009 and

Table 6.22 Annual wind and wave power portfolio capacity values

	FOR	2007	2008	2009	2010
System wind	n/a	234 MW	246 MW	224 MW	180 MW
Portfolio 1	10%	103 MW	146 MW	131 MW	53 MW
	20%	96 MW	134 MW	121 MW	48 MW
	30%	85 MW	119 MW	108 MW	43 MW
Portfolio 2	10%	97 MW	n/a	124 MW	50 MW
	20%	90 MW	n/a	114 MW	46 MW
	30%	81 MW	n/a	102 MW	41 MW
Portfolio 3	10%	98 MW	n/a	115 MW	50 MW
	20%	91 MW	n/a	106 MW	46 MW
	30%	80 MW	n/a	93 MW	40 MW
Portfolio 4	10%	113 MW	149 MW	134 MW	58 MW
	20%	106 MW	137 MW	124 MW	53 MW
	30%	93 MW	121 MW	110 MW	47 MW
Portfolio 5	10%	117 MW	150 MW	134 MW	59 MW
	20%	108 MW	137 MW	123 MW	54 MW
	30%	95 MW	122 MW	110 MW	48 MW

2010. Three portfolios (portfolios 1, 4 and 5) are examined for the full 4 years, while the two remaining portfolios are examined for 2007, 2009 and 2010. Given L4's low capacity value, there would be little impact on the overall results. Table 6.22 illustrates a large variation across the 4 years for the capacity values of each portfolio being examined, with peak values of up to 150 MW (30%) occurring in 2008 for portfolio 5, and minimum values as low as 40 MW (8%) occurring in 2010 for portfolio 3.

ELCC figures for a representative FOR of 10% are illustrated in Figure 6.41 along with the variation in the capacity value of system-wide wind power over the same period. It is important to note that the installed wind power generation increased substantially over the study period (almost doubling from 805 to 1406 MW over the 4 years), and hence it is more appropriate to view these figures as a percentage of installed capacity. The wind capacity value increased by 12 MW in absolute terms from 2007 to 2008, but decreased as a percentage of installed capacity from 29% to 25%. Such a decrease can be attributed both to underlying inter-year trends in wind power and to the diminishing marginal capacity value of increased variable energy resources on the power system. The wave energy resource experienced off the coast of Ireland has a significant capacity value (up to 30% of installed capacity in 2008). The capacity value is enhanced slightly through spatial diversity, with a number of sites providing the potential for the deployment of wave energy farms. Should substantial levels of installed capacity be reached, it will provide an important contribution to the generation adequacy of the power system.

The overall trend over the 4 years is an initial rise (peaking at 150 MW in 2008 in Table 6.22), followed by a decline in the capacity value of wave power (reaching

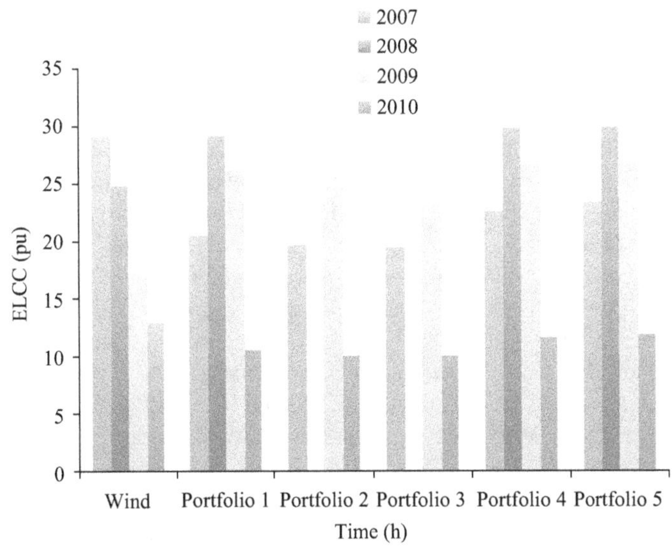

Figure 6.41 Inter-year variation for wind and wave (for a 10% wave FOR) (data unavailable in 2008 for portfolios 2 and 3)

as low as 40 MW in 2010), indicative of a decline in the available resource over that period. The wind power capacity value (as a percentage of installed capacity) appears to have declined continuously over the 4 years. Wave power could have experienced a higher capacity value in 2008 and 2009, with the year-on-year drop in 2008/2009 being lower for wave power than for wind. However, 2010 was a much worse year for wave power with its capacity value falling below that of wind. While the capacity value of wave power appeared consistent across the period 2007–2009, its low value in 2010 highlights the likelihood of large inter-annual variations, reducing its reliability as a source of firm generation capacity on a power system. While the wave energy resource off the Irish coast can exceed that experienced in many other parts of the world [52, 53], it could also provide a significant contribution to the capacity value of the power system, hence contributing to the generation adequacy of the entire system [54].

6.5 References

[1] T. Thiringer, J. MacEnri and M. Reed, 'Flicker evaluation of the SeaGen tidal power plant', *IEEE Transactions on Sustainable Energy*, vol. 2, no. 4, pp. 414–422, October 2011.

[2] J. MacEnri, M. Reed and T. Thiringer, 'Influence of tidal parameters on SeaGen flicker performance', Philosophical Transactions Series A, Mathematical Physical and Engineering Sciences, vol. 371, no. 1985. doi: 10.1098/rsta.2012.0247, January 2013.

[3] P. Ricci, J.B. Saulnier, F.O. António and M.T. Pontes, 'Time domain models and wave energy converters performance assessment'. *Proceedings of the ASMET 27th International Conference on Offshore Mechanics and Artic Engineering, 15–20 June, 2008*. Estoril, Portugal, 2008.
[4] EN50160:2010: Voltage Characteristics in Public Distribution Systems.
[5] J. MacEnri, M. Reed and T. Thiringer, 'Power quality performance of the tidal energy converter, SeaGen', *ASME 30th International Conference on Ocean, Offshore & Arctic Engineering*. Paper No. 49549, June 2011.
[6] F. Salcedo J. L. Villate, Y. Torre-Enciso, M. Santos and D. Ben Haim. *Grid Integration of Wave Energy Farms: Basque Country Study*. ICOE 2010, October 2010.
[7] M. Egan, R. Alcorn, T. Lewis, A. Blavette and D. O'Sullivan, *Wave Energy Grid Integration in Ireland – a Case Study*. ICOE 2010, October 2010.
[8] bimep (Biscay Marine Energy Platform). Available at: http://www.eve.es/bimep/
[9] EirGrid, 'Transmission forecast statement'. Available at: http://www.eirgrid.com/transmission/transmissionforecaststatement/
[10] W. H. Michel, 'Sea spectra revisited'. *Marine Technology*, vol. 36, no. 4, pp. 211–227, 1999.
[11] N. Daratha, H. Polinder and M. de Sousa Prado, 'A first-order energy storage requirements estimation for an Archimedes wave swing park', *Proceedings of the International Conference on Sustainable Energy Technologies 2008 (ICSET08)*, 24–27 November 2008, Singapore, pp. 1161.C–1165.C, 2008.
[12] E. Tedeschi, M. Santos, P. Ricci, M. Molinas and J.L. Villate, 'Control strategies for the grid integration of wave energy converters at the Biscay Marine Energy Platform'. *European Wave and Tidal Energy Conference*, Southampton, 2011.
[13] D. O'Sullivan, D. Mollaghan, A. Blavette and R. Alcorn, Dynamic characteristics of wave and tidal energy converters and a recommended structure for development of a generic model for grid connection, 2011. Available at: www.iea-oceans.org
[14] E. Tedeschi, M. Molinas, M. Carraro and P. Mattavelli, 'Analysis of power extraction from irregular waves by all-electric power take off', *IEEE Energy Conversion Congress and Exposition (ECCE)*, 2010 IEEE, pp. 2370–2377, 12–16 September 2010. doi: 10.1109/ECCE.2010.5617893.
[15] IEC, *IEC 61400-21: Wind turbines – Part 21: Measurement and Assessment of Power Quality Characteristics of Grid Connected Wind Turbines*, TC/SC 88, 2008.
[16] R. Alcorn. *Wave Station Modelling Based on the Islay Prototype Plant*. PhD thesis, Queen's University Belfast, UK, 2000.
[17] A. J. Nambiar, A. E. Kiprakis, D. I. M. Forehand and A. Wallace. 'Effects of array configuration, network impacts and mitigation of arrays of wave energy converters connected to weak, rural electricity networks'. *Proceedings of the 3rd International Conference on Ocean Energy*, Bilbao, Spain, 2010.

[18] D. O'Sullivan, F. Salcedo, A. Blavette, M. Santos and A.W. Lewis. 'Case studies on the benefits of energy storage for power quality enhancement: Oscillating water column arrays'. *Proceedings of the 4th International Conference on Ocean Energy*, Dublin, Ireland, 2012.

[19] M. Santos, A. Blavette, E. Tedeschi, D. O'Sullivan and F. Salcedo. 'Case study on the benefits of energy storage for power quality enhancement: Point absorber arrays'. *Proceedings of the 4rd International Conference on Ocean Energy*, Dublin, Ireland, 2012.

[20] F. Sharkey, J. MacEnri, Bannon E., M Conlon and K. Gaughan. 'Voltage flicker evaluation for wave energy converters: Assessment guidelines'. *Proceedings of the 4th International Conference on Ocean Energy*, Dublin, Ireland, 2012.

[21] A. Blavette, D. O'Sullivan, T. Lewis and M. Egan. 'Impact of a wave farm on its local grid: Voltage limits, flicker level and power fluctuations'. *Proceedings of the IEEE/MTS OCEANS12 Conference*, Yeosu, South Korea, 2012.

[22] J. Aubry. *Optimisation du dimensionnement d'une chaîne de conversion électrique directe incluant un système de lissage de production par supercondensateurs: application au houlogénérateur SEAREV*. PhD thesis, Ecole normale supérieure de Cachan, France, 2011.

[23] A. Blavette. *Grid Integration of Wave Energy & Generic Modelling of Ocean Devices for Power System Studies*. PhD thesis, University College Cork, Ireland, 2013.

[24] F. Thiebaut, D. O'Sullivan, P. Kracht, S. Ceballos, J. Lopez, C. Boake, J. Bard, ... A. W. Lewis. 'Testing of a floating OWC device with movable guide vane impulse turbine power take-off'. *Proceedings of the 9th European Wave and Tidal Energy Conference*, 2011.

[25] Website of the Sustainable Energy Authority of Ireland (SEAI): http://www.seai.ie

[26] A. Babarit, 'Impact of long separating distances on the energy production of two interacting wave energy converters', *Ocean Engineering*, vol. 37, no. 8–9, pp. 718–729, June 2010.

[27] J. Falnes. *Ocean Waves and Oscillating Systems: Linear Interactions Including Wave-Energy Extraction*. Cambridge University Press, Cambridge, the United Kingdom, 2002.

[28] B. G. Cahill and A. W. Lewis. 'Wave energy resource characterization of the Atlantic Marine Energy Test Site'. *Proceedings of the 9th European Wave and Tidal Energy Conference Series*, Southampton, UK, 2011.

[29] National Grid. *National Electricity Transmission System (NETS) Seven Year Statement*. Technical Report, National Grid, 2011.

[30] WestWave consortium: www.westwave.ie/

[31] IEC. *IEC Standard 61000-4-15, TC/SC 77A: Electromagnetic Compatibility (EMC) – Part 4–15: Testing and Measurement Techniques – Flickermeter – Functional and Design Specifications,* ed2.0, 2010.

[32] NORDEL. Nordic Grid Code, 2007.

[33] IEC. *IEC Standard 61000-3-7: Electromagnetic Compatibility (EMC) Part 3–7: Limits Assessment of Emission Limits for the Connection of Fluctuating Installations to MV, HV and EHV Power Systems*, 2008.

[34] T. Ackermann. *Wind Power in Power Systems*. Wiley, 2012.

[35] S. Lundberg. *Electrical Limiting Factors for Wind Energy Installations*. Master's thesis, Chalmers University, Sweden, 2000.

[36] US Navy. 'Navy issues new shore energy policy to achieve energy security goals'. Published on 7 October 2012. Available at: http://www.navy.mil/. Accessed on 24 October 2012.

[37] WavePlam. *Wave Energy: A Guide for Investors and Policy Makers*. Technical Report, 2009 [Project funded under the Intelligent Energy Europe Programme].

[38] IEC Technical Committee 114. Marine Energy – Wave, Tidal and Other Water Current Converters. Available at: http://www.iec.ch/dyn/www/f?p=103:7:0::::FSP_ORG_ID:1316

[39] EirGrid, *Winter Outlook 2011–12*, 2011.

[40] L. L. Garver, 'Effective load carrying capability of generating units', *IEEE Transactions on Power Apparatus and Systems*, vol. PAS-85, no. 8, pp. 910–919, 1966.

[41] B. Hasche, A. Keane and M. O'Malley, 'Capacity value of wind power, calculation and data requirements: The Irish power system case', *IEEE Transactions on Power Systems*, vol. 26, no. 1, pp. 420–430, 2010.

[42] C. J. Dent, A. Keane and J. W. Bialek, 'Simplified methods for renewable generation capacity credit calculation: A critical review', *IEEE Power and Energy Society General Meeting*, Minneapolis, 2010.

[43] F. Fusco, G. Nolan and J. V. Ringwood, 'Variability reduction through optimal combination of wind/wave resources – an Irish case study', *Energy*, vol. 35, no. 1, pp. 314–325, 2009.

[44] E. D. Stoutenburg, N. Jenkins and M. Z. Jacobson, 'Power output variations of co-located offshore wind turbines and wave energy converters in California', *Renewable Energy*, vol. 35, no. 12, pp. 2781–2791, 2010.

[45] D. E. Reeve, Y. Chen, S. Pan, V. Magar, D. J. Simmonds, A. Zacharioudaki, 'An investigation of the impacts of climate change on wave energy generation: The Wave Hub, Cornwall, UK', *Renewable Energy,* vol. 36, no. 9, pp. 2404–2413, 2011.

[46] E. B. L. Mackay, A. S. Bahaj, and P. G. Challenor, 'Uncertainty in wave energy resource assessment. Part 2: Variability and predictability', *Renewable Energy*, vol. 35, no. 8, pp. 1809–1819, 2009.

[47] R. Billinton and R. N. Allan, *Reliability Evaluation of Power Systems*, 2nd edn., New York, NY: Plenum Press, 1996.

[48] C. Ensslin, M. Milligan, H. Holttinen, M. O'Malley and A. Keane, 'Current methods to calculate capacity credit of wind power, IEA collaboration', *IEEE Power and Energy Society General Meeting*, Pittsburgh, 2008.

[49] Pelamis Wave Power, 'Pelamis P-750 wave energy converter'. Available at: http://www.pelamiswave.com/, 2010.

[50] D. Kavanagh, A. Keane and D. Flynn, 'Challenges posed by the integration of wave power onto the Irish power system', *European Wave and Tidal Energy Conference*, Southampton, 2011.

[51] Single Electricity Market Operator (SEMO), *The Single Electricity Market (SEM) Trading and Settlement Code Version 10*, 2011.

[52] *Options for the Development of Wave Energy in Ireland – A Public Consultation Document*, Marine Institute, Sustainable Energy Ireland, 2002.

[53] SEAI. *Ocean Energy – Roadmap to 2050*. Sustainable Energy Authority of Ireland, 2010.

[54] D. Kavanagh, A. Keane and D. Flynn, 'Capacity value of wave power', *IEEE Transactions on Power Systems*, vol. 28, no. 1, pp. 412–420, 2013.

Chapter 7
Electrical energy storage systems

D. Murray, J. Aubry, B. Multon and H. B. Ahmed

7.1 Introduction

This chapter examines electrical energy storage systems (ESSs) for wave energy converters (WECs). The motivations for including on-board energy storage are outlined in terms of power smoothing, low-voltage ride-through (LVRT) and ancillary services. Various wave energy converter technologies are explored as well as their inherent energy storage mechanisms. Electrical energy storage technologies, including batteries, capacitors, and supercapacitors, are then compared.

Two case studies are described where electrical energy storage, namely supercapacitor energy storage, is applied to help smooth the output power from a WEC. The first case study examines power smoothing for a SEAREV WEC and the second case studies explores smoothing the output power from an offshore oscillating water column (OWC) WEC operated at variable speed.

This chapter highlights the issue of ensuring a long lifetime for any employed energy storage device so as to match the desired intervals for non-routine disruptive maintenance. It also describes long lifetime testing that validates supercapacitor cycle lifetime, the temperature effect on cycle lifetime and cycle lifetime with an application power profile. An ageing model for supercapacitor devices is then presented.

The final section of the chapter details the power smoothing quality criteria and the contribution of supercapacitor-based energy storage to power smoothing performance. A trade-off between power quality and the cost and size of the ESS is found.

7.2 Motivations for energy storage

7.2.1 Power smoothing

A plot of wind speed versus time and wind power versus time is shown in Figure 7.1. Power fluctuations are seen to occur rapidly as divergence around a mean value, and as wind power is proportional to the cube of wind speed, there is a dramatic increase in the standard deviation compared to the wind speed profile.

During the transfer of energy from wind to waves, many of these rapid power fluctuations are smoothened out. As waves are effectively an integration of wind power, their available power and direction is steadier and can be forecasted and predicted to a better degree than wind; this is one of the principal advantages of wave power. A sample plot of normalised wave elevation over time is shown in Figure 7.2. Wave power is a function of the wave height and comparing Figure 7.1 and Figure 7.2 it is seen that the energy transfer of wind to wave smooths out much of the rapid power changes.

Ocean wave periods typically vary from 1 to 20 s, and without some form of energy storage or power smoothing, the output grid power from the device will display large power fluctuations over this time period.

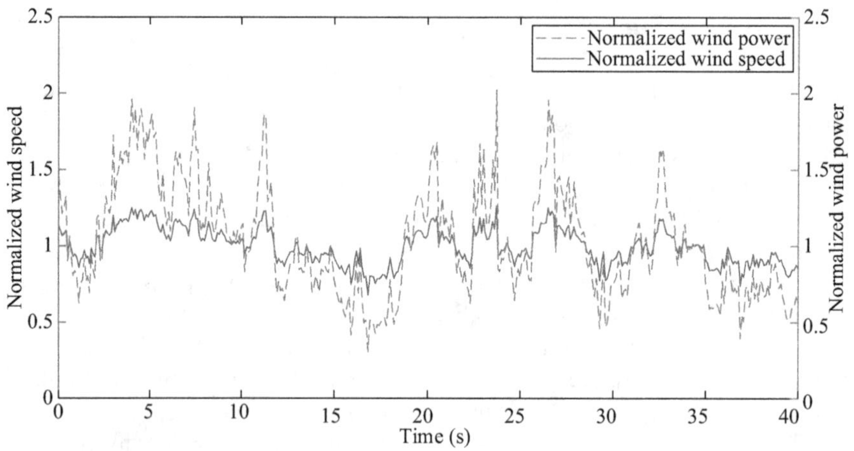

Figure 7.1 Wind speed/power short-term variability [1]

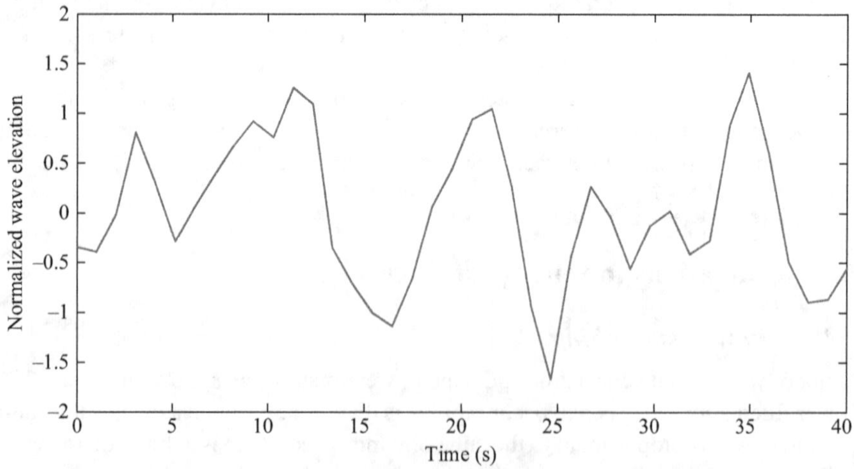

Figure 7.2 Wave elevation short-term variability [1]

A power source with large variations will have increased cost and decreased system lifetime due to elevated power losses and larger ratings required for equipment. If the WEC is connected to a weak grid, flicker and voltage and frequency deviation issues arise, as the varying current may interact with the grid impedance to affect local voltage levels [2]. Flicker is related to the voltage changes in the supply that result in variations to the light output from lighting sources. The human perceptibility to this varying light intensity prompted the definition of a mechanism for calculating flicker severity on which IEC standards are based. Humans are most sensitive at a frequency of 8.8 Hz. At lower WEC input power frequencies and with WECs' large input power variations, it is expected that flicker and local voltage levels will still vary at a noticeable rate to the human eye if power smoothing is not implemented.

The issue of equipment rating is related to the peak-to-average power output ratio as opposed to fault conditions. Larger power rated equipment clearly costs more. For example, increased conductor area is required to handle the increased currents, and increased insulation is required to handle the increased voltage rating. A peak-to-average power output ratio of 1:1 allows the equipment to be used optimally and most cost effectively. Equipment like generators, transformers, cables etc. are able to handle transient peak ratios of up to 3:1 to 5:1, where the time at which the equipment is operated at peak power is limited by the thermal time constant. Power electronic equipment usually has a much lower time constant, which effectively results in rated power and peak power being very close. To achieve full controllability, power electronic converters are typically utilized to control the electrical power flows. The power electronic converters will experience this input peak power and need to be rated accordingly unless power smoothing equipment is inherent in the power take-off system.

Due to power fluctuations, the system losses fluctuate and dissipate heat cyclically. This thermal cycling for equipment with different temperature coefficients and different coefficients of thermal expansion degrades interconnections throughout the system, for example between wire bonds and silicon in power electronic converter modules. Component, and hence system, lifetime is directly related to the amplitude and frequency of the power fluctuations, most especially for power electronics devices with their very short thermal time constants.

Comparing two power sources with the same average power, but one with a constant power output and the other with a fluctuating power output, will show that there are increased power losses in the fluctuating power output case. This additional power loss component is described as

$$P_{loss} = R_{sys}\left[\frac{1}{T}\int_0^T {I_{rms}}^2(t)dt - {I_{avg}}^2\right] \quad (7.1)$$

It should be noted that using an energy storage device to smooth output power and remove this power loss component is unlikely to increase the overall efficiency of the system, due to the losses in the ESS.

7.2.2 Low-voltage ride-through (LVRT)

As well as power smoothing, energy storage can help with low-voltage ride-through (LVRT) in grid-connected WECs [3]. In a grid low-voltage event (short-circuit conditions), the ability to transfer power to the grid is limited as it is a function of the grid voltage. If this occurs when the WEC is experiencing high input power, there will be a transient power imbalance unless energy storage is employed. If the input wave power is exported to the grid without an energy storage device via a power converter, the DC-link voltage of this converter will rise dangerously unless controlled. As well as using energy storage to ride through this fault, another option would be to burn off excess energy in a controlled manner using a power electronic converter and a dissipative load bank circuit.

7.2.3 Ancillary services

Energy storage might also help with the operation of specific WECs. For example in offshore WECs employing Wells turbines, energy storage can provide the energy to accelerate these non-self-starting, high inertia devices from rest [3]. While the power converter used in the power take-off mechanism is often bi-directional and could theoretically be employed to start the machine, it is preferable to allow power flow in one direction only to help with ratings and minimize cost of safety and protective equipment. This also has advantages from a grid operator perspective, by limiting starting current surge from the grid as well as in terms of the import capacity of the grid connection.

Electrical power is needed in offshore WECs for lighting, communications, equipment monitoring and control purposes. Heating and ventilation equipment may also need to be powered. These operations need power, even in times of no/low input wave energy. Thus, some form of energy storage is needed. Due to the high energy density values of batteries, these seem ideal devices for this application.

7.3 Approaches to implementation of energy storage

There are many different technologies and approaches to extracting power from the ocean waves and there are many methods of employing energy storage in the system. Energy storage implementation can be examined for WECs on an individual basis, where installation would probably take place on-board the individual WEC, as well as on a collective basis, where the power output from a wave farm is connected to an ESS on or offshore.

7.3.1 Approaches to the implementation of energy storage for a farm of WECs

As outlined in section 7.2, a motivation for including energy storage in a WEC system is to help with power smoothing. Aggregating the output power from many WECs in a wave farm has been examined in [4] where three WECs (SEAREV point absorbers) were operated over 20 days, and the standard deviation of power reduced by 80% compared to single device operation. According to a study, power smoothing is independent of array layout and sea-state conditions, but instead depends on the number of systems in the array [5]. Simulations also conclude that less electrical energy storage is needed for a centralized energy storage device for a wave farm, compared to individual WEC energy storage [6].

It was simulated that the power variation is reduced for a farm with 64 uncorrelated devices [7], although it is possible that power peaks may occur simultaneously in many WECs resulting in large export power fluctuations. This effect is partly shown in [6]. Clearly, a smoother output from an individual WEC would improve this smoothing by aggregation process.

If an ESS is sized for a farm of WEC devices, it will be required to handle a large power throughput. Presently, no commercial wave farms are in existence and the industry is still in the development stage. It is likely that grid code requirements would need to be fulfilled for a wave farm, as power quality to the grid will be of greater importance compared to a single device being considered.

Energy storage options for the scenario where the storage system is placed onshore are not limited by size and space constraints. The power quality to the grid can be greatly enhanced by large-scale systems that provide long-term energy storage. These long-term options include pumped hydro, compressed air energy storage (CAES) and large-scale battery installations. This chapter focuses on short-term energy storage options and on-board ESS devices.

7.3.2 Approaches to the implementation of energy storage for individual WECs

Many different types of WECs are currently in development, and full-scale prototypes are rare. Recent reviews identified about 100 projects at various stages of development and this number seems to be increasing as new concepts outnumber those that are being abandoned [8].

This large number of WECs encompasses devices with a wide range of technologies utilizing a variety of methods to extract energy from the ocean waves and some with inherent energy storage built into the power take-off mechanism. There are several methods of subdividing the devices according to various criteria. In [8] the devices are classified under working principle and a select number of devices that reached the prototype stage or were the object of extensive development effort are given as examples. This classification is:

- Oscillating water column (with air turbine).
- Oscillating bodies (with hydraulic motor, hydraulic turbine, direct-drive generators).
- Overtopping (with low-head hydraulic turbine).

Energy storage may be inherent and built into the power take-off systems of these WECs. The following three devices are used to examine these ESSs.

- Oscillating water column (OWC) – inertial energy storage (flywheel effect of turbine).
- Attenuator – hydraulic accumulator energy storage.
- Overtopping device hydro energy storage.

The energy contained in ocean waves is greatest offshore where wave interaction and friction with the local geometry is minimized. This chapter primarily focuses on offshore WECs. The level of energy storage utilized, as well as the efficiency of the device, is dependent on the control system implemented. For oscillating body and OWC converters, if the device is to be an efficient absorber its own frequency of oscillation should match the frequency of the incoming wave; that is, it should operate at near-resonance conditions [8].Clearly a control scheme utilizing energy storage will likely reduce the output power fluctuation. However, some control methods aimed at maximizing the power output can result in significantly larger power fluctuation, e.g. the 'latching' control scheme implemented in some point absorbers.

7.3.2.1 Oscillating water column (OWC) – inertial energy storage

In an OWC, incident waves compress and expand the air within a chamber causing airflow across a turbine. These turbines are generally Wells turbines or impulse turbines and they convert the bi-directional pneumatic power to uni-directional, but pulsating, mechanical power from which a generator produces electricity.

In terms of offshore OWCs, an example of an offshore floating device that has proven successful at scaled testing is the OE Buoy. This is based on the backward bent duct buoy (BBDB) principle and is developed by OceanEnergy (www.oceanenergy.ie). A quarter-scale device has been successfully operating in Galway Bay off Ireland for over 2 years and it also employs a Wells turbine [9, 10]. In the BBDB, the OWC inlet is oriented away from the wave direction, which was found to be an improvement over the inlets that are oriented into the wave direction.

An overview of the BBDB OWC WEC is shown in Figure 7.3. The power transfer overview is displayed in Figure 7.4. This reveals where the flywheel energy storage effect of the turbine inertia is experienced in the power take-off system. The feedback interaction of the power take-off is also noted.

Fixed shoreline structures were built in Portugal (Pico, Azores, 1999, [11]) and Scotland (LIMPET, island of Islay, 2000, [12]). The device in Portugal is rated at

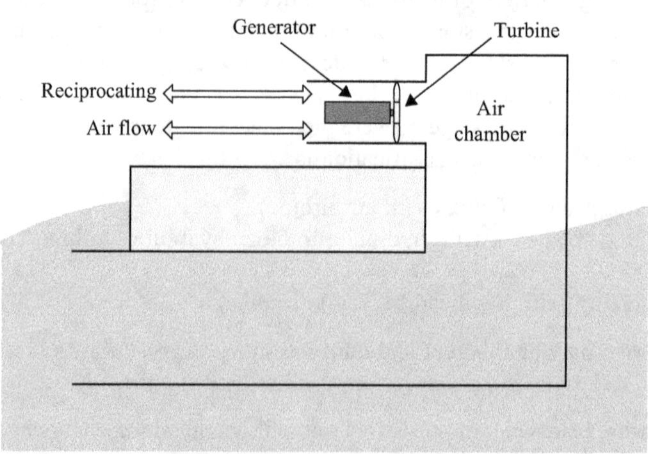

Figure 7.3 BBDB OWC WEC overview

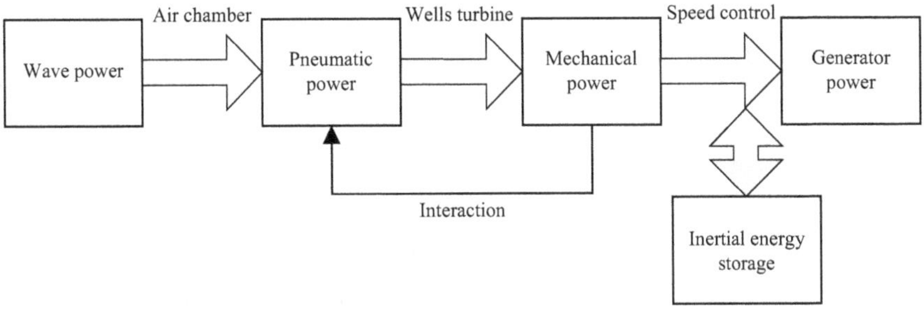

Figure 7.4 OWC WEC power transfer overview

400 kW. It uses a Wells turbine with an inertia of 595 kg m^2 and a speed range of 750–1,500 rpm [12]. The LIMPET contains two 250 kW Wells turbines, each with a speed range of 700–1,500 rpm and inertia of 1,250 kgm^2 per unit [14].

Utilizing the Wells turbines in a variable speed strategy offers significant energy storage due to the flywheel effect of the turbine inertia. For exampe, the difference in energy stored in a rotating turbine with inertia J, between two speeds ω_1 and ω_2, is

$$W_{turbine} = \frac{1}{2}J\left(\omega_1{}^2 - \omega_2{}^2\right) \tag{7.2}$$

This gives a figure of 1.5 kWh for the Pico device, which equates to almost 14 s of rated power. For the LIMPET device, the energy stored per turbine is 3.2 kWh, giving over 46 s of rated power.

It is noted from the power transfer overview diagram in Figure 7.4 that if there is no mechanical power produced on the turbine, the control scheme may allow the generator to continue exporting power to the grid. This input to output power differential is balanced utilizing the inertial energy storage of the Wells turbine and this will cause the turbine speed to reduce. Conversely, if there is excess mechanical power produced on the Wells turbine of which the generator exports a fraction to the grid, the excess mechanical power will cause the turbine speed to increase. These power flows are described as

$$P_{mech}(t) = P_{gen}(t) + P_{loss}(t) + J\omega(t)\frac{d\omega(t)}{dt} \tag{7.3}$$

7.3.2.2 Attenuator – hydraulic accumulator energy storage

Attenuators are multi-segment floating devices that align with the wave direction. The differing heights of waves along the length of the device cause the floating segments to move relative to each other. The resulting motion at the joints between segments is converted into electrical power through hydraulic or other means.

An example of an attenuator is the Pelamis device developed by Pelamis Wave Power (www.pelamiswave.com). This semi-submerged device consists of four or five cylindrical segments where hydraulic rams at the segment joints pump oil through hydraulic motors driving three electrical generators. This power take-off includes high pressure storage gas accumulators providing some in-built energy storage.

A Pelamis P2 machine was installed offshore at Orkney in Scotland in 2010. Also, three Pelamis machines were installed and operated at Agucadoura in Portugal, 5 km offshore in 2008 to become the world's first wave farm; they have since been removed. Each device had a capacity of 750 kW in total from its three 250 kW generators.

A schematic of an attenuator WEC is shown in Figure 7.5. The power transfer conversion process shown in Figure 7.6 gives a graphical representation of where the high pressure accumulator energy storage is in the power take-off system. Again it should be noted, the power take-off is a highly coupled system where the control strategy greatly influences the feedback and interaction between stages of this simplified power transfer process as well as extractable power.

The short-term energy storage of the accumulator pressure in each joint is set by the difference between the primary transmission energy intake and the secondary transmission outlet. The primary transmission consists of a hydraulic system that converts the wave power into stored energy in the accumulators, while the secondary transmission consists of hydraulic motors coupled to three-phase asynchronous generators that convert this stored energy into electricity exported to shore [15, 16]. The hydraulic control system is achieved through electronically controlled valves that control fluid flow between the hydraulic cylinders in the movable joints, and the high pressure accumulator and low pressure reservoir.

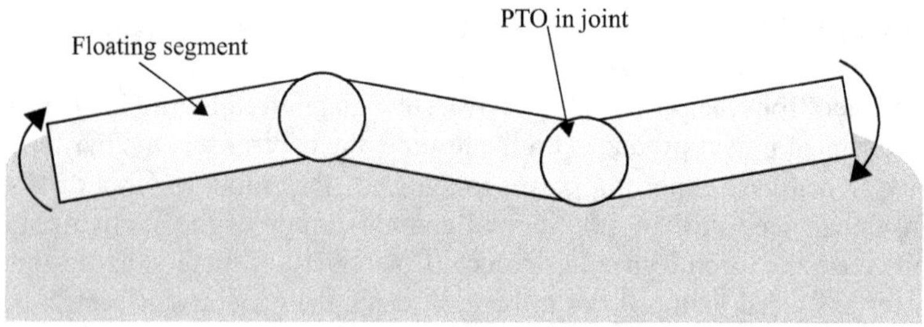

Figure 7.5 Attenuator WEC overview

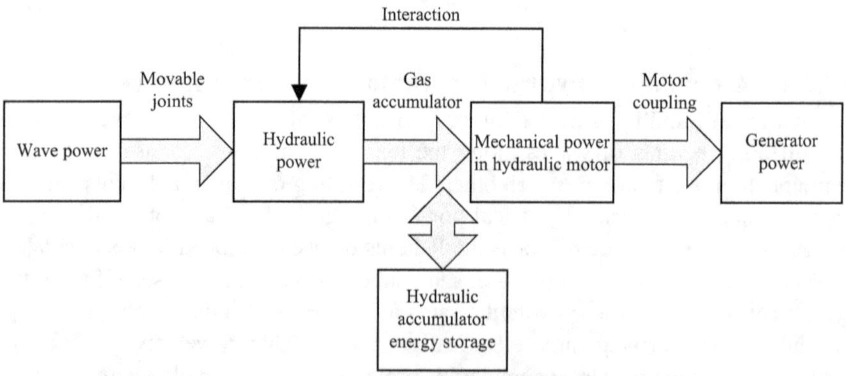

Figure 7.6 Attenuator WEC power transfer overview

The energy transfer during gas expansion in an accumulator, assuming the ideal gas law is obeyed and the process is isentropic (constant entropy) [17], is given as

$$W_{gas} = mc_v(T_1 - T_2) \tag{7.4}$$

where m is the mass of the gas, c_v is the specific heat at constant volume, T_1 is the absolute temperature in Kelvin before expansion, and T_2 is the absolute temperature in Kelvin after expansion.

Further information on gas accumulators as an energy storage device in WECs is found in [17].

7.3.2.3 Overtopping device – hydro energy storage

Overtopping devices direct waves and the sea water up over a structure and store this water above the level of the ocean in a reservoir. This potential energy of the water is converted first into kinetic and then into electrical energy using a conventional low-head hydro turbine when releasing the water back into the sea. The reservoir itself forms a large energy storage mechanism that allows for the smoothing of the short-term power variability of the waves.

An overview of an overtopping WEC is shown in Figure 7.7, and the power transfer overview in Figure 7.8 shows where the energy storage mechanism is in the overall power take-off system.

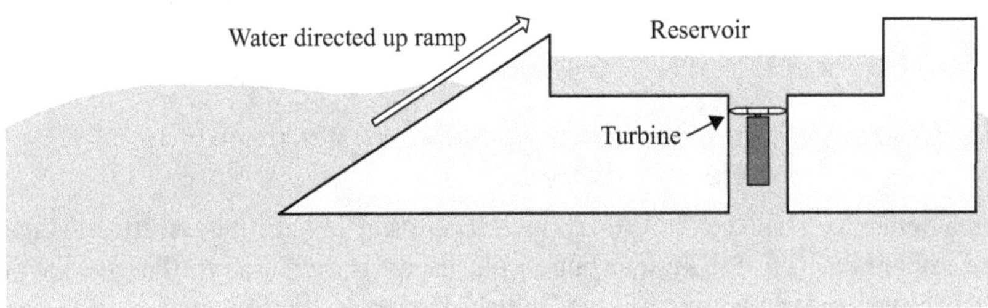

Figure 7.7 Overtopping device WEC overview

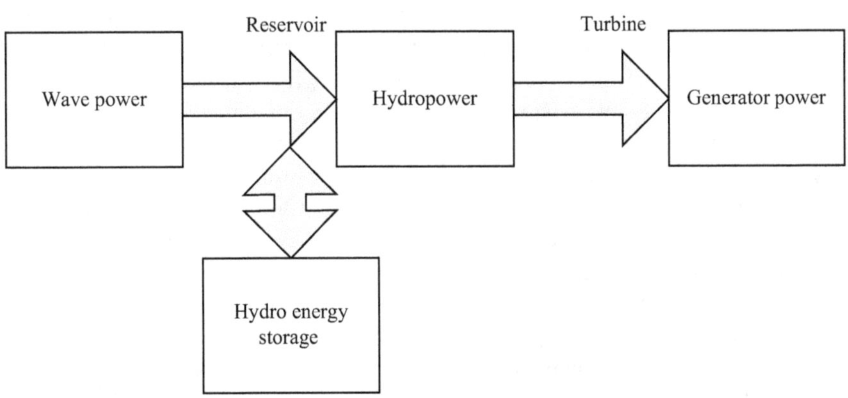

Figure 7.8 Overtopping device WEC power transfer overview

An example of an overtopping device is the Wave Dragon developed by Wave Dragon Ltd (www.wavedragon.net). This offshore device employs ramps and reflectors to help focus the waves into the reservoir. A 1:4.5 scaled prototype is operating in Nissum Bredning, Denmark and is rated at 20 kW. It contains seven low-head Kaplan turbines each connected to 2.3 kW permanent magnet generators. It is designed for a site with a low power wave climate of 0.4 kW/m. This has a reservoir capacity of 55 m^3 with a ramp height of 3.6 m above sea level. The largest wave dragon specified would be designed for a wave climate of 48 kW/m, having a reservoir of 14,000 m^3 and a ramp height of 19 m. These are very large devices. The 0.4 kW/m device has a total length of 33 m, a width of 55 m and a weight of 237 tonnes. The 48 kW/m device has a total length of 220 m, a width of 390 m and a weight of 54,000 tonnes. These large devices contain significant energy storage in their reservoirs above sea level.

For example the potential energy stored in a reservoir of water of density ρ (1,000 kg/m^3), volume V and falling height of h is

$$W_{reservoir} = \rho g h V \tag{7.5}$$

where g is the gravitational constant of 9.81 m/s^2. Assuming the total volume of water is at the ramp heights specified above (a large over approximation), i.e. 3.6 m for the 0.4 kW/m device and 19 m for the 48 kW/m device, the total energy contained in the full reservoirs is evaluated as 0.54 and 725 kWh respectively for these two systems.

7.3.3 *Energy storage strategies in the electrical power take-off systems of offshore WECs*

After the generator stage of the WEC power take-off (PTO), further electrical stages may be implemented in the system before electricity is delivered to the grid. Typically a back-to-back converter is used, which decouples the frequency of the generator from the fixed voltage and fixed (50 or 60 Hz) frequency of the electrical grid, and allows the generator to operate at variable speeds. This back-to-back converter contains a DC-link whose DC voltage is sustained by a capacitor bank. In a standard control scheme the grid coupled converter delivers power to the grid to maintain this DC-link voltage close to its set-point. The capacitor operational voltage range as well as its capacitance determines the energy stored in the DC-link stage of the PTO.

Further electrical energy storage control may be implemented using a DC–DC converter connected between the DC-link and the electrical storage device, typically a battery, supercapacitor (SC) module or capacitor bank. The DC–DC converter allows control of the power flows to and from the electrical ESS. The electrical PTO schematic after the generator is shown in Figure 7.9, similar to a schematic given in [18], and also similar to the full power converter topology for a wind turbine [19]. This power electronic topology is also given for a battery ESS in [20]. This layout is valid for most generator types, except for a doubly-fed induction generator, where the converter is placed on the rotor windings allowing for converter de-rating [21].

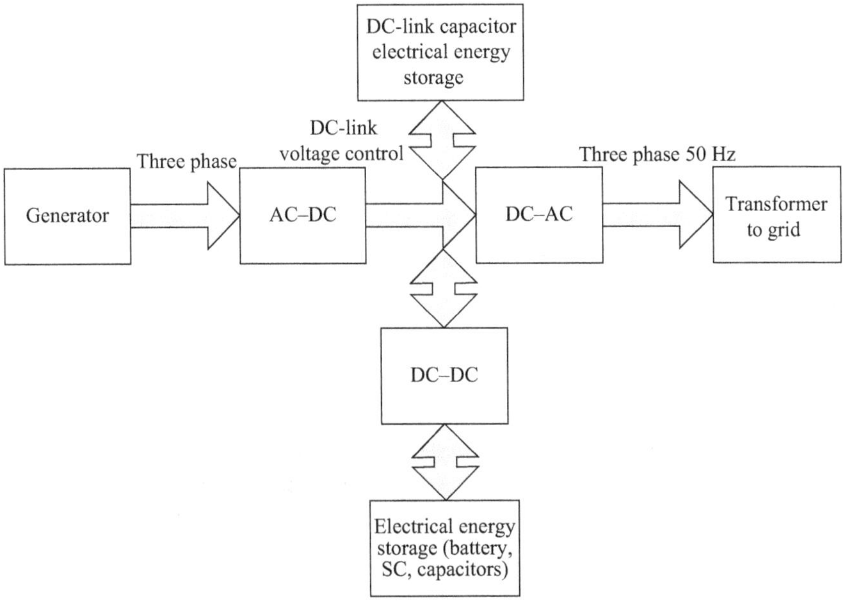

Figure 7.9 Electrical power transfer overview

7.4 Electrical energy storage – technology description

Viable electrical energy storage technologies for offshore wave energy applications include batteries, supercapacitors (SCs), capacitors and superconducting magnetic energy storage (SMES) devices. A summary of each technology is given, with a focus on the application of utilization on-board an offshore WEC.

7.4.1 Superconducting magnetic energy storage (SMES)

SMES systems store energy in the magnetic field around a superconducting coil created by the flow of DC current. The coil achieves superconductivity when cooled to a temperature below its superconducting critical temperature. Once charged, the current is sustained with very low losses and the magnetic field energy can be stored according to

$$W_{SMES} = \frac{1}{2}LI^2 \tag{7.6}$$

where I is the magnitude of the current and L is the associated inductance at this current level.

SMES is currently costly and hasn't fully emerged from the development stage [21, 22]. It consists of many essential parts, including a cryogenically cooled refrigerator that increases breakdown vulnerability in the harsh offshore wave climate as well as increases the necessary space and mechanical support. For these reasons SMES has not yet been considered for offshore ocean energy applications.

7.4.2 Batteries

Batteries are high energy density electrical storage devices that have undergone significant development in recent times. With the increased research into electric vehicles, rechargeable batteries are undergoing continuous development. Currently lithium-ion (Li-ion) batteries are the technology of choice being installed in new electric vehicles as their improved performance over NiMH batteries are now being realized as production costs decrease. Some Li-ion batteries for electric and hybrid electric vehicles have energy densities as high as 140 Wh/kg and power densities of up to 745 W/kg [23].

7.4.3 Supercapacitors (SCs)

Supercapacitors (SCs) are also known as electric double layer capacitors (EDLC), ultracapacitors and electrochemical double layer capacitors. This variety of names comes from a number of market leaders, or from the physical composition that effectively contains a double layer that increases capacitance. They have a very high energy density and are governed by the same equations as all capacitors. The value of capacitance, C, and the energy stored at a particular voltage V, W_{SC}, are

$$C = \frac{\varepsilon A}{d} \tag{7.7}$$

$$W_{SC} = \frac{1}{2} C V^2 \tag{7.8}$$

SCs use a porous carbon-based electrode with a large surface area typically between 1 and 2 million m^2/kg. The charge separation distance, less than 10 Å (10×10^{-10} m), is much smaller than what can be accomplished using conventional dielectric materials. These properties give SCs their extremely high capacitance in accordance with (7.7), with values ranging from a few farads up to 5,000 farads. However, critically, due to the very small charge separation distance in the 'double layer', voltage ratings are low, typically close to 2.7 V. To achieve higher voltages, strings of series connected SCs are created. Consequently, voltage balancing circuits are usually added due to the relatively large capacitance variations of individual SCs.

SCs have demonstrated robustness in applications with photovoltaics, where SCs complemented battery storage devices and improved system performance and battery lifetime [25, 26]. They can also be operated at sea for long periods of time [25]. SCs have been used in wind turbine pitch systems, hybrid vehicles, trains, buses and lift trucks. The time constant of SCs is typically around 1 s. Their small energy density but large power density suggest they are ideal short-term energy storage options, especially for ocean energy applications if their lifetimes can be shown to be compatible with the required service life of such equipment in an offshore WEC.

7.4.4 Capacitors

The three main capacitor technologies are ceramic, electrolytic and film capacitors.

Ceramic capacitors are typically utilized for high-frequency applications and have very low equivalent series resistance (ESR) ratings but have poor aging characteristics.

Electrolytic capacitors are typically used to store large amounts of energy with a relatively high capacitance compared to other capacitors (not SCs). Electrolytics are utilized in DC-link applications in power converters to help maintain the bus voltage during any large power deviations. Drawbacks include poor tolerances and poor high-frequency characteristics.

Film capacitors are larger and more expensive than electrolytic capacitors, but they have higher surge or pulse load capabilities, have high rated voltages (up to kV range), lower aging and higher ripple current capability. They have very low ESR and equivalent series inductance (ESL) values.

As capacitors are typically situated very close to the power electronics in power converters to help maintain the DC-bus voltage, reduce wire lengths and minimize parasitic inductance, they are not offered in modules normally but are integrated into the power converter during design and construction. For this reason the technology comparison below doesn't include a capacitor module.

7.4.5 Technology comparison

It is difficult to make a comparison of technologies that encompass many manufacturers' products and their slight differences. For this reason, leading products of each technology are chosen. A large SC module from Maxwell Technologies is compared with a high energy Li-ion battery module from SAFT batteries, a large electrolytic capacitor from EPCOS and a high performance lead-acid battery from Enersys. Comparisons are made typically using modules to take into account any control, safety and cooling equipment that might also be needed when using the technology, and allow a comparison of these standalone modular devices.

Li-ion batteries cycle life is in the range of several thousand cycles at present, although SAFT (www.saftbatteries.com) show cycle life increasing for a lower depth of discharge (DOD) for their Synerion 48E module. They specify over 1 million cycles for 3.2% to 6.0% DOD (determined by extrapolation) depending on charge rates.

Their lifecycle analysis is determined using an end of life definition of 70% of the initial capacity remaining. The nominal 2,200 kWh Saft Synerion 48E module would only have an effective usable energy of 71 Wh at a DOD of 3.2% corresponding to 1 million cycles. This would produce effective energy densities and specific energies in the ranges of 4 Wh/l and 4 Wh/kg. While charging rates are not given for the above figure, using the lower limit of 3.2% DOD (71 Wh) at the maximum continuous discharge capacity of 1,150 W gives a cycle time of almost 450 s. This would take over 14 years to complete 1 million cycles if operated continuously at this charge/discharge rate.

While SCs cannot compete with batteries in terms of energy density, their much longer cycle life, power density, operational temperature range and ability to fully discharge make them an energy storage option that must be considered in many applications. A typical SC has an energy density of over 5 Wh/kg, a power density of over 6,000 W/kg and a rated lifetime of 1 million cycles (www.maxwell.com). Table 7.1 demonstrates that SCs' power and energy densities lie in a range between capacitors and batteries. Coupled with this, SCs have charge/discharge efficiencies ranging from 0.85 to 0.98 [26]. These characteristics are very complementary to those of batteries and thus appear ideal for a WEC application.

Table 7.1 Electrical energy storage technology comparison

	Maxwell Technologies BMOD0063 P125	**SAFT Synerion 48E**		**Epcos B41456B8150M**	**Enersys genesis EP 12V G12V70EP**
Technology	SC module	Li-ion battery module		Aluminium Electrolytic capacitor	Lead acid battery
Voltage (V)	125	48		63	12
Capacitance (F)	63			0.15	
Energy (Wh)	137	2,200 (70% DOD)	71 (3.2% DOD)	0.083	644 (1C)
Max cont power (W)	15,000	1,150		1,575	644 (1C)
Volume (V)	70	17		1.03	10
Weight (kg)	61	19		1.3	24
Cycle life (cycles)	1,000,000 (75% DOD)	10,000 (70% DOD)	1,000,000 (3.2% DOD)		400 (80% DOD)
Energy density (Wh/L)	1.96	129	4.2	0.0806	64.4
Specific energy (Wh/kg)	2.25	116	3.7	0.0638	26.8
Power density (W/L)	214	67.6	67.6	1,529	64.4
Specific power (W/kg)	246	60.5	60.5	1,212	26.8

7.5 EES case studies

This section provides case studies for two different device types, outlining methodologies for implementing ESSs in both cases. The first example is a direct-drive pendular wheel type device, SEAREV; and the second is an OWC device with a Well's turbine.

Case study: SEAREV

Introduction

The SEAREV WEC is composed of a pendular wheel (weighing several hundred tons), which is totally enclosed in a floating buoy [27]. This wheel is free to oscillate in the buoy. The excitation forces of the swell, acting on the buoy, generate an oscillating rotation between the inner wheel and the buoy.

The power take-off (PTO) of the SEAREV can be either a hydraulic system or an all-electric system. The hydraulic system incorporates hydraulic

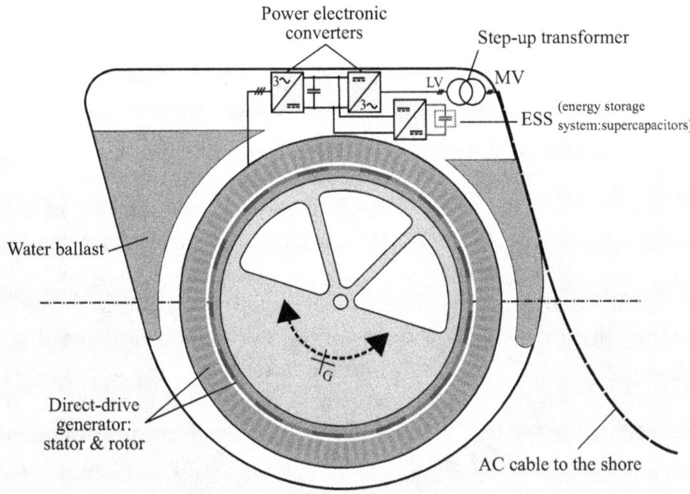

Figure 7.10 Direct-drive version of the SEAREV wave energy converter with power smoothing system

rams, accumulators and a hydraulic motor driving a classical induction generator. In the all-electric version, a direct-drive generator with a fully controlled power electronic converter is driven by the pendular wheel. As this all-electric version does not contain an inherent means of energy storage (the hydraulic version contains hydraulic accumulators for energy storage), an electrical ESS is required to smooth the produced power (Figure 7.10). By applying a damping torque T_{damp} on the wheel as in (7.9), the PTO can produce electrical energy from this motion.

One of the advantages of the direct-drive solution is the possibility to control the applied torque instantaneously. It is then possible not only to apply a viscous damping torque, i.e. proportional to the angular velocity of the pendular wheel relative to the float, denoted $\Omega(t)$, but also to easily limit the produced power to a maximum (levelling) value, denoted P_{lev}. So the corresponding damping torque relation is the following [28]:

$$T_{damp}(t) = \begin{cases} \beta\Omega(t) & \textit{if } \beta\Omega(t)^2 \leq P_{lev} \\ \dfrac{P_{lev}}{\Omega(t)} & \textit{if } \beta\Omega(t)^2 > P_{lev} \end{cases} \tag{7.9}$$

where β is the viscous damping coefficient adjustable in relation with the sea-state.

It is interesting to note that this capability to level the converted produced power would not be so easy in the case of a hydraulic PTO and/or in the case of a linear motion of the PTO. This production levelling strongly reduces the peak-to-average ratio of the produced power, without significant loss of harvested wave energy. The main advantage of this strategy is to enable rated power optimization of the power electronic converter [29].

In order to compute instantaneous power time-series over various sea-states, we rely on the work of researchers of Laboratoire de mécanique des fluides (CNRS – Ecole Centrale de Nantes). For many years, they have been developing hydrodynamical and mechanical models to simulate the SEAREV WEC behaviour [26, 30]. The former is based on linear potential theory, assuming that the excitation forces of the swell can be expressed as a sum of monochromatic components, while the latter takes into account the non-linearity of the inner pendulum movement equation. These numerical models have been packaged in a 'white-box', and made us possible to carry out various studies, in the field of electrical engineering, around the direct-drive conversion chain for this WEC [28, 29, 31, 32].

From our point of view, the considered parameters of the SEAREV numerical model are those defining the sea-state: H_s and T_p and also those relative to the damping torque control parameters: β and P_{lev} (cf. Figure 7.11). The geometry of the buoy and the inner wheel will not be modified in the following study.

In the following case, in order to evaluate the yearly (and over the life-time) performances, we consider wave data statistics corresponding to Yeu island in Vendée, France. These wave data statistics are presented in Figure 7.12, where four scatter plots are shown:

- the yearly sea-state percentage of occurrence (or probability)
- the average mechanical power harvested on different sea-states with $\beta = 4$MNms and $P_{lev} = 1$MW (which are constants whatever the sea-state is, even if it is possible to tune β)

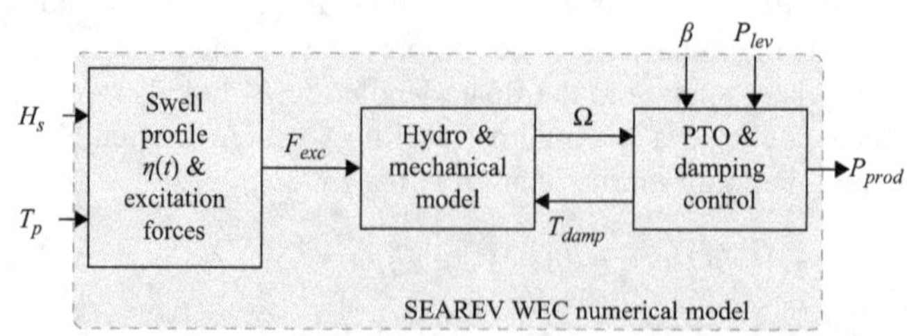

Figure 7.11 Block diagram of the numerical model of the SEAREV (from the sea-state to the mechanical power profile)

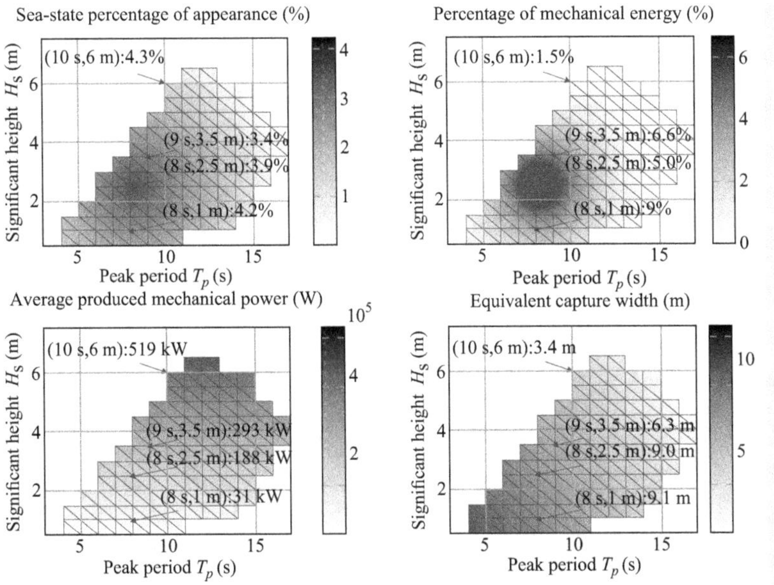

Figure 7.12 Wave data statistics of Yeu Island, Vendée, France, average mechanical power absorbed by the SEAREV wave energy converter (30 m wide) and equivalent capture width

- the sea-state percentage of mechanical energy, which is obtained with the sea-state probability weighted by the average mechanical power harvested on different sea-states
- the equivalent capture width, which is the ratio of average mechanical power produced by the SEAREV to the power per metre of wave front, considering that the power per metre of wave front is obtained with $P_w = 420 H_s T_p (\text{W/m})$

Figure 7.13 gives a particular example of time-series that can be obtained after simulation of the SEAREV behaviour corresponding to a ($T_p = 9$ s; $H_s = 3.5$ m) sea-state (ISSC spectrum) and damping parameters values of $\beta = 4$ MN ms and $P_{lev} = 1$ MW.

The EES energy management

The EES of the SEAREV is placed on the DC-bus of the SEAREV direct wave energy converter (DWEC) and is controlled with a reversible (half-bridge) DC–DC converter (see Figure 7.14). The control scheme given in Figure 7.15 is the most obvious for the determination of the storage power

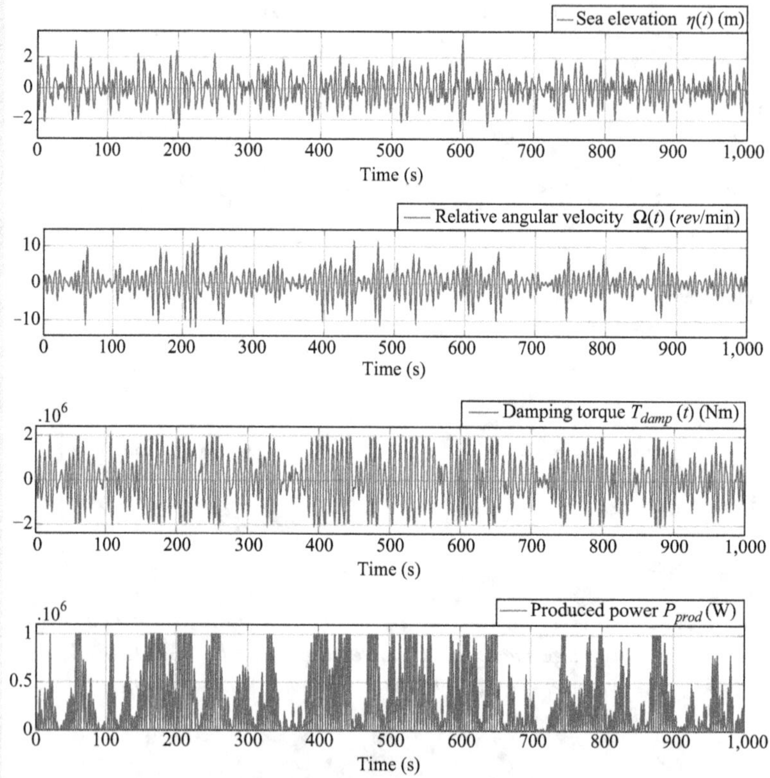

Figure 7.13 Example of computed 1,000 s time-series corresponding to a ($T_p = 9$ s; $H_s = 3.5$ m) sea-state (ISSC spectrum) and damping parameters values of $\beta = 4$ MN ms and $P_{lev} = 1$ MW

set-point $P_{sto}{}^*$ and the smoothing control of the incoming mechanical power P_{prod}. However, this storage control scheme does not ensure that the State of Energy (*SoE*) will be kept between 0% and 100% anytime (i.e. in our case, for the SC, the voltage should be kept in the considered voltage range). That is why many authors have enhanced this control scheme with fuzzy logic rules depending on the *SoE* [33–38]. The objective of these rules is to lowering the storage power set-point when the *SoE* is high and vice versa in order to keep the *SoE* within its limits.

Taking this idea further, we can easily demonstrate that by making the grid power set-point $P_{grid}{}^*$ a linear function of the *SoE* (cf. Figure 7.16), the control is also equivalent to a low-pass filtering of P_{prod}, but, while ensuring the *SoE* constraints with a wise choice of P_{max} and P_{min}.

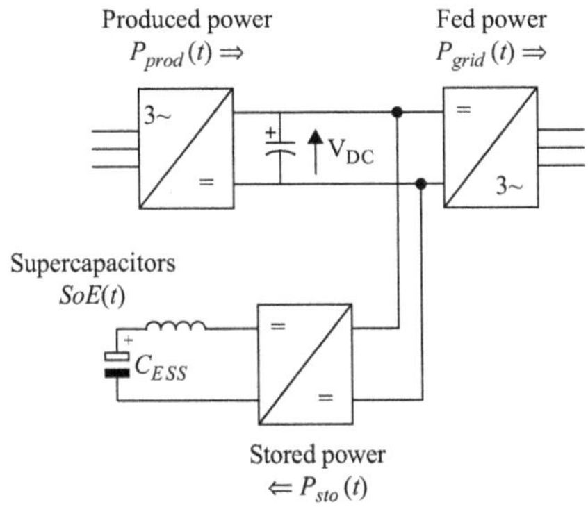

Figure 7.14 Configuration of the EES system on the DC-bus of the SEAREV wave energy converter

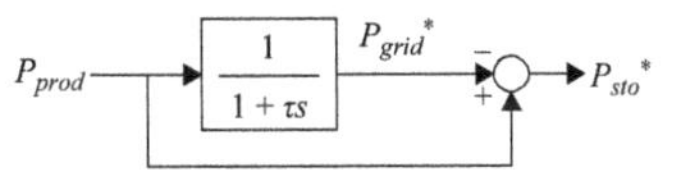

Figure 7.15 A classical control scheme for the determination of the storage power set-point

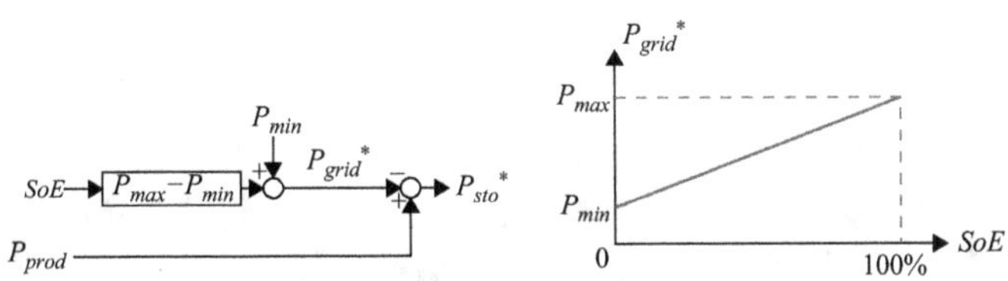

Figure 7.16 The proposed storage power set-point control ensuring SoE limits

Demonstration

If $SoE = 0$ and P_{min} is chosen so as $P_{min} \leq \min P_{prod}$, then P_{sto} is necessarily positive and then SoE increases. In the same way, if $SoE = 1$ and P_{max} is chosen so as $P_{min} \geq \max P_{prod}$, then P_{sto} is necessarily negative and SoE decreases. It demonstrates that, with this control of the storage set-point, the SoE of the EES is guaranteed to stay within the range $[0; 1]$.

Equations (7.10) and (7.11) demonstrate that this linear relation between P_{grid} and SoE is equivalent to a low-pass filtering of P_{prod}:

$$\begin{aligned}\frac{dSoE}{dt} &= \frac{1}{E_{rated}}\frac{dE_{EES}}{dt} \\ &= \frac{1}{E_{rated}}\left(P_{prod} - P_{grid}\right)\end{aligned} \tag{7.10}$$

Assuming that $P_{grid} = (P_{max} - P_{min})SoE + P_{min}$ (cf. Figure 7.16) we can rewrite:

$$\left(\frac{E_{rated}}{P_{max} - P_{min}}\right)\frac{dP_{grid}}{dt} + P_{grid} = P_{prod} \tag{7.11}$$

$$\tau_{smooth}\frac{dP_{grid}}{dt} + P_{grid} = P_{prod} \tag{7.12}$$

Equation (7.12) corresponds to a first order linear differential equation that reflects the fact that P_{grid} corresponds to a low-pass first order filtering of the power production P_{prod} with a filtering time constant $\tau_{smooth} = \frac{E_{rated}}{P_{max} - P_{min}}$.

Example

Figure 7.17 gives an example of power and SoE time-series that can be obtained. Here, the rated energy of ESS is equal to 3kWh(10.8MJ). P_{max} and P_{min} are chosen respectively equal to 1MW (the maximum of power production for a power leveling at 1 MW) and 0MW (the minimum of

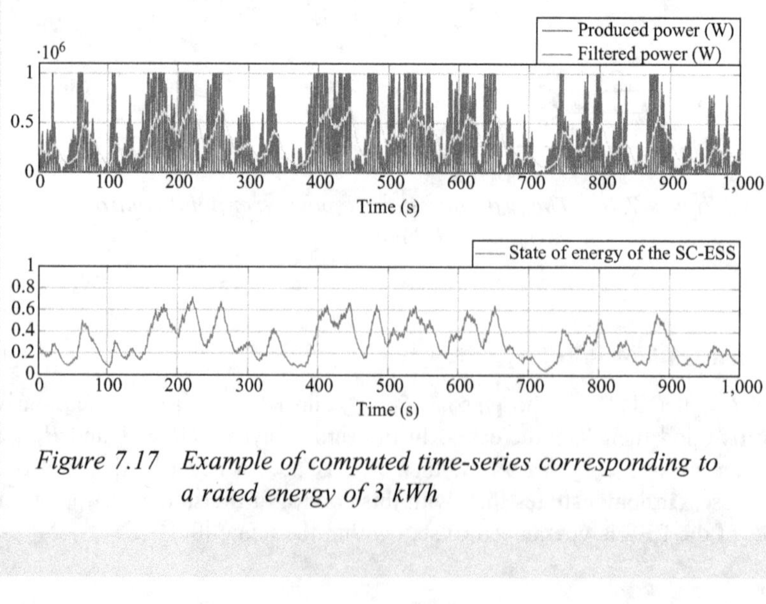

Figure 7.17 Example of computed time-series corresponding to a rated energy of 3 kWh

the power production for a viscous damping). The equivalent time constant of filtering is then equal to 10.8 s. Other simulation parameters are ($T_p = 9$ s; $H_s = 3.5$ m) sea-state (ISSC spectrum) and damping parameters values of $\beta = 4$ MN ms and $P_{lev} = 1$ MW.

Case study: OWC turbine

This case study examines power smoothing in a single offshore WEC by developing an ESS utilizing the turbine inertia in a variable speed strategy and smoothing the generator output power with supercapacitors (SCs). Actual sea-state data is used with an experimentally derived WEC model [39] to obtain the full-scale power flows and system speed response. From this, an SC system is sized and integrated into the Simulink model representing a full-scale 500 kW prototype WEC system.

The oscillating water column (OWC) WEC considered employs a Wells turbine without blow-off valves. A simplified schematic of the WEC is shown in Figure 7.3. To extract realistic power take-off (PTO) information from the device at variable speed, the Simulink model created in [39] is used. This was based on experimental data from a quarter-scale prototype operated offshore in an Atlantic test site. The simulations take account of the impact of varying turbine speed on the pneumatic power production. The model inputs are pneumatic power and turbine speed, and the output is turbine torque. Non-dimensional quantities are used to allow scaling to full size.

With so many WECs in development, any variable speed strategy is unique to each device and its location. Factors to be considered when devising a control strategy are discussed in [40] and consist of

- Remaining within speed limits.
- Efficient performance.
- Power quality to the grid.
- Utilizing a realistic control procedure where measureable quantities are used.

Numerous variable speed strategies [13, 40, 41] were examined and compared using the turbine model described above with sea-state data. The strategy that produced optimum performance was developed in [13] where generator torque is evaluated from a measure of turbine speed.

The control scheme consists of two parts. The first part was developed by measuring the average mechanical power produced at a fixed machine speed. The fixed machine speed maximizing the average power for each of the 13 sea-states was found, and these speeds and powers were plotted. Results from two mid-power sea-states are shown in Figure 7.18.

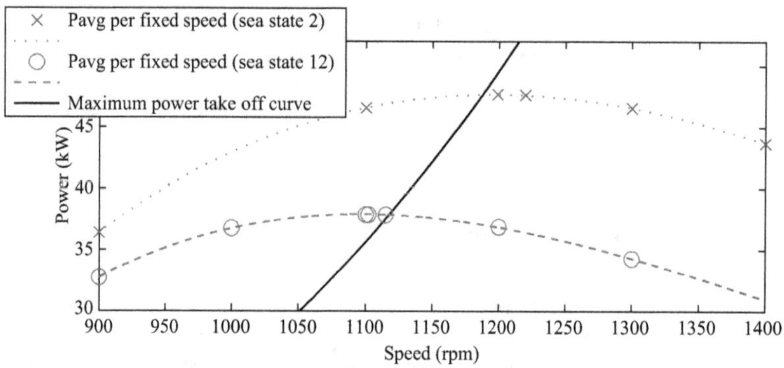

Figure 7.18 Average mechanical power versus fixed speed for two sea-states and maximum power curve

Using the curve fitting tool in Matlab with these maximum average powers and corresponding fixed speeds for each sea-state, the empirical power coefficients in (7.13) were derived (producing an *R*-square value of 0.9996). This curve is also shown in Figure 7.18.

$$P_{gen} = 0.0005307\omega^{3.797} \tag{7.13}$$

The second part of the control scheme limits the generator power as shown in (7.14) and ensures that the turbine does not over-speed to avoid mechanical stress and possible failure.

$$P_{gen} = \left[{P_{max}}^2 - J\left|\frac{dP_{gen}}{dt}\right|\left({\omega_{max}}^2 - \omega^2\right)\right]^{\frac{1}{2}} \tag{7.14}$$

where $J|dP_{gen}/dt| = 100$ MW s^{-2} kg m^2 as in [13], and $\omega_{max} = 1,500$ rpm and turbine inertia were set at 595 kg m^2 in line with other full-scale OWC Wells turbines [13].

The control algorithm sets the generator power to the maximum value evaluated from (7.13) and (7.14) according to the turbine speed as shown in Figure 7.19.

Simulated plots of input pneumatic power, electrical power and speed are shown in Figure 7.20(a), (b), and (c). Chattering of the generator power occurs around the speed where (7.13) comes into effect. To prevent this chattering, a switched controller is used where the local maximum generator torque achieved is maintained until the speed drops by a predetermined level (this hysteresis value was set at 80 rpm). The resulting power profile and turbine speed are shown in Figure 7.20(d) and (e).

As shown in Figure 7.20 the generator electrical power contains large peaks that occur only occasionally. It is proposed to further smooth this

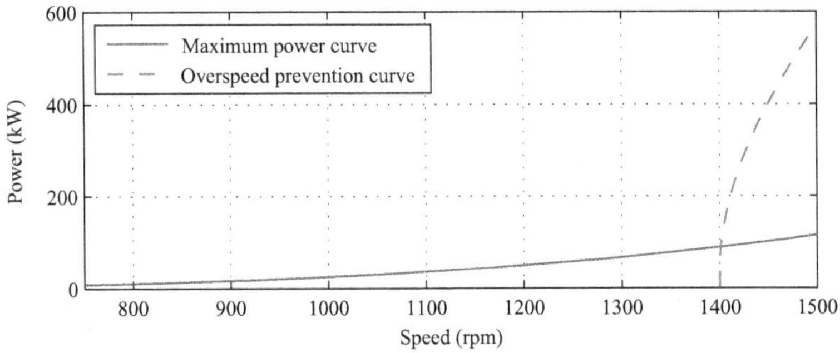

Figure 7.19 Speed control power curve of generator power versus speed

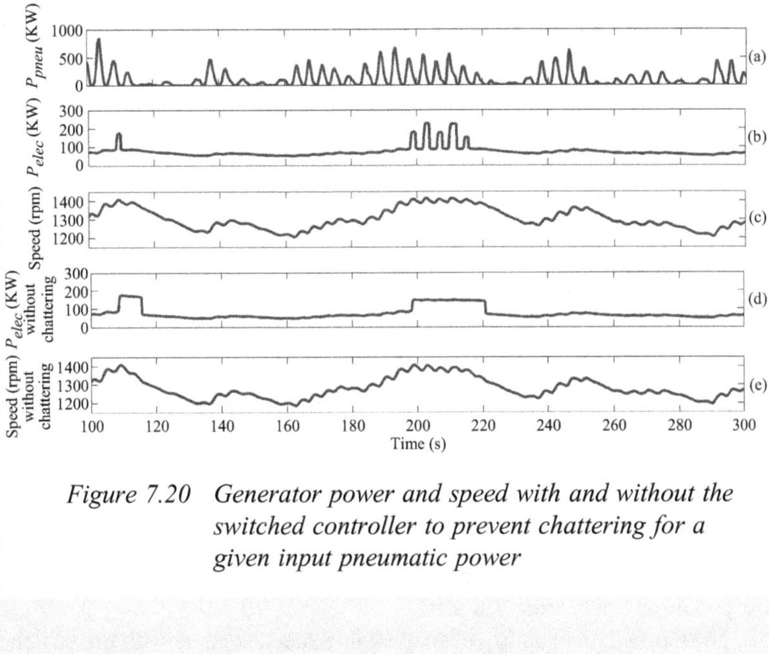

Figure 7.20 Generator power and speed with and without the switched controller to prevent chattering for a given input pneumatic power

power with SCs connected to the DC-bus of the back-to-back converter that couples the generator to the grid.

The number of generator power peaks for each sea-state was measured, and multiplied by occurrence values to evaluate the total number of peaks over the 5 year WEC maintenance interval. It was assumed that the WEC would not be operational in very low or high energy sea-states. Therefore, the device would be functioning over 70% of the time, with approximately 981,000 peaks of electrical power to be smoothed. This number of peaks is within the specified cycle lifetime of some SC modules.

The discharge strategy attempts to maintain the SCs at their lowest operational voltage (half rated voltage) to make available the SC energy capacity for absorbing power peaks. Once the generator power exceeds a predetermined value (dependent on the sea-state), the SCs prevent any excess power flowing to the grid and absorb the difference. Once the input power drops below this value, the SCs maintain this power to the grid until the minimum voltage is achieved. A voltage hysteresis band prevents discharge of the SCs until the band is exceeded to ensure no rapid charging and discharging cycles occur.

The SCs are sized for the maximum energy sea-state of the WEC that produces 294 kW of pneumatic power on average. Sizing was based on multiples of the BMOD0063 P125 63 F module from Maxwell Technologies (utilizing SCs of the same technology as the SCs under test), with parameters given in Table 7.2. Two modules in series give a voltage range of 125–250 V, and five of these parallel strings satisfy all module ratings and limit the maximum output grid power to 175 kW for the sea-state producing maximum energy. Due to the physical size and significant ballast of the WEC, the SC system size is not necessarily a constraint.

Table 7.2 Parameters of the SC module [42]

Maxwell Technologies BMOD0063 P125	
Capacitance (F)	63
Max continuous current (A)	150
Max peak current, 1 s (A)	750
R_{esr} DC (mΩ)	18
Cycle life (cycles)	1,000,000
Mass (kg)	59.5

The level of power smoothing was indicated by measuring the standard deviations of the different powers in the system over the simulation time using the mechanical power on the turbine as a reference. These results were: 1 pu for mechanical power, 0.43 pu for electrical power and 0.33 pu for grid power. The modelling work produced the results given in Table 7.3 for the most commonly occurring sea-state.

The smoothing capability of the variable speed scheme with the switched controller is not captured in the standard deviation measure of 0.43 pu for electrical power, which was an increase on the standard deviation of 0.37 before the switched controller was added, although it is observed in Figure 7.20. The major advantage of integrating SCs and turbine inertia in this power smoothing scheme is the reduction of the peak-to-average grid power.

Table 7.3 *Variable speed strategy with SCs simulation performance data*

Strategy	From [13]	From [13] with switched controller	From [13] with switched controller + SCs
Efficiency (%)	54.9	55	55
Peak to avg P_{mech}	6.9	6.8	6.8
Peak to avg P_{elec}	5.5	4.6	4.6
Peak to avg P_{grid}	5.5	4.6	2.2
SD of P_{mech} (pu)	1	1	1
SD of P_{elec} (pu)	0.37	0.43	0.43
SD of P_{grid} (pu)	0.37	0.43	0.33
Measured ω_{max} (rpm)	1,440	1,428	1,428
Measured ω_{in} (rpm)	941	941	941

7.6 Issues associated with electrical energy storage

The problem of precisely evaluating the lifetime of supercapacitors is an area still undergoing research. We can find in the literature two approaches: one based on the counting of charge–discharge cycles performed by the ESS, and a second one based on an ageing model where the two main factors are the cell voltage and cell temperature.

The number of cycles that a supercapacitor can perform is always specified in the manufacturer's datasheets, though these long lifetimes are difficult to validate and test. Through tests an ageing model is developed for a small capacitance supercapacitor based on temperature and constant current cycling.

But for some applications the number of charge–discharge cycles is quite difficult to define because the stored power profile is complex, it is more interesting to dispose of a generic model of ageing.

This section presents results from these two approaches.

7.6.1 Cycling

Maintenance intervals in offshore WECs are necessarily long with a typical desired interval for non-routine, disruptive maintenance of 5–10 years. This is due to the high cost and difficulty in carrying out maintenance [43], where working in an unstable environment for floating WECs, docking issues, and the availability of equipment and labour on days with low enough sea swell are key concerns. This maintenance interval gives the desired minimum lifetime of any employed energy storage element.

An indication of the operational time of a WEC in that time is given in [44], where Harrison and Wallace state that their device is idle for about one third of the year though it is difficult to define this value at the early stage of development for many WEC technologies.

Typical ocean wave periods can vary from 1 to 20 s or even higher. Using a wave period of about 10 s for a typical full-scale WEC gives a frequency of 0.1 Hz. If a short-term energy storage device is to be utilized for power smoothing over each wave period, it will need to survive power cycling for 5 years at close to this rate. The number of power cycles expected with these figures over this maintenance interval is

Power cycles $= 6$ cycle/min $= (6)(60)(24)(365.25)(5)(2/3) \approx 10.5$ million cycles/5 years

For a WEC that rectifies the input wave power, the power frequency to the energy storage device will be doubled. This results in 21 million cycles over 5 years if power smoothing is to take place over each wave period. This poses serious lifetime issues for any energy storage equipment that is likely to be cycled every wave cycle.

7.6.1.1 Supercapacitor (SC) cycling

Comparing the given 1 million lifecycle value for SCs with this lifecycle specification, it appears that SCs cannot help attenuate normal operating output power fluctuations. While it can be inferred that reducing the operating voltage will extend cycle lifetime, as yet no data on this topic is available, and it is unknown if 21 million power cycles is attainable. Some SCs require voltage balancing due to capacitive tolerances between cells in order to ensure no over-voltage takes place. The reliability of each of these balancing circuits, as well as each series SC cell in a module, is of concern when investigating continued power cycling operation over 5 years.

The case study given in section 7.5.2 describes the use of SCs in an integrated ESS with turbine inertia. In this case study, the SCs are expected to undergo just under 1 million power cycles.

SCs do not have a hard failure point to indicate end of life, but rather a maximum parameter deviation, typically a reduction of capacitance by 20% and an increase in the equivalent series resistance (ESR) by 100%. While SC lifetime has been tested before, it has typically been accelerated testing, where elevated voltages and temperatures were used. Based on changes in lifetime at small deviations of voltage and temperature at elevated values, typical lifetimes at normal conditions were determined from extrapolations [45–47]. Maxwell Technologies provide some results from their lifetime testing but only up to 150,000 cycles, and then extrapolate to 1 million [34]. Also, their testing procedure provided 15 s of rest between every cycle.

Test setups are built to establish the SC cycle lifetimes under the standard and application test conditions. SC standard testing at rated temperature is also carried out to determine the extent to which elevated temperatures affect lifetime. This life testing will help determine whether SC cycle lifetime is a limiting factor in the application of power smoothing in offshore WECs.

7.6.1.2 Standard cycle lifetime testing

In this case study, to decrease cycle testing time, the smallest capacitance SC from Maxwell Technologies was chosen for lifecycle testing. Thirty BCAP0005 P270 cells were characterized. Each SC, which is charged at the rated current of 1.6 A to the rated voltage of 2.7 V, undergoes a 5 s rest period (approximately five time constants) and the voltage and time are measured. The SC is then discharged at

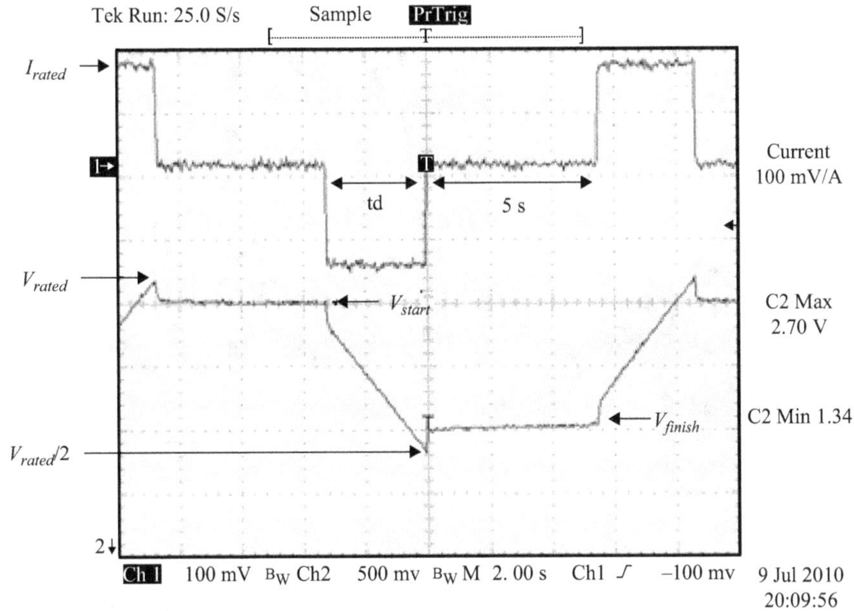

Figure 7.21 SC current and voltage during characterization

rated current to half rated voltage, and another 5 s rest period takes place before measuring the final voltage. Plots of this characterization profile are shown in Figure 7.21, and the capacitance and ESR are evaluated according to

$$C = \frac{I_{rated} t_d}{V_{start} - V_{finish}} \tag{7.15}$$

and

$$R = \frac{V_{finish} - V_{rated}/2}{I_{rated}} \tag{7.16}$$

An SC with close to average specification was chosen for testing. The test setup, shown in Figure 7.22, consists of a power supply to charge the SC, an electronic load, a high precision voltmeter, and a thermocouple monitor taking temperature readings of the top, body and leg of the SC, as well as the ambient temperature. These devices are operated using GPIB hardware under the control of a Matlab file. The testing is carried out at ambient temperature with continuous rated current.

Constant current cycling between rated and half rated voltage is carried out continuously. Characterization tests occur every 100 cycles and are performed over five consecutive cycles, from which average values are obtained giving more accurate readings. The BCAP0005 P270 SC has a specified cycle lifetime of 500,000, where end of life is specified as a 30% reduction of capacitance, or a 100% increase in ESR. The degradation is shown in the results section.

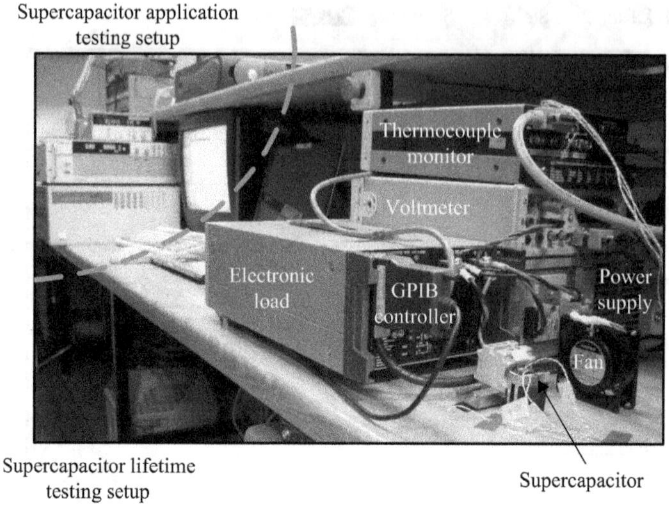

Figure 7.22 SC lifecycle test setup

7.6.1.3 Standard cycle lifetime testing at rated temperatures

While high temperature testing has been performed before, it has typically been carried out in conjunction with high voltage testing. In an unmanned offshore WEC, the SCs will be utilized in modules that may also be placed inside on-board control rooms with poor thermal management, leading to operating temperatures in excess of the tested 25°C room temperature. To test the extent to which elevated temperatures will adversely affect the SCs' cycle lifetime, a BCAP0005 P270 SC was placed inside a thermal chamber that was maintained at 64°C. Constant current cycling testing was performed on the device continuously, in accordance to the standard cycle lifetime specifications described above. The SC was placed in a heat sink under the airflow of a fan inside the chamber. This induced a temperature rise of 1°C at the continuous current rating, allowing operation at the SC rated temperature.

7.6.1.4 Application testing

From the modelling work described in section 7.5.2, the full-scale SC power profile is obtained. Using Froude scaling [48], these powers are scaled down to values relevant to the BCAP0005 P270 SC under test. To scale time down, divide by scale to the power of 0.5, and to scale power down, divide by scale to the power of 3.5. A scale factor of 21.135 was chosen to match the continuous powers of the module scaled with the tested SC. The other relevant resultant scaled values closely match the tested SC ratings as shown in Table 7.4. As the resultant usable energy of the SC is lower than the scaled value, the maximum voltage limit is expected to be reached during testing.

The model SC power profile is developed from the most commonly occurring sea-state before voltage limits are encountered, with the grid power limited to 145 kW. This sea-state contains over 30 min of data and produces 10 power peaks; this is also close to the average power peak rate over yearly operation.

The application SC test utilizes similar equipment as outlined in the lifecycle testing. Due to Froude scaling the applied power profile lasts 395 s. This is looped three times before characterization tests are carried out. Testing is performed continuously and the SC selected for testing has the closest specifications to the SC in the lifecycle apparatus.

7.6.1.5 Results

Almost 4 million cycles have been tested on the SC under standard test. The initial parameters and testing conditions of the SCs under test are shown in Table 7.5. Testing runs continuously for a few days at a time, after which data files are collected, equipment checked, and parameters noted before the computer-controlled testing is restarted. The degradation of capacitance and ESR are seen in Figures 7.23 and 7.24 respectively.

The SC itself is rated for 500,000 cycles and almost 4 million cycles are shown tested in Figures 7.23 and 7.24. Previous available testing data demonstrated operation

Table 7.4 SC module scaled to values relevant to tested SC

	SC module scaled	Tested SC
Continuous power (W)	4.32	4.32
1 sec power (W)	10.17	9.18
Usable energy (J)	18.5	13.7

Table 7.5 Initial parameters of the SCs under test conditions

	Cycle life SC	Thermal SC	Application SC
Initial C (F)	5.794	5.747	5.759
Initial R_{esr}, DC (mΩ)	0.108	0.105	0.103
Average relative humidity (at start of testing run) (%)	35.0	41.0	34.1
Average ambient temperature (°C)	25.1	64.1	23.8

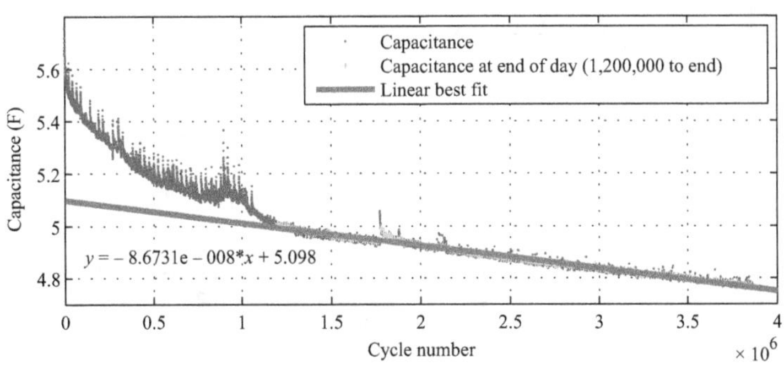

Figure 7.23 Capacitance versus cycle number during cycle lifetime testing at 25°C ambient

up to 150,000 cycles. Figures 7.23 and 7.24 validate SC performance discussed in [49], where there is an exponential decrease of capacitance initially before capacitance degradation is more linear. It is expected that near end of life an exponential fall off of capacitance will occur.

A linear best-fit line is applied to the capacitance graph using points obtained at the end of each testing day, and from 1.2 million cycles onward. This prevents life estimation errors from the initial exponential decrease and from capacitance recovery that occurs while testing stops. Initially no testing occurs overnight, or on weekends (this weekend recovery is clearly seen up to 1 million cycles in Figure 7.23), but eventually testing is carried out continuously. If these trends continue, the capacitance value will reach end of life first, in line with current literature [49]: capacitance will reach 70% of its nominal value of 5 F after 20.5 million cycles, although it is stated by Maxwell that there is an exponential decrease of capacitance near end of life that might affect this estimate.

The variations of capacitance and ESR for the SC over one day of testing are shown in Figures 7.25 and 7.26, and the variation of ambient temperature and the SC

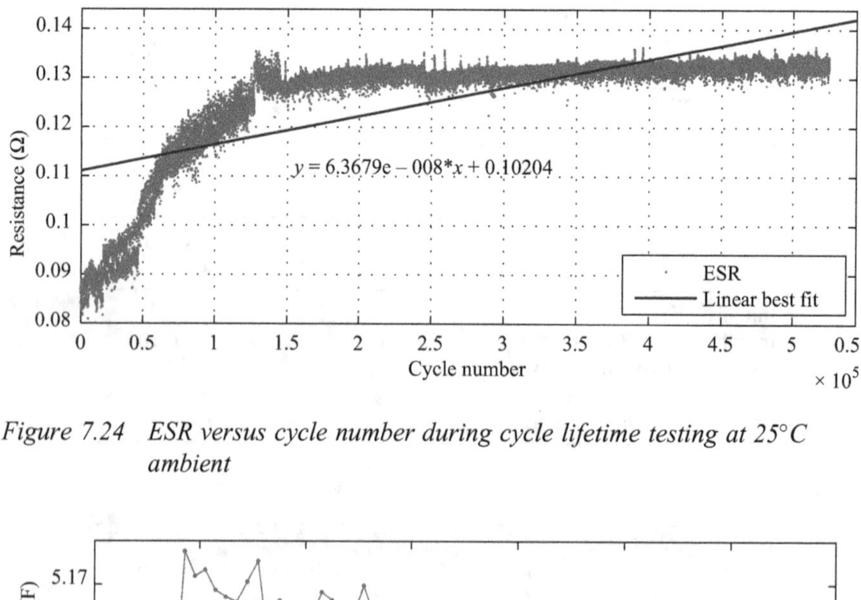

Figure 7.24 ESR versus cycle number during cycle lifetime testing at 25°C ambient

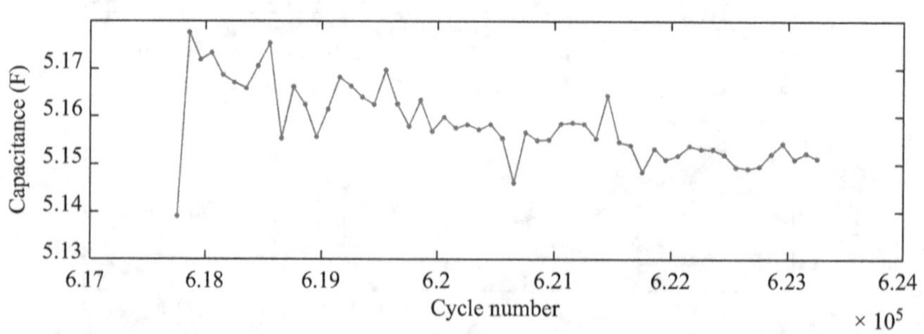

Figure 7.25 Cycle lifetime testing capacitance over one day at 25°C ambient

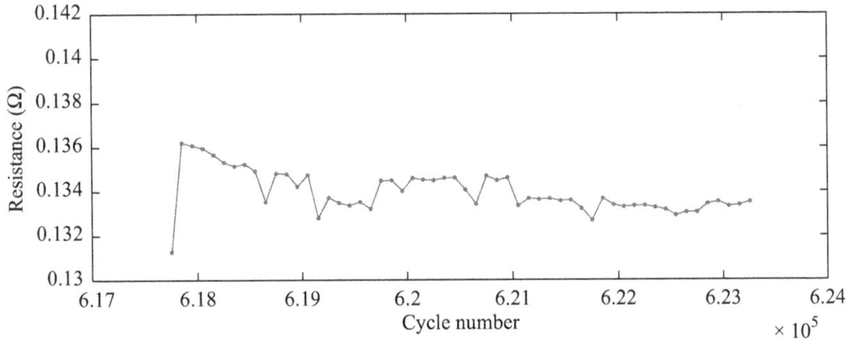

Figure 7.26 Cycle lifetime testing ESR over one day at 25°C ambient

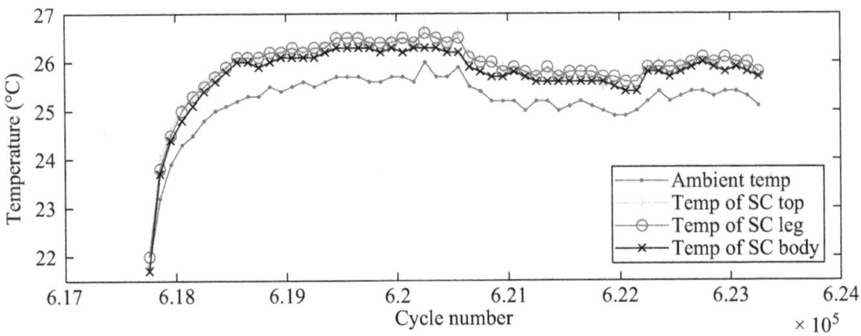

Figure 7.27 Ambient and SC temperature for the cycle lifetime setup over one day

temperature measured at the top, body and leg is shown in Figure 7.27. SC temperature rises only slightly above ambient, due to the action of the heat sink and fan. Also, the first characterization point in Figures 7.25–7.27 is taken before the SC has undergone any power cycles and after an overnight rest.

Temperature has a significant effect on SC lifetime, and a typical reference is that an increase of 10°C will halve the lifetime of the SC [50, 51]. From this it is expected that operation at the rated temperature of 65°C will lead to capacitance degradation at 16 times that which is experienced at a room temperature of 25°C.

Standard cycle lifetime testing at rated temperature has achieved over 700,000 cycles. The degradation of capacitance and ESR are seen in Figures 7.28 and 7.29 respectively. These plots are much smoother than the other tests' plots, as due to added safety checks the testing could be carried out continuously from the start. One day of rest occurred during testing and the resulting capacitance recovery is clearly seen at around 610,000 cycles. A linear best-fit line is applied to the capacitance graph using points obtained at the end of each testing day, and from 200,000 cycles onward. If these trends continue, capacitance will reach end of life first after 3.75 million cycles.

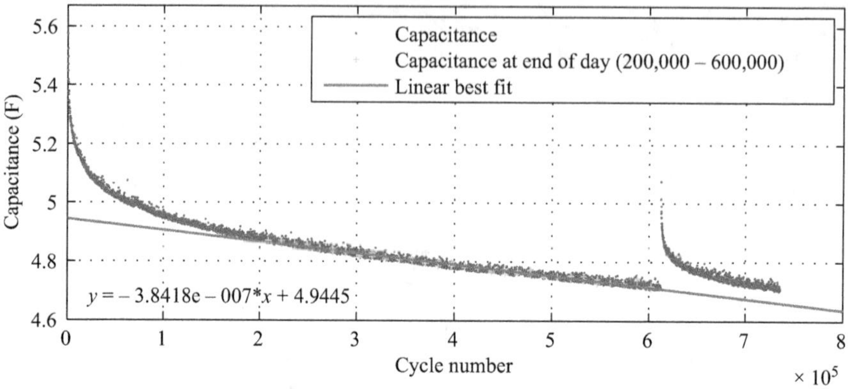

Figure 7.28 Capacitance versus cycle number during standard testing at 64°C ambient

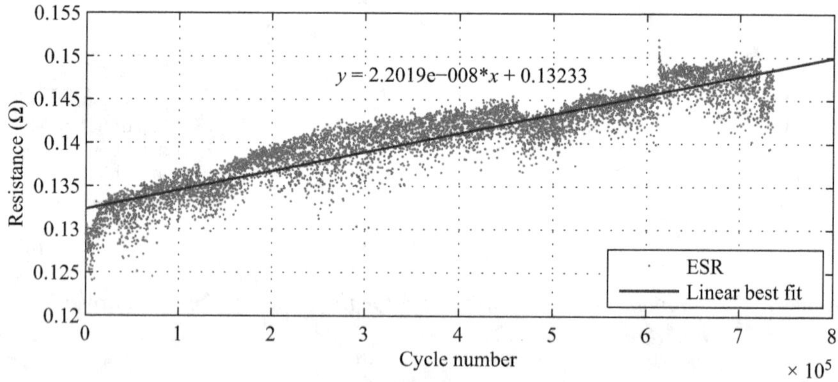

Figure 7.29 ESR versus cycle number during standard testing at 64°C ambient

To investigate the effect of temperature on the SC cycle lifetime, an Arrhenius equation that looks at the temperature dependence on the rate of a chemical reaction was applied to the trends of the two standard cycle lifetime tests [52].

$$t_i = Be^{\frac{-E_a}{kT_i}} \tag{7.17}$$

where T_i is the absolute temperature of interest in Kelvin, t_i is the reaction time in hours, k is Boltzmann's constant, B is a parameter to be determined, and E_a is the activation energy in eV given by

$$E_a = \frac{k\ln\left(\frac{t_1}{t_2}\right)}{\frac{1}{T_1} - \frac{1}{T_2}} \tag{7.18}$$

The time to end of life for the two test points are represented by t_1 and t_2, and the corresponding absolute temperatures are T_1 and T_2.

Applying these equations to the work in this case study produces an E_a of 0.38 eV and 0.015 h for B. These parameters were obtained from single sample testing at 65°C and 26°C, and are only applicable at constant current cycling between rated and half rated voltage. From this it is found that SC cycle lifetime is halved for an increase in temperature of around 15°C for the temperature range of interest.

Application lifetime testing has achieved over 500,000 cycles. This corresponds to two and a half years of operation at full scale. A sample of the SC applied current and resultant voltage is presented in Figure 7.30 and the degradation of capacitance and ESR are seen in Figures 7.31 and 7.32 respectively. A linear best-fit line is applied to the capacitance graph using points obtained at the end of testing each day, and from 240,000 cycles onward. Up until about 200,000 cycles, testing is stopped overnight and at weekends, and in accordance with [49] a lifetime estimate from a linearization should only be applied once the capacitance recovery resulting from breaks in testing has diminished. If these trends continue, capacitance will reach end of life first after 4.25 million cycles corresponding to over 20 years of operation at sea.

The application test appears to reach end of life after 4.25 million cycles compared to 20.5 million cycles for the standard test. While it is difficult to compare projected lifetime degradations for the plots, it is interesting to note the testing time required. The application test would require 1,780 days of non-stop testing to achieve 4.25 million cycles, whereas the standard test would require 1,550 days of non-stop testing to achieve 20.5 million cycles.

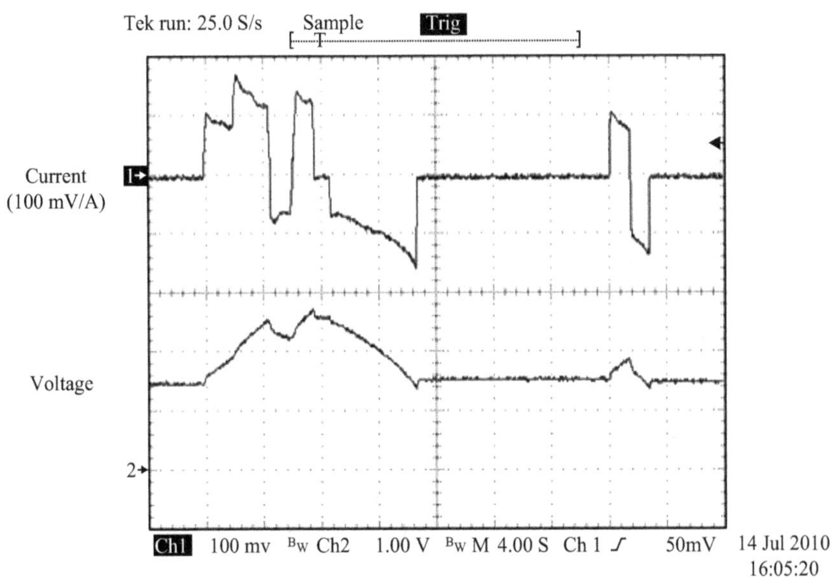

Figure 7.30 SC current and voltage from application testing

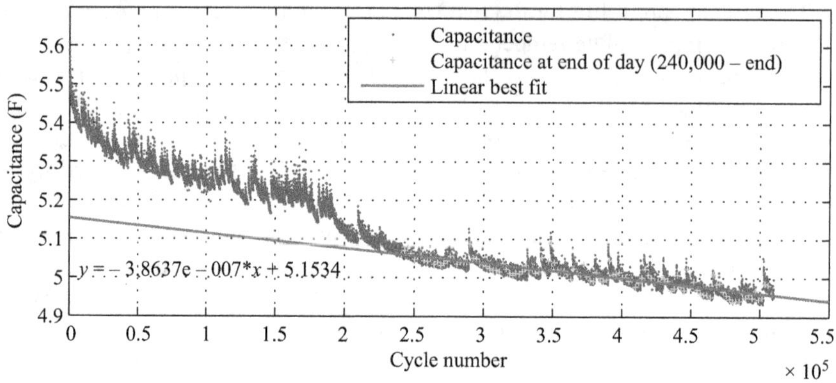

Figure 7.31 Capacitance versus cycle number during application testing at 24°C ambient

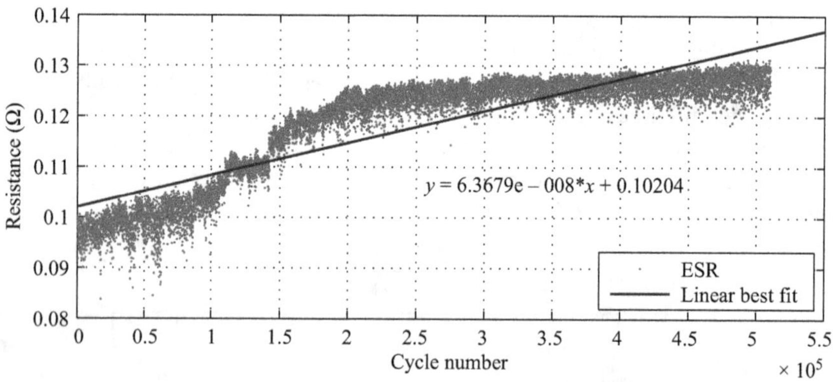

Figure 7.32 ESR versus cycle number during application testing at 24°C ambient

7.6.1.6 Conclusions

As long component lifetime is a requirement for offshore WECs, computer-controlled lifetime testing setups have been built using GPIB equipment to validate SC lifetimes and determine application lifetime. Almost 4 million cycles have been tested on the standard setup, 700,000 cycles at rated temperature, and 500,000 cycles on the application setup equivalent to 2.5 years of full-scale operation. Previous results of available documented cycle testing reached 150,000 at most [50]. A summary of these results are shown in Table 7.6. An Arrhenius degradation was assumed to create an equation for expected time of life based on operating temperature and trends of single sample testing at 65°C and 26°C on small capacitance SCs; a temperature increase of approximately 15°C halves lifetime.

Table 7.6 Summary of BCAP0005 P270 SC testing results

Rated cycle lifetime at 25°C	**500,000**		
	Cycle life SC	**Thermal SC**	**Application SC**
Cycles tested	4 million	700,000	500,000
Equivalent full-scale time of operation			2.5 years
Average ambient temperature (°C)	25.1	64.1	23.8
Expected cycle lifetime	20.5 million	3.75 million	4.25 million
Equivalent full-scale time of operation			21.5 years
Testing time required for expected cycle lifetime	1,551 days	284 days	1,778 days

Also, based on initial trends of single sample tests, the application tested SC appears to have a much smaller cycle life compared to the standard lifecycle tested SC, but it is interesting to note that the application test would require slightly more time to reach this end of life estimate.

This case study has demonstrated that if SCs are utilized with turbine inertia for power smoothing in an offshore WEC with the algorithm described, then cycle lifetime of SCs appears not to be a limiting factor in the long maintenance intervals (typically 5–10 years) that would be experienced, even if placed in relative on-board hot spots.

7.6.2 Ageing model

An ageing model is necessary in order to take into account the degradation in time of the main parameters of an SC-ESS in sizing optimization procedure and to define a state of ageing (SoA). During ageing, capacitance and equivalent series resistance (ESR) of supercapacitors vary. Assuming that when $SoA = 1$, ESR ESR_{EES} increases by 100% and capacitance C_{EES} decreases by 20%, their evolution according to SoA can be expressed as

$$ESR_{EES}(t) = ESR_{EES,init} \times (1 + SoA(t)) \tag{7.19}$$

$$C_{EES} = C_{EES,init} \times (1 - 0,2SoA(t)) \tag{7.20}$$

The ageing process of a supercapacitor depends essentially upon two factors: temperature and voltage. It was found that the lifetime (defined by the time to reach $SoA = 1$) of an SC is reduced by a factor of 2 when voltage increases by 200mV or temperature increases by 10K [53]. Therefore, the lifetime at the temperature θ and voltage V, denoted $\mathscr{L}(\theta, V)$, can be expressed according to the lifetime $\mathscr{L}(\theta_0, V_0)$ as in (7.21), measured in temperature and voltage reference conditions, and can be deduced from manufacturer data (Maxwell data are considered here) :

$$\mathscr{L}(\theta, V) = \mathscr{L}(\theta_0, V_0) \times 2^{\frac{\theta_0 - \theta}{10}} \times 2^{\frac{V_0 - V}{0.2}} \tag{7.21}$$

This expected lifetime is plotted in Figure 7.33.

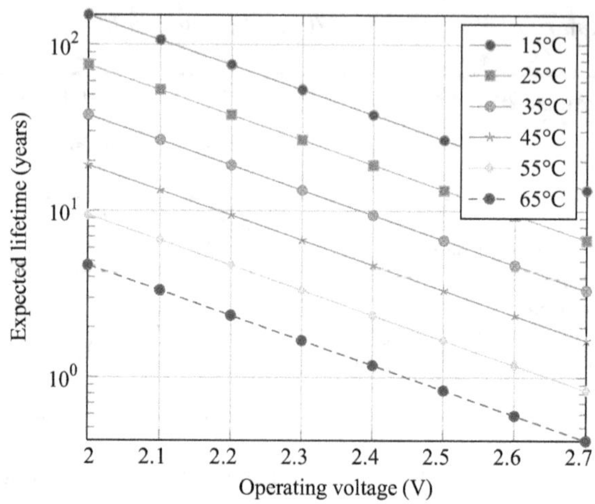

Figure 7.33 Expected lifetime of a supercapacitor according to operating voltage and temperature

Taking into account varying conditions of temperature and voltage according to the time, the state of ageing (SoA) can be defined as

$$SoA(t) - SoA(t = 0) = \int_0^t \frac{1}{\mathscr{L}(\theta(t), V(t))} dt \tag{7.22}$$

According to this definition, the SoA becomes 1, when the end of life of the SC can be considered as reached. Voltage and temperature operating conditions can be calculated from SoE and P_{sto} as follows:

$$V(t) = \frac{V_{rated}}{2}\sqrt{1 + 3\ SoE} \text{ (assuming voltage range between } V_{rated} \text{ and } \tfrac{V_{rated}}{2}) \tag{7.23}$$

$$\theta(t) = \theta_{ambient} + R_{th}ESR_{EES}\overline{\left(\frac{P_{sto}(t)}{V(t)}\right)^2} \tag{7.24}$$

In practice, the thermal time constant of manufactured supercapacitor cells and modules is relatively high compared to the power fluctuations period (1,000 s vs. 4 s). Our computed power profiles are 1 h long, which is why the temperature rise can be considered as a constant over a sea-state and calculated with the average value of losses. We can note that this thermal problem simplification would not have been possible if the simulated profiles would have been longer than few times the thermal time constant of the supercapacitor module. Finally, voltage conditions are essentially related to the SoE and temperature conditions are related to the ambient plus self-heating caused by ESR losses. A strong hypothesis is the uniformity of the temperature in the supercapacitor module. We made other hypotheses also like the independence of the capacitance and ESR according to the temperature and voltage over the duration of a power profile. However, the evolution of ESR and capacitance over greater time length due to ageing is taken into account. The eventual non-uniformity can affect the ageing evolution but this is not considered in this study.

Through the increase in ESR_{EES} and indirectly through the decrease in C_{EES}, ageing causes a supplementary heating and can shoot up. Then, the evolution of SoA can be highly non-linear, and the lifetime estimation requires an iterative calculation scheme. The time step of this calculation has to be in the range of months or even years. In practice, this value is chosen so as the variation of SoA is kept below 10%. Therefore, a variable time step scheme of integration is considered. During each step (representing months or years), the increase in SoA is computed on the basis of many simulations: one for each sea-state corresponding to the considered wave data statistic. For each sea-state, several random draws of initial phases for the sea surface elevation profile generation should be considered.

7.7 Sizing of the capacity according to performance criterions

The role of an ESS in direct-drive WECs is to smooth the power produced by the PTO operating in direct conversion. With this type of conversion, unlike those involving a hydraulic stage, the electrical power output becomes zero approximately every 3–5 s (for 6 s to 10 s sea-states) and then has a naturally high peak-to-average ratio.

7.7.1 Smoothing quality criteria

In order to evaluate the sizing of this ESS, we have to define smoothing quality criteria.

7.7.1.1 Percentage of extra losses due to fluctuating grid power

As D. O'Sullivan *et al.* mention in [54], and for a constant (or quasi-constant) voltage level, the RMS-to-average ratio of the grid power gives an indication of the extra losses due to the transport of fluctuating power rather than its average:

$$\Delta P_{loss} = \frac{\overline{P_{grid}^{\,2}}}{\overline{P_{grid}}^{\,2}} \tag{7.25}$$

However, the sole absolute value of this criterion doesn't make sense and only a comparison between two values (e.g. with and without smoothing) gives a first image of the smoothing benefit provided by an ESS.

7.7.1.2 Criteria of power dispersion

Various classical criteria of dispersion can be used to quantify the smoothing quality. We consider here the mean absolute deviation (MAD) and the standard deviation (STD). We apply these criteria both on the smoothed power, $P_{grid}(t)$, and on its time-derivative, $\frac{dP_{grid}}{dt}$.

The MAD of a time-dependant variable $x(t)$ is defined as

$$\mathrm{MAD}(x(t)) = \overline{\left|x(t) - \overline{x}\right|} \tag{7.26}$$

where $\bar{x}$ is the average of $x(t)$. The STD definition is

$$\mathrm{STD}(x(t)) = \sqrt{\overline{(x(t) - \bar{x})^2}} \tag{7.27}$$

7.7.1.3 Flicker severity: definition of a modified flicker coefficient

The flicker severity levels (instantaneous $s(t)$, short-term P_{st} or long-term P_{lt}) depend on short-circuit apparent power S_k at the considered point of common coupling of the grid and will be all the more important so as the grid will be weak at the PCC (low value of S_k) [55]. These values depend also on the grid impedance angle ϕ_k. Indeed, flicker is calculated from the RMS voltage variations profile $\Delta V(t)$, which itself results from the power profiles fed into the grid (active $P_{grid}(t)$ and reactive $Q_{grid}(t)$) through the following expression:

$$\Delta V(t) = \frac{R_k P_{grid}(t) + X_k Q_{grid}(t)}{3V_n} \tag{7.28}$$

where V_n is the nominal RMS voltage of the grid (phase to neutral), and R_k and X_k are respectively the resistive and reactive parts of the grid impedance, which are related to short-circuit apparent power S_k and the grid impedance angle ϕ_k through

$$Z_k = \frac{3{V_n}^2}{S_k} \tag{7.29}$$

$$R_k = Z_k \cos \phi_k \tag{7.30}$$

$$X_k = Z_k \sin \phi_k \tag{7.31}$$

It can be demonstrated that the product of the flicker severity and the short-circuit apparent power no more depends on short-circuit apparent power but solely on the grid impedance angle and on the power profiles. On the basis of this property, a flicker coefficient has been defined for wind turbines [56–58]. The flicker coefficient is a normalized measure of the flicker severity emitted from a wind turbine during continuous operation and is defined as follows:

$$c(\phi_k, v_w) = P_{st} \frac{S_k}{S_n} \tag{7.32}$$

where v_w is the annual average wind speed defined in [58], P_{st} is the flicker emission from the wind turbine on a grid with a short-circuit apparent power S_k, while S_n is the rated apparent power of the wind turbine. The flicker coefficient of a wind turbine is interesting because it permits to deduce the flicker emission of a wind turbine on any grid. However, the flicker coefficient must be given with the corresponding rated power S_n, which is not easy to define in the context of wave energy conversion [59]. Then, in the following text, we do not want to normalize by S_n, and will only give the product of the flicker severity and the short-circuit apparent power S_k of the grid on which the flicker has been evaluated. This modified flicker coefficient is denoted c^*:

$$c^*(\phi_k) = P_{st}S_k \tag{7.33}$$

In order to evaluate the flicker emission of a WEC on a grid showing a particular value of short-circuit apparent power, this modified flicker coefficient has just to be divided by the particular value of short-circuit apparent power.

For example, in France, the flicker severity is limited to 0.25 for any short-circuit apparent power above 40MVA (otherwise, this limit have to be multiplied by the ratio of 40MVA to the effective short-circuit power). Therefore, the modified flicker coefficient must be below 10MVA to comply with regulatory limits whatever the strength of the grid.

Example: Estimation of the flicker emission of the SEAREV without energy storage means

In order to justify the use of energy storage means, we can first evaluate the level of flicker severity without active power smoothing. For wind turbines, the flicker must be given at the 99th annual percentile and, by extension, we will choose this recommendation for WECs. For wind turbines, flicker severity increases with one parameter, i.e. the speed of wind, and then the 99th flicker percentile corresponds to the wind speed 99th percentile, whereas for WECs, the evolution of the flicker severity according to the two values H_s and T_p is not so obvious.

In Figure 7.34, the modified flicker level coefficient c^* is plotted as a function of the sea-state parameters, for a grid impedance angle ϕ_k of 30°. It appears that flicker severity rises not only with the increase in the significant height H_s but also with the decrease in eT_p. As a result, flicker severity does not vary like wave power resource (which is proportional to $H_s^2T_p$) but rather like the average mechanical power (see Figure 7.12). In Figure 7.34, we can see that this regulatory limit is exceeded for many sea-states (grey areas, the area where the flicker complies with regulatory limits has been whitened).

In view of particular sea-state statistics of a sea site, here Yeu Island (cf. Figures 7.12), it is then possible to estimate the 99th percentile of the short-term and long-term flicker severity. In this case, we have

Short term $c^*(30°) = 34.5$ MVA
Long term $c^*(30°) = 33.3$ MVA

These values correspond to the flicker emitted by only one SEAREV DWEC. It can be noted that the flicker emitted by n SEAREV, even judiciously arranged, will not be n times the flicker emitted by 1 SEAREV. For wind farms, it has been found that the flicker emitted by n wind turbines is approximately the product of the square root of n and the flicker emitted by one wind turbine. These values have been computed on the basis of the 106 sea-states from the wave data statistics (year 1999) of Yeu Island taking into account their probability of occurrence. For each sea-state, eighty 1 h wave elevation profiles were generated.

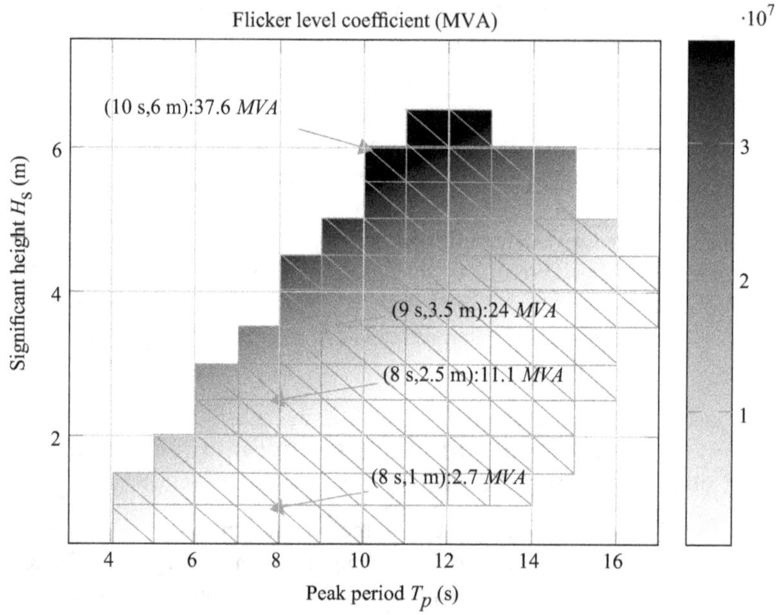

Figure 7.34 Modified flicker level coefficient $\mathbf{c}^*$ *without energy storage and active power smoothing, based on the 120 min average of* P_{st} *values (as prescribed in the CEI standard), for a grid impedance angle* $\phi_k = 30°$ *and damping parameters values of* $\beta = 4$ *MNms and* $P_{lev} = 1$ *MW. The area with values below the 10 MVA regulatory limit has been filled with white*

It means that for a grid impedance angle ϕ_k of 30°, the short-circuit apparent power at the point of common coupling must be greater than 138 MVA$(= 34.5/0.25)$ in order to comply with the regulatory limit without need of ESS. Below this value of short-circuit apparent power, the smoothing of the active power production becomes mandatory. To comply with the regulatory limit whatever the strength of the grid, this flicker coefficient has to be reduced by 71%.

Example: Contribution of supercapacitor-based ESS to smoothing performances

In this last example, we will see the influence of the EES energy rating E_{rated} on different performances smoothing and cost criterion.

MAXWELL, a supercapacitor manufacturer, proposes 63F125V modules that can be connected up to 12 in series: the BMOD0063 P125 (Table 7.7) [60]. Each of these modules is composed by a string of 48 supercapacitor cells 3, 000F2.5V: the BCAP3000 [61]. An active control system ensures the balancing between the cells (cf. Figure 7.35).

Table 7.7 Specifications of the module BMOD0063 P125

Description	Value	Unit
Rated capacitance	63	F
Rated voltage	125	V
Equivalent series resistance	223	214
Thermal resistance	0.04	°C/W
Thermal capacitance	33,370	J/°C

*Figure 7.35 Maxwell Technologies® supercapacitors module BMOD0063 P125 (left, L*W*H (mm) = 619*425*265, 60.5 kg) and cell BCAP3000 (right, L*D (mm) = 138*60, 510 g)*

We consider a DC-bus voltage of 1,300V (for a 690V AC connection) and a half-bridge IGBT DC–DC converter. The studied SC-ESS will then be composed by $N_{parallel}$ strings in parallel, each composed by N_{series} modules BMOD0063 P125 in series. In order to reach a rated voltage value of the stack near but below the DC-bus voltage, N_{series} must be lower or equal to 10. Each BMOD0063 P125 module corresponds approximately to an energy content of 100Wh assuming a voltage working range between V_{rated} and $V_{rated}/2$.

Results are presented in Figure 7.36. For this computation, the number of random draws for initial phases of the sea elevation profile has been limited to 30 because of computation time (80 previously). Then it still means that for each sea-state, we consider 30 different hour-profiles of produced power. Finally, taking into account all the sea-states of the wave data statistics, 3,200 h of realistic power time-series data are generated and used (Table 7.8).

As expected, all the power quality criteria decrease with E_{rated} decreasing (see the first plot in Figure 7.36). They are given as the percentage of their values obtained without storage in order to estimate the benefit of energy storage means. The value obtained without storage is specified in the legend. We can note that the 99th percentile of flicker coefficient has been reduced by more than 71%, which has been previously estimated as the needed reduction to comply with the regulatory limit whatever the strength of the grid.

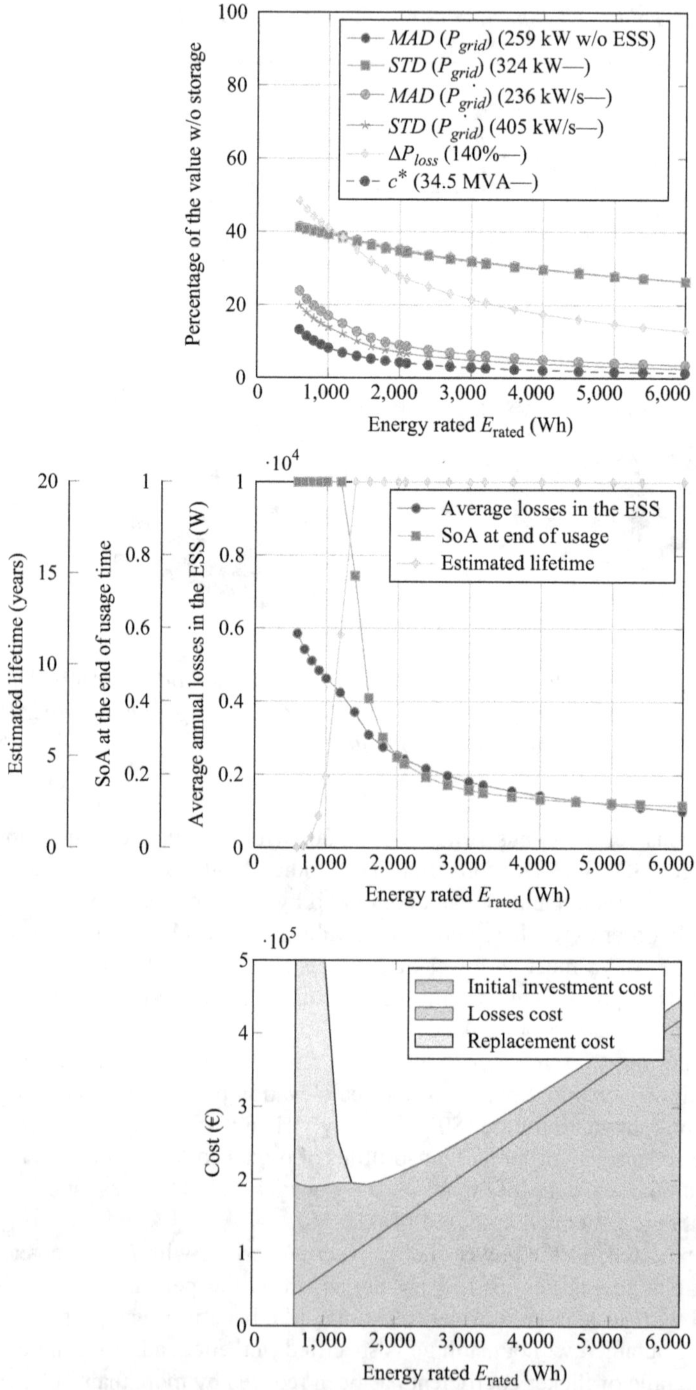

Figure 7.36 Evolution of smoothing and cost criteria according to the rated energy of the supercapacitor energy storage system

Table 7.8 Supercapacitor ESS sizing problem parameters

Description	Symbol	Value	Unit
Temperature of the ambience	$\theta_{ambient}$	30	°C
Rated voltage of the ESS	V_{rated}	$125 \times N_{series}$	V
Maximal grid power	P_{max}	1	MW
Minimal grid power	P_{min}	0	MW

Thanks to power smoothing, losses in electric equipment placed downstream of the EES (grid converter, step-up transformer, power cable etc.) can be significantly reduced (see ΔP_{loss}). For the highest values of E_{rated}, these losses are even only increased by 20%–30% compared with those caused by an ideal constant power. One can note that the losses in the EES are not taken into account in this criterion. Another advantage for the electric components placed downstream is the diminution of the cycling fatigue that can improve the global lifetime of the system.

On the basis of the considered ageing model, we can see in the second plot in Figure 7.36 that the estimated lifetime of the EES reaches 20 years easily (the ageing calculation is stopped at this age), except for the lowest values of E_{rated} (below 1.6kWh). The state of ageing (SoA) at the end of the usage time (considered here as 20 years) gives an indication of the remaining health of supercapacitors when the ageing calculation is stopped. The cost of offshore replacements of the EES can be very high, while the yearly window of possible maintenance can be very short. It is therefore crucial for the ESS to have a high lifetime. These estimations show that it is possible to stand a reasonable period (20 years) without needing to intervene for replacement.

The estimated lifetime increases with E_{rated} and $N_{modules}$ because of two main effects:

1. P_{sto}, the power in the storage system, is distributed over the modules. Then, for a given storage power profile, the losses due to the equivalent serial resistance (ESR) are inversely proportional to the square of the number of modules.
2. A tendency of the probability of high values of *SoE* to decrease and. Indeed, because of the *SoE* control strategy, the value of the *SoE* corresponds to the ratio of the output (smoothed) power to the maximal power P_{max} and focuses around a level corresponding to the ratio of average power to maximum power.

The cost of the ESS can be estimated taking into account an initial cost of about 70€ per Wh of supercapacitor (which is slightly overestimated), a cost for the internal losses assuming an income shortfall of 15c€ per kWh of losses and a replacement cost calculated on the basis of the ESS lifetime and the needed replacements of the system (the cost of maintenance is not considered and should be added). We can see in the third plot in Figure 7.36 that an optimum value of E_{rated} minimizes this cost. This value is around $1,600$Wh and $1,800$Wh taking account the flatness of the cost function and the uncertainties of the ageing modelization. Lowest values of energy rating are not interesting because they have higher cost and lower power quality criterion. But higher values of E_{rated} permit to have a higher power quality. Therefore, there is a trade-off between the power quality and the cost of the ESS.

7.8 References

[1] D. Murray, J. Hayes, M. G. Egan, A. W. Lewis and D. O'Sullivan, 'The benefits of device level short term energy storage in ocean wave energy converters', in Rosario Carbone, Ed., *Energy Storage in the Emerging Era of Smart Grids*, 2011.

[2] R. G. Alcorn and W. C. Beattie, 'Power quality assessment from the LIMPET wave-power station', *Proceedings of the 11th International Offshore and Polar Engineering Conference*, 2001.

[3] D. B. Murray, M. G. Egan, J. G. Hayes and D. L. O'Sullivan, 'Applications of supercapacitor energy storage for a wave energy converter system', *Proceedings of the 8th European Wave and Tidal Energy Conference*. Uppsala, Sweden, pp. 786–795, Uppsala, Sweden, 2009.

[4] M. Rahm, O. Svensson, C. Boström, R. Waters and M. Leijon, 'Experimental results from the operation of aggregated wave energy converters', *Renewable Power Generation, IET,* 2012, pp. 149–160.

[5] J. Tissandier, A. Babarit and A. H. Clément, 'Study of the smoothing effect on the power production in an array of SEAREV wave energy converters', The International Society of Offshore and Polar Engineers (ISOPE), 6–11 July 2008. Vols. *ISOPE-2008: Eighteenth (2008) International Offshore and Offshore and Polar Engineering Conference Proceedings*. ISBN 1-880653-68-0.

[6] M. Molinas, O. Skjervheim, B. Sørby, P. Andreasen, S. Lundberg and T. Undeland, 'Power smoothing by aggregation of wave energy converters for minimizing electrical energy storage requirements', *Proceedings of the 7th European Wave and Tidal Energy Conference*. Porto, Portugal, pp. 3–8, Porto, Portugal, 2007.

[7] S. H. Salter, 'World progress in wave energy–1988', *International Journal of Ambient Energy*, vol. 10, pp. 3–24, 1989.

[8] A. F. d. O. Falcão, 'Wave energy utilization: A review of the technologies', *Renewable and Sustainable Energy Reviews*, vol. 14, pp. 899–918, 2010.

[9] HMRC, UCC, 'State of the art analysis, a cautiously optimistic review of the technical status of wave energy technology', WAVEPLAM 2009. Available at: http://www.waveplam.eu/files/downloads/SoA.pdf

[10] D. O'Sullivan, J. Griffiths, M. G. Egan and A. W. Lewis, 'Development of an electrical power take off system for a sea-test scaled offshore wave energy device', *Renewable Energy*, vol. 36, pp. 1236–1244.

[11] A. F. d. O. Falcão, 'The shoreline OWC wave power plant at the Azores', *Proceedings of the Fourth European Wave Energy Conference*. Aalborg, Denmark, pp. 42–48, Aalborg, Denmark, 2000.

[12] T. Heath, T. J. T Whittaker and C. B. Boake, 'The design, construction and operation of the LIMPET wave energy converter (Islay, Scotland)', *Proceedings of the 4th European Wave Energy Conference*. Aalborg, Denmark, pp. 49–55, Aalborg, Denmark 2000.

[13] A. F. d. O. Falcão, 'Control of an oscillating-water-column wave power plant for maximum energy production', *Applied Ocean Research*, vol. 24, pp. 73–82, 2002.

[14] C. Boake, *Islay LIMPET Wave Power Plant, Publishable Report, 1 November 1998 to 30 April 2002.* The Queen's University of Belfast. 2003. Contract JOR3-CT98-0312.

[15] B. Czech and P. Bauer, 'Wave energy converter concepts: Design challenges and classification', *IEEE Industrial Electronics Magazine*, vol. 6, pp. 4–16, 15 June 2012.

[16] R. Henderson, 'Design, simulation, and testing of a novel hydraulic power take-off system for the Pelamis wave energy converter', *Marine Energy*, vol. 31, pp. 271–283, February 2006.

[17] A. F. d. O. Falcão, 'Modelling and control of oscillating-body wave energy converters with hydraulic power take-off and gas accumulator', *Ocean Engineering*, vol. 34, pp. 2021–2032, October 2007.

[18] M. Molinas, O. Skjervheim, P. Andreasen, T. Undeland, J. Hals, T. Moan and B. Sorby, 'Power electronics as grid interface for actively controlled wave energy converters', *International conference on clean electrical power*. ICCEP'07. IEEE, pp. 188–195, 2007.

[19] J. M. Carrasco, L. G. Franquelo, J. T. Bialasiewicz, E. Galván, R. C. P. Guisado, M. A. M. Prats, J. I. León and N. Moreno-Alfonso, 'Power-electronic systems for the grid integration of renewable energy sources: A survey', *IEEE Transactions on Industrial Electronics*, vol. 53, pp. 1002–1016, 2006.

[20] F. Wu, X. P. Zhang and P. Ju, 'Application of the battery energy storage in wave energy conversion system', *International Conference on Sustainable Power Generation and Supply*. SUPERGEN'09. IEEE, pp. 1–4, 2009.

[21] C. A. Luongo, 'Superconducting storage systems: An overview', *IEEE Transactions on Magnetics*, vol. 32, pp. 2214–2223, 1996.

[22] M. H. Ali, B. Wu and R. A. Dougal, 'An overview of SMES applications in power and energy systems', *IEEE Transactions on Sustainable Energy*, vol. 1, pp. 38–47, 2010.

[23] S. Amjad, S. Neelarkrishnan and R. Rudramoorthy, 'Review of design considerations and technological challenges for successful development and deployment of plug-in hybrid electric vehicles', *Renewable and Sustainable Energy Reviews*, vol. 14, pp. 1104–1110, 2010.

[24] M. E. Glavin, P. K. W. Chan, S. Armstrong and W. G. Hurley, 'A stand-alone photovoltaic supercapacitor battery hybrid energy storage system', *13th Power Electronics and Motion Control Conference*. EPE-PEMC 2008. IEEE, pp. 1688–1695, 2008.

[25] D. V. Weeren, H. Joosten, R. Scrivens and A. Schneuwly, 'Ultracapacitors double operational life of wave measurement buoys'. Available at: www.altenergymag.com. Accessed 21 July 2012.

[26] H. Douglas and P. Pillay, 'Sizing ultracapacitors for hybrid electric vehicles', in *The 31st Annual Conference of the IEEE Industrial Electronics Society*, 2005,. IEEE, pp. 1599–1604, 2005.

[27] A. Babarit, A. H. Clément and J.-C. Gilloteaux, 'Optimization and time-domain simulation of the Searev wave energy converter', *The 24th International Conference on Offshore Mechanics and Arctic Engineering*, vol. 2. OMAE, pp. 703–712, 2005.

[28] M. Ruellan, H. Ben Ahmed, B. Multon, C. Josset, A. Babarit and A. H. Clément, 'Design methodology for a SEAREV wave energy converter', *IEEE Transactions on Energy Conversion*, vol. 25, no. 3, pp. 760–767, 2010.

[29] J. Aubry, H. Ben Ahmed and B. Multon, 'Sizing optimization methodology of a surface permanent magnet machine-converter system over a torque-speed operating profile: Application to a wave energy converter', *IEEE Transactions on Industrial Electronics*, vol. 59, no. 5, pp. 2116–2125, 2012.

[30] C. Josset, A. Babarit and A. H. Clément, 'A wave-to-wire model of the SEAREV wave energy converter', *Journal of Engineering for the Maritime Environment*, vol. 221, no. 2, pp. 81–93, 2007.

[31] J. Aubry, M. Ruellan, H. Ben Ahmed and B. Multon, 'Minimization of the kWh cost by optimization of an all-electric chain for the SEAREV Wave Energy Converter', *The 2nd International Conference on Ocean Energy*, 2008.

[32] J. Aubry, P. Bydlowski, B. Multon, H. Ben Ahmed and B. Borgarino, 'Energy storage system sizing for smoothing power generation of direct wave energy converters', in *The 3rd International Conference on Ocean Energy*, pp. 1–7, 2010.

[33] L. Leclercq, 'Control based on fuzzy logic of a flywheel energy storage system associated with wind and diesel generators', *Mathematics and Computers in Simulation*, vol. 63, no. 3–5, pp. 271–280, 2003.

[34] A. B. Marques, G. N. Taranto and D. M. Falcao, 'A knowledge-based system for supervision and control of regional voltage profile and security', *IEEE Transactions on Power Systems*, vol. 20, no. 1, pp. 400–407, 2005.

[35] C. Abbey and G. Joos, 'Supercapacitor energy storage for wind energy applications', *IEEE Transactions on Industry Applications*, vol. 43, no. 3, pp. 769–776, 2007.

[36] S. M. Muyeen, R. Takahashi, T. Murata and J. Tamura, 'Integration of an energy capacitor system with a variable-speed wind generator', *IEEE Transactions on Energy Conversion*, vol. 24, no. 3, pp. 740–749, 2009.

[37] L. Jerbi, L. Krichen and A. Ouali, 'A fuzzy logic supervisor for active and reactive power control of a variable speed wind energy conversion system associated to a flywheel storage system', *Electric Power Systems Research*, vol. 79, no. 6, pp. 919–925, 2009.

[38] G. O. Suvire and P. E. Mercado, 'Active power control of a flywheel energy storage system for wind energy applications', *IET Renewable Power Generation*, vol. 6, no. 1, p. 9, 2012.

[39] D. P. Cashman, D. L. O'Sullivan, M. G. Egan and J. G. Hayes, 'Modelling and analysis of an offshore oscillating water column wave energy converter', *Proceedings of the 8th European Wave and Tidal Energy Conference*, pp. 924933, Uppsala, Sweden, pp. 924–933, 7–10 September 2009.

[40] P. Justino and A. F. d. O. Falcão, 'Rotational speed control of an OWC wave power plant', *Journal of Offshore Mechanics and Arctic Engineering*, vol. 121, pp. 65–70, May 1999.

[41] J. Duquette, D. O'Sullivan, S. Ceballos and R. Alcorn, 'Design and construction of an experimental wave energy device emulator test rig', *Proceedings of the 8th European Wave and Tidal Energy Conference*. Uppsala, Sweden, pp. 443–452, 7–10 September 2009.

[42] Technologies, Maxwell. BMOD0063 P125 ultracapacitor datasheet. Available at: www.maxwell.com. Accessed 23 July 2012.

[43] D. L. O'Sullivan and A. W. Lewis, 'Generator selection for offshore oscillating water column wave energy converters', *13th Power Electronics and Motion Control Conference*. EPE-PEMC 2008. IEEE, pp. 1790–1797, 2008.

[44] G. P. Harrison and A. Robin Wallace, 'Sensitivity of wave energy to climate change', *IEEE Transactions on Energy Conversion*, vol. 20, pp. 870–877, December 2005.

[45] W. Lajnef, J. M. Vinassa, O. Briat, H. El Brouji and E. Woirgard, 'Monitoring fading rate of ultracapacitors using online characterization during power cycling', *Microelectronics Reliability*, vol. 47, pp. 1751–1755, 2007.

[46] K. Paul, M. Christian, V. Pascal, C. Guy, R. Gerard and Z. Younes, 'Constant power cycling for accelerated ageing of supercapacitors', *13th European Conference on Power Electronics and Applications. EPE*' 09. IEEE, pp. 1–10, 2009.

[47] H. El Brouji, O. Briat, J. M. Vinassa, N. Bertrand and E. Woirgard, 'Comparison between changes of ultracapacitors model parameters during calendar life and power cycling ageing tests', *Microelectronics Reliability*, vol. 48, 2008.

[48] Wavenet. *Results from the work on the European thematic network on wave energy*. March 2003. ERK5-CT-1999-20001.

[49] Technologies, Maxwell. BOOSTCAP energy storage modules life duration estimation, application note 1012839, 2007. Available at: www.maxwell.com. Accessed 25th July 2012.

[50] Product Guide, Maxwell Technologies BOOSTCAP ultracapacitors. Available at: www.maxwell.com. Accessed 23rd July 2012.

[51] D. Linzen, S. Buller, E. Karden and R. W. De Doncker, 'Analysis and evaluation of charge-balancing circuits on performance, reliability, and lifetime of supercapacitor systems', *IEEE Transactions on Industry Applications*, vol. 41, pp. 1135–1141, 2005.

[52] G. Alcicek, H. Gualous, P. Venet, R. Gallay and A. Miraoui, 'Experimental study of temperature effect on ultracapacitor ageing', *European Conference on Power Electronics and Applications*. IEEE, pp. 1–7, 2007.

[53] P. Kreczanik, C. Martin, P. Venet, G. Clerc, G. Rojat and Y. Zitouni, 'Constant power cycling for accelerated ageing of supercapacitors', *EPE*, pp. 1–10, 2009.

[54] D. L. O'Sullivan, D. Murray, J. G. Hayes and M. G. Egan, 'The benefits of device level short term energy storage in ocean wave energy converters', in R. Carbone, Ed., *Energy Storage in the Emerging Era of Smart Grids*, 2011.

[55] IEC, 'IEC 61000: Electromagnetic compatibility (EMC) – Part 4-15: Testing and measurement techniques – Flickermeter – Functional and design specifications', 2010.

[56] A. Larsson, 'Flicker emission of wind turbines during continuous operation', *IEEE Transactions on Energy Conversion*, vol. 17, no. 1, pp. 114–118, 2002.

[57] T. Ackermann, *Wind Power in Power Systems*, Chichester, UK, John Wiley & Sons, Ltd, 2005.

[58] IEC, 'IEC 61400-21: Wind turbines – Part 21: Measurement and assessment of power quality characteristics of grid connected wind turbines', 2008.

[59] A. Babarit, J. Hals, M. J. Muliawan, A. Kurniawan, T. Moan and J. Krokstad, 'Numerical benchmarking study of a selection of wave energy converters', *Renewable Energy*, vol. 41, pp. 44–63, 2012.

[60] Maxwell Technologies, 'BMOD0063 P125 Datasheet'; Document number: 1014696.3. Accessed on line in August 2012.

[61] Maxwell Technologies, 'K2 SERIES ULTRACAPACITORS'; Document number: 1015370.3. Accessed on line in August 2012.

Chapter 8

Control systems – design and implementation

8.1 Overview of control scheme

J. V. Ringwood

8.1.1 Introduction

Control systems, despite often being 'invisibly' incorporated within products, devices and vehicles, are ubiquitous. They are prevalent within the automotive and aerospace industries and form part of the vanguard of technologies in increasing performance, improving fuel economy and increasing safety. One of the most appealing aspects of incorporating control technology in many systems is that the addition of extra control functionality can usually be achieved merely through the addition of extra software code though, in many cases, additional sensors and actuators may be required.

This relatively simple implementation modality masks both the capability of control systems and the high level of engineering underpinning the development of a suitable control algorithm. For example, many high-performance model-based control design methods require an accurate mathematical model of the system to be controlled and a significant number of man-hours can be absorbed in modelling. Nevertheless, there is usually a good case to be made for the incorporation of control technology to improve the performance (both technical and economic), reliability and safety of a system. In this chapter, we will examine the role that control engineering can play in making ocean energy technology more competitive.

In an ideal world, one should consider the design of a complete system from the top down. However, convention has it that physical systems are usually designed by the discipline-specific experts and the control problem is then addressed in a subsequent step by control engineers, working in collaboration with the discipline-specific experts. Such an approach, though prevalent in the bulk of industrial applications of control, is non-optimal. There are some notable exceptions, though, with a notable one being in the design of flight control systems. In the 1970s, aircraft were designed to be open-loop unstable with the result that, in the absence of a closed-loop control system, such aircraft could not be flown by human pilots. While this put complete reliance on the control system, the advantage was that significant gains in aircraft manoeuvreability and economy of flight could be achieved. Some preliminary studies [1] have been carried out which examine the interaction between the optimal design (geometry) of a wave energy device and an accompanying control system. The results suggest a strong interaction between the type of control system used and the optimal device geometry, with optimal energy capture as an objective.

For ocean energy applications, control systems can offer performance benefits. Assuming that the prime energy converter is designed first, the addition of control can offer significant energy capture enhancement. While the area of tidal turbine control is not so well established, there are close similarities with wind turbine control, one notable difference being that the density of the medium is about 1000 times greater in the case of tidal turbines. A general study on wind turbine control [2] suggests that a variable-speed turbine, requiring torque and speed control, can absorb 2.3% more energy than a fixed-speed counterpart, where the speed is fixed by the electrical grid frequency. In the case of wave energy, the numbers are more dramatic. A study on latching control [3], which is a relatively simple implementation of the more ideal complex-conjugate control, suggests that energy capture can increase by as much as a *factor* of 2 with control in irregular waves and by up to a *factor* of 4 in regular waves.

We also need to consider if the addition of a control system may drive the system more aggressively in an attempt to increase energy capture, perhaps leading to shortened device lifetimes. While the addition of control to a (tidal) turbine is likely to be relatively benign, the use of motion-exaggerating control for a reciprocating wave energy device can have a dramatic effect on device motion. Consequently, the balance between increased energy capture (income) and increased device wear (cost) needs to be carefully considered. Some formulations are now appearing which attempt to balance such quantities [4], though explicit enumeration of the cost of increased wear remains an open issue.

While potentially affecting more aggressive device motion, there are some redeeming features of control which may help the designer in practical application. For example, physical constraints can be explicitly included in many control formulations [5], resulting in a control action that respects (and is optimal within) the physical system constraints. In addition, most optimal control formulations allow some explicit trade-off between control action and the main objective (e.g. setpoint tracking, energy maximisation, etc), which provides a design handle on the level of aggressiveness of the control [6]. Control science also provides a body of knowledge relating to the design of control systems which are tolerant (in some respect, but usually with reduced performance) to device, actuator or sensor failure [7].

Finally, one might consider the various 'levels' of control that might be required in an ocean energy application. Clearly, there is a top level of supervisory control which assesses the incident energy resource and may curtail the operation of the device in the face of extreme conditions. Such curtailment may be a requirement in order to preserve the device integrity, ensure safe operation, or be required by legislation, as in the case of wind turbines. For the tidal energy case (unlike wind energy), the extremes of current flow are known and the tidal current device will be designed to operate in energy capture model over this relatively narrow operating range. Wave energy devices, however, will frequently encounter sea states which are outside their normal operational envelope and some

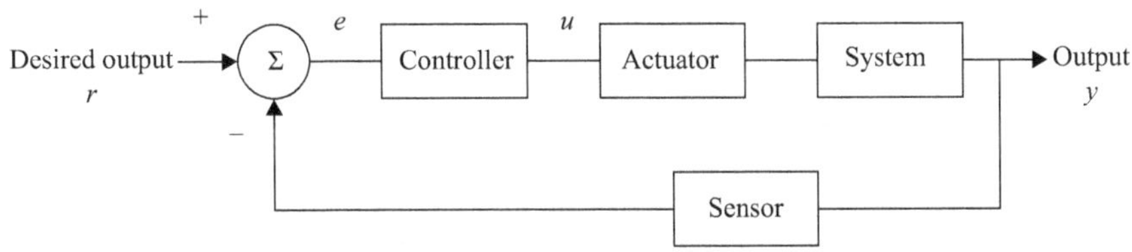

Figure 8.1 Typical tracking problem for a control system

supervisory strategy may be necessary to ensure that device integrity is retained. While such supervisory control is important, it is beyond the scope of this chapter.

8.1.2 What is control?

While the reader might like to peruse the 1548 page excellent 'encyclopedia' of [8] in an effort to understand the essence of control, it is pertinent to try to articulate the fundamental utility of control that makes it useful in the context of ocean energy. In general, control attempts to devise algorithms that force a system to follow a desired path, objective or behaviour modality. Traditionally, the control problem is defined by a tracking problem, as shown in Figure 8.1, where the objective is for the system output to follow the reference input. While problems of these type do occur in ocean energy applications, for example speed control of a tidal turbine, it is more useful to broaden the set of problem descriptions and potential solutions a little in order to assess the potential of control engineering in an ocean energy context.

In general, control problems are usually defined by some subset of the following:

Definition 1: General control problem definition

`Maximize/minimize Performance objective (max. energy, min. error) subject to: System constraints (amplitudes, rates, forces, etc)`

i.e. a constrained optimization problem. The definition above is not inconsistent with the purpose of the system in Figure 8.1 where the objective function is usually some measure (e.g. a quadratic measure) of the difference between the output and its desired value, i.e. the tracking error.

The desired performance of the tracking system in Figure 8.1 can be specified in a variety of ways:

1. Desired transient response, e.g. [9].
2. Desired steady-state response, e.g. [10].
3. Desired closed-loop poles (roots of the closed-loop transfer function), e.g. [11].
4. Trade-off between control energy (u^2) and tracking error (e^2) [12].
5. Minimization of the sensitivity of the closed-loop system to variations in the system description [13].
6. Minimization of the sensitivity of the closed-loop system to external disturbances [13].

Items 5 and 6 in the list above relate to the system *robustness*, and specific control methodologies to address these objectives have been developed since the late 1970s.

In most cases, control design methods provide an explicit solution for the controller in Figure 8.1, while some methods solve the more general optimization problem defined above at each time step. In the next section, we will see how such specific or general solutions can be useful in the control of ocean energy devices.

Finally, some control methods require a mathematical model of the system in order to determine the control algorithm and such methods are termed *model-based*. The requirement for an accurate mathematical system model often involves considerably more work than the calculation of the controller itself, though *system identification techniques* [14] can be employed to determine a blackbox model, i.e. a model which has no structural relationship to the physical system. The combination of system identification techniques with a mathematical procedure for controller determination can be used to develop adaptive controllers, which have the capability to adapt to unknown (in 'self-tuning mode') or time-varying systems. Adaptive control schemes based on linear system models also have the capability to track variations in a linear model due to the presence of non-linearity, though non-linear systems are best controlled with a dedicated fixed-parameter non-linear controller. Significant care and attention must also be paid to adaptive schemes to ensure stability and convergence over all operating regimes [15].

For ocean energy systems, the modelling effort can be considerable, since hydrodynamic modelling is involved. While a variety of comprehensive non-linear modelling methodologies are available for hydrodynamic modelling, including smooth particle hydrodynamics (SPH) or computational fluid dynamics (CFD) approaches, the difficult of incorporating such models into a control formulation requires the use of more compact and structurally simple models. In addition, the very significant computational complexity of SPH or CFD models preclude their direct use for real-time controller implementation. Instead, model-based control strategies usually use compact linear models, which are based on either local linearization about an operating point (see, e.g., [16, 17]) for the turbine case, or linear boundary-element models [18, 19] for the wave energy case. Even modest non-linear extensions to linear boundary element methods can result in models which are computationally intractable for real-time control [20], while some specific parameterizations (e.g. to include viscosity effects [21])give non-linear parametric forms that may be possible to incorporate in model-based control schemes.

8.1.2.1 Varieties of control algorithms

With a view to an examination of the spectrum of control methodologies available, the diagram of Figure 8.1 is now slightly reconfigured to that of Figure 8.2 for convenience of notation. The operators K, G and H are purposely free of continuous-time/discrete-time or linear/non-linear classification, with the intention of keeping the discussion as broad as possible. It is hoped that any loss of mathematical rigour is compensated by the increased scope of the diagram's use.

In Figure 8.2, the 'system' and 'actuator' blocks of Figure 8.1 have been combined into G while the additional inputs, d and ξ, represent disturbances

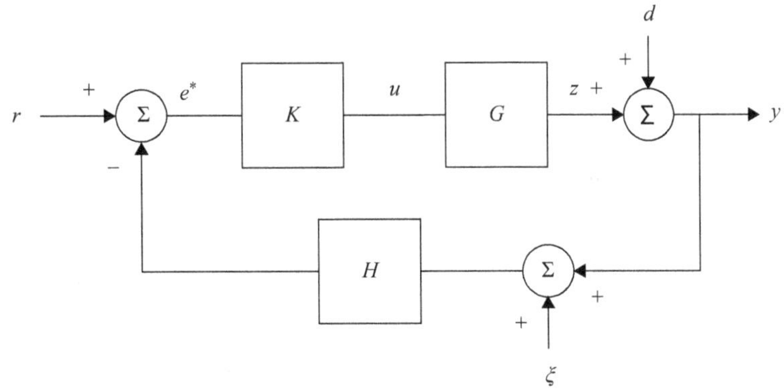

Figure 8.2 Feedback system specification

(deterministic and/or stochastic) and measurement noise respectively. The '*' designation is used to denote that e^* is not necessarily an exact calculation of $e = r - y$, since H may be non-unitary.

Proportional, Integral, Derivative (PID) Control

One of the most ubiquitous control methods has been a mainstay of process control and many other application areas for over seven decades. In this paradigm, the control signal is formed as a linear combination of error, integral of error (which aids good steady-state control) and first derivative of error (which aids good transient response), expressed in continuous-time form as

$$u(t) = k_p e^*(t) + k_i \int e^*(t)dt + k_d de^*(t)/dt \tag{8.1}$$

Various other forms, including PI (Proportional-integral) and PD (proportional-derivative) forms, are also possible, while a number of embellishments, including velocity forms, bumpless transfer, derivative on output only and derivative filtering make the PID controller more useful in many practical applications. The proportional, integral and derivative gains (k_p, k_i and k_d respectively) can be determined from a variety of model-based design rules, or can be tuned by hand. The strong intuitive relationship between the controller parameters and (particularly) the closed-loop step response gives the opportunity to tune, or fine-tune, the parameters of the PID controller by observing the resulting changes to the closed-loop response and has been largely responsible for the widespread appeal of PID control. A comprehensive treatment of PID tuning rules is given in [22]. PID controllers are best suited to systems with predominantly linear dynamics, though gain-scheduled versions can be used to control non-linear systems [23].

Pole placement

The controller K can also be determined by defining the positions of the desired closed-loop poles, which are strongly related to the closed-loop transient response. Such calculations can be easily done in either discrete- or continuous-time, though transient response specifications are more easily specified in the (continuous-time) s-plane. Designs performed in continuous-time can be easily transformed to discrete-time versions using discretization based on pole-zero mapping [24]. Pole placement required a model of the system, but is easily performed for both models described by transfer functions [25] or in state-space form [26]. For systems described in transfer function form, some design freedom (choice of observer polynomial) is available beyond achievement of the closed-loop poles, while pole placement for multivariable systems (systems with multiple inputs and outputs) normally results in a nonunique solution for the feedback gains, allowing, e.g., some extra design freedom which can be used to address robustness issues.

Optimal control

There are many varieties of optimal control, which has its origins in the birth of the 'modern control' era of the 1960s. It can be used with both state-space [12] and transfer function [27] models and is equally applicable to continuous- and discrete-time models. Traditionally, the (continuous-time) linear quadratic regulator (LQR) optimal control problem is formulated as

$$\min_{u} J = \int_{0}^{t^*} (e^T Qe + u^T Ru)dt \tag{8.2}$$

where $e = r - y$ and Q and R are weighting matrices which define the trade-off between error minimization and reduction of the control energy used. For example, in high-accuracy servomechanism problems, the choice of $Q \gg R$ is usually made, while in satellite control applications, where fuel is limited, a choice of $R \gg Q$ is more appropriate. Since there is an implicit relationship between Q, R and the closed-loop poles, traditional optimal control can be seen as an alternative way of determining the closed-loop poles. The solution to the problem in (8.2), however, is not simple, requiring the solution to a Ricatti equation for systems described in state-space form [12] and a Diophantine equation for the transfer function form [27]. However, a range of computer-based tools are now available to assist with the solution to these equations.

In general, the formulation in (8.2) is for linear models and can be specified either as a finite-horizon (t^* is finite) problem or as an infinite horizon ($t^* \to \infty$). A particular evolution of the finite horizon optimal control problem which is formulated in discrete time and uses a future prediction of the system output is model-based predictive control (MPC) [28], which has seen significant adoption in the process, and other, industries, rivalling PID control in popularity.

The basic problem formulation for MPC is

$$\min_{u(0),\ldots,u(h_c)} \sum_{k=0}^{h_p} e(k)^T Qe(k) + \sum_{k=0}^{h_c} \Delta u(k)^T R\Delta u(k) \tag{8.3}$$

where $\Delta u(k) = u(k) - u(k-1)$, $e(k) = r(k) - y(k)$ and h_p and h_c are the prediction and control horizons respectively. Using a model of the system, the future output is predicted up to h_p steps ahead and the optimal control sequence, $u(0),\ldots,u(h_c)$, calculated so that the system output follows a reference trajectory. The next control input, $u(0)$ is then applied to the system and the optimal future control sequence recalculated. This 'receeding horizon' procedure allows the system to effectively counteract output disturbances (d in Figure 8.2) and gives the control paradigm good robustness properties.

MPC is flexible and can deal with a number of practical system characteristics. For example, if the system contains pure delay (of, say, n_d steps) then the performance function in (8.3) is recast with n_d as the lower limit on the first summation term. This effectively implements a Smith predictor [29], placing the delay component outside the control loop. An extension which is very useful in many applications is the possibility to implement constraints on the system inputs, outputs and states. This can be used to implement an optimal control which respects physical limits on system variables, such as velocities, positions, forces, currents and voltages. The constrained formulation [30] includes (8.3) together with a set of constraints:

$$\begin{aligned} y_{\min} \leq &\quad y \quad \leq y_{\max} \\ u_{\min} \leq &\quad u \quad \leq u_{\max} \\ &|\Delta u| \leq u_{\max} \end{aligned}$$

etc.

MPC can use both linear and non-linear models. For the back MPC formulation, an algebraic solution $u(k)$, in terms of $y(k)$ and $r(k)$, is possible. For MPC formulations involving constraints and/or a non-linear system model [31], the general (numerical) optimization problem in Definition 1 above must be solved. However, the subject of efficient optimization algorithms for constrained/non-linear MPC has received considerable attention [32].

Finally, one very useful exploitation of the general MPC framework is the possible substitution of the quadratic regulation penalty (e.g. $e^T e$) in (8.3) with an energy-related term, such as `current x voltage` or `force x velocity`, coupled with a change from minimization to maximization of the performance function. Such a possibility has particular relevance to ocean energy and will be further discussed in section 8.1.3.

Robust control

Robust control theory developed rapidly from the 1970s as a paradigm that explicity tried to synthesize controllers which were optimally robust to variations in the system (due to time variance, non-linearity etc.) and disturbances. For robust control design, both a system model *and* a measure of the uncertainty in the model are required. The uncertainty can be expressed in both structured (leading to μ-synthesis [33]) or unstructured (leading to H_∞ design [34]) forms. For illustration, the robust control formulation for H_∞ design will be briefly outlined. The performance function for the classical H_∞ control design problem can be specified as

$$J_\infty = \left\| \begin{matrix} W_1(s)S(s) \\ W_2(s)T(s) \end{matrix} \right\|_\infty \tag{8.4}$$

where

$$S(s) = (1 + GKH(s))^{-1}, \qquad T(s) = GKH(s)(1 + GKH(s))^{-1} \tag{8.5}$$

are the sensitivity and complementary sensitivity functions (with reference to Figure 8.2) respectively, and we note that $S(s) + T(s) = 1$. $S(s)$ specifies the closed-loop sensitivity to variations in $G(s)$ and $K(s)$ and also is the transfer function relating the disturbance, d, to the system output, y. $T(s)$ specifies the relationship between the measurement noise, ξ, and the output, y, and also determines robust stability.

Robust stability is guaranteed by ensuring that the weight $W_2(s)$ overbounds the plant (multiplicative) perturbation in the maximum singular value sense as

$$\bar{\sigma}[W_2(j\omega)] \geq \bar{\sigma}[\Delta(j\omega)] \forall \omega \geq 0 \tag{8.6}$$

where

$$G(s) = G_o(s)(1 + \Delta(s)) \tag{8.7}$$

with G_o being the nominal system model. In general, the weighting function $W_1(s)$ is chosen to:

- Penalize sensitivity, $S(s)$, at low frequency, giving good low-frequency disturbance rejection, and
- Ensure that system performance (dynamic response) is maintained in spite of parameter variations in $G(s)$ at low frequency.

W_2 is chosen to:

- Ensure robust stability by covering $\Delta(s)$, i.e. that condition (8.6) is met, and
- Attenuate high-frequency measurement noise in ξ, by driving $T(s)$ down at high frequency.

In addition, the relative positions of $W_1(s)$ and $W_2(s)$ determine the closed-loop bandwidth, controlling the dynamic response to setpoints (r) and disturbances (d).

One of the issues with early robust control was that, while robustness was explicitly handled, no specifications on performance (e.g. tracking) could be explicitly made. One solution is to design an LQR regulator and then reinforce controller robustness using the loop transfer recovery (LTR) method [35]. Robust versions of predictive control, including the possibility to deal with constraints, are also available [36] and H_∞ design methods have also been extended to non-linear systems [37]. While robust control presents the possibility to deal with non-linear systems using a linear robust control approach, where the non-linearity is reflected in the uncertainty in the linear model, this is not a recommended approach, since the performance will be degraded across the operational spectrum (in order to ensure robust stability), compared to a dedicated non-linear design.

In general, the optimization problems resulting from minimization of cost functions such as (8.4) are complex, but formulation of robust control problems as a set of linear matrix inequalities (LMIs) allows the application of efficient numerical tools [38].

Non-linear control

While control algorithms developed for linear systems apply to the vast bulk of such systems (with some exceptions relating to open-loop stability/instability, non-minimum phase zeros, presence of time delay etc.), there is no general theory of non-linear systems and most non-linear control design methods are for specific non-linear forms, e.g. Hammerstein and Volterra. A number of generic non-linear control formulations, and using artificial neural network (ANN) system models, are presented in [39] and [31] and include methods inspired by model-reference adaptive control, internal-model control (IMC) and MPC. An alternative is to try to apply feedback linearization [40] to the non-linear system, followed by linear design. Other popular non-linear control approaches include backstepping [41] and sliding-mode control [42], which is a form of variable strucutre control using high gain and a switching control strategy to implement robust feedback control.

8.1.3 Control systems for ocean energy

While section 8.1.2 has focussed mainly on the classical control problem of regulation of some variable to a desired value, and indeed such problems are encountered in ocean energy applications, there is a broader set of problems which can also be addressed by control system technology. The purpose of this section is to present this broad problem definition and examine how this problem may be addressed, or broken down into smaller parts which may be more easily solved.

8.1.3.1 Problem specification

In the case of both tidal and wave energy, the general problem is to maximize energy capture, subject to grid and environmental constraints. However, we might modify the objective of energy capture maximization to that of maximization of economic return [4],which requires a balance to be achieved between maximizing energy capture and minimizing wear on components. However, the move to an economic performance function also requires the accurate articulation of capital and operational costs, which is quite onerous for the relatively immature field of ocean energy, and significantly complicates the optimization problem. Instead, for the current analysis, in order to retain a focus on the fundamental control issues, we will limit ourselves to the general problem of energy capture maximization.

There are two broad approaches which may be taken to solve the energy maximization problem:

(a) Overall extremum seeking control [43], with little use of a detailed model of the system, or
(b) Determination of an optimal setpoint for the system, which gives maximum energy capture, followed by a regulatory to make sure this setpoint is achieved.

Option (a) is attractive from the point of view of the lack of requirement for a detailed model, but may have dynamic performance limitations in convergence rates and may have difficulty finding a global maximum over a non-convex performance surface. For example, in a wave energy application, the controller may not converge to the appropriate setting before the instantaneous wave frequency changes.

Interestingly, a common framework for both wave and tidal energy may be adopted for option (b), as shown in Figure 8.3. The particulars for tidal and wave

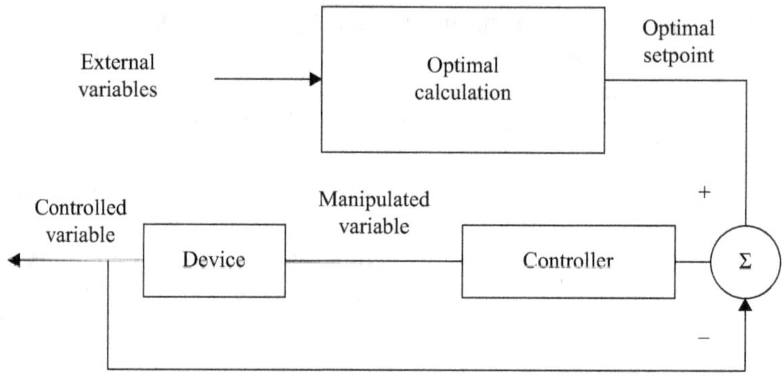

Figure 8.3 Sequential optimal calculation and regulation

devices are detailed in sections 8.1.3.2 and 8.1.3.3. For the standard feedback regulation section of Figure 8.3, any one of the techniques mentioned in section 8.1.2 can be chosen, based on the particular system description, the level of control fidelity required and the appetite for computational complexity. Since both tidal turbine and wave energy device dynamics are relatively slow (with the possible exception of the power converter section), there is much scope for the implementation of complex control strategies.

8.1.3.2 Tidal energy

In the case of a tidal turbine, optimal blade pitch, β, and rotor velocity (via the tip–speed ratio (TSR), λ) are set based on the incident flow velocity in order to maximize the power coefficient, C_p.Standard feedback loops are then used to control the pitch actuators and the torque in the rotor in order to achieve the desired pitch angle and rotor speed. The manipulated variable for the pitch control is the power to the pitch actuators (voltage and/or current). For torque control, either the back-to-back (B2B) power converter (where one is used) or the generator excitation can be used as control actuators.

It is important to note that the relationship between β, λ and C_p is specific to each tidal turbine and must be determined for each particular case. However, this relationship is then fixed, though some slight variation may occur due to component wear, or possible water density variations. We also note that when a tidal turbine reaches its rated power, the turbine must be 'depowered' in order to avoid exceeding any rated specifications. In this situation, it is not required to maximize power conversion and, for variable pitch turbines, blade pitch can be adjusted in order to limit power converted.

Control system for wind turbines are now well developed [2, 16, 17] and many of these schemes can be successfully exploited in tidal turbine control. However, some important differences between wind and tidal turbines are articulated in [44]. Section 8.2 of this chapter examines the detail of control strategies for marine current turbines.

8.1.3.3 Wave energy

For the wave energy case, the optimal velocity profile needs to be determined from the excitation force experienced by the device and the power take-off (PTO) system is then manipulated by the feedback controller to achieve this velocity profile. The optimization problem in Figure 8.3, for the wave energy case, is effectively an impedance matching problem, since the input to the system is broadly sinusoidal and the wave energy device can be represented by its (complex) intrinsic impedance, as represented in Figure 8.4. For optimum power transfer from the device to the PTO, we can use the well-known result:

$$Z_{PTO} = Z_i^* \tag{8.8}$$

where * denotes the complex conjugate. In order to see the typical form of Z_i we can use the simple wave energy device model, originally proposed in [18]:

$$m\ddot{x}(t) + m_\infty \ddot{x}(t) + \int_0^{+\infty} h_r(\tau)\dot{x}(t-\tau)d\tau + K_v\dot{x}(t) + k_b x(t) = f_{ex}(t) \tag{8.9}$$

where

$$f_{ex}(t) = \int_{-\infty}^{t} h_e(\tau)\eta(t-\tau)d\tau \tag{8.10}$$

with $\eta(t)$ being the free surface elevation, $h_e(\tau)$ and $h_r(\tau)$ are the kernels associated with the excitation force and radiation damping convolutions, respectively, m is the device mass, m_∞ is infinite frequency component of added mass and K_v and K_b are the coefficients of viscous loss and restoring force (buoyancy/gravity) respectively. With:

$$B(\omega) = \mathcal{F}\left[\int_0^{+\infty} h_r(\tau)\dot{x}(t-\tau)d\tau\right] \tag{8.11}$$

we can identify the *intrinsic impedance*, Z_i, of the device as

$$Z_i(\omega) = F_{ex}(\omega)/V(\omega) = B(\omega) + k_v + j\omega\left[m + m_\infty - \frac{k_b}{\omega^2}\right] \tag{8.12}$$

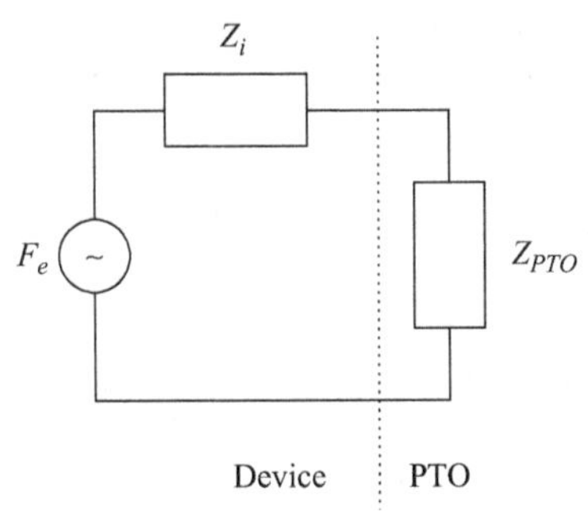

Figure 8.4 Impedance matching problem for wave energy device

If the wave excitation is monochromatic, i.e. ω has a single value and is known, then (8.8) is straightforward to calculate, with the optimal velocity profile:

$$v_{opt} = \int_{-\infty}^{t} h_v(\tau) f_{ex}(t-\tau) d\tau \tag{8.13}$$

with $h_v(\tau) = \mathcal{F}^{-1}\{1/2b(\omega)\}$. However, real (irregular) sea conditions present some difficulties:

- Reactive power must be supplied by the PTO.
- The calculation in (8.13) is non-causal.
- There are no constraints on the motion.

With regard to constraints, Budal [45] and Evans [46] considered optimal power absorbtion under motion constraints. As discussed previously, MPC provides a mechanism to allow motion constraints to be explicitly specified in the control problem formulation. The obvious difficulty with MPC is that it's normally used for quadratic regulation problems (see (8.3)), but reformulation of the cost function to address a suitable energy maximisation problem, as

$$\min_{u(0),\ldots,u(h_c)} \quad T_s \sum_{k=0}^{h_p} [v(k) f_{ex}(k) - v(k) f_r(k)] \tag{8.14}$$

where available mechanical energy (i.e. the difference between excitation and radiation energy, assuming no losses) is maximized and gives the opportunity of using standard MPC tools. In (8.3), T_s is the sampling period, noting that power is the product of force and velocity, while energy is the integral (or discrete-time summation) of power over time. Relatively standard MPC formulations for wave-energy device control have been considered in [47] and [48], while an MPC problem which parameterizes the system variables in terms of basis functions [49] results in the computationally straightforward solution of a set of linear equations. It would be reasonable to suggest that MPC implicitly implements complex-conjugate control, since the performance function is the same as that for impedance matching, albeit with the optimum obtained within the envelope of constraints. Finally, a particularly simple sub-optimal controller, but with the capability of implementing amplitude constraints, is reported in [50].

With regard to the causality problem, a number of causal approximations to complex-conjugate control are available, including latching [3, 51], declutching [52] and other methods, e.g. [53]. One causal approach, which uses the broad LQR strategy of section 8.1.2.1 is reported in [54]. An alternative way of dealing with non-causal control is to attempt to predict the future free-surface elevation or the excitation force, as required in (8.13). While (8.13) suggests that a forecast of f_{ex} for $t \to \infty$ is required, [55] evaluates the realistic forecasting requirements in terms of the device and the typical seas in which such a device might operate find that, in general, there is a strong positive correlation between forecasting requirements and likely ease of forecasting. Some forecasting methods for excitation force and free surface elevation are described in [56].

Section 8.3 of this chapter examines the detail of control strategies for wave energy devices.

8.1.4 Conclusions

Control technology, which includes significant components of system identification (blackbox modelling) and optimisation, has a significant role to play in increasing the functionality, performance and economic viability of ocean energy systems. While the more traditional control problem of output regulation appears to a significant extent in ocean energy application, some of the primary issues (such as converted energy maximization) are more difficult to address using standard control methods. However, some control methods which solve more general optimization problems (as articulated in Definition 1), such as MPC, can be adapted to address the energy maximization objectives.

Other aspects of control technology which, to date, have not seen application in the ocean energy area include fault-tolerant control [57]which could be important in ocean energy systems which typically have very limited access for maintenance. Fault-tolerant control could also be applied to arrays of devices in order to maintain grid compliance and smooth energy output of an ocean energy farm if one or more devices develop faults.

8.2 Implications of control schemes for electrical system design in tidal energy converters

S. Benelghali, *M.E.H. Benbouzid* and *J.F. Charpentier*

Nomenclature

ϕ	Flux
β	Pitch angle
σ	Total leakage coefficient, $\sigma = 1 - M^2/L_sL_r$
Ω_{gen}	Mechanical speed ($\Omega = \omega/p$)
C_p	Power coefficient
d,q	Synchronous reference frame index
h	Viscosity coefficient
J	Rotor inertia
$L(M)$	Inductance (mutual inductance)
$MPPT$	Maximum Power Point Tracking
$P(Q)$	Active (Reactive) power

PI	Proportional-Integral
PID	Proportional-Integral-Derivative
p	Pole pair number
R	Resistance
s, (r)	Stator (rotor) index (superscripts)
$T_{em}(T_{gen})$	Electromagnetic torque (Mechanical torque)
$V(I)$	Voltage (current)
V_{tides}	Tidal current speed
η	Mechanical efficiency of the turbine power train
θ_r	Rotor position
λ *TSR*	Tip-Speed-Ratio
$\omega(\omega_s)$	Rotor electrical speed (electrical synchronous speed)

The tidal current turbulences and the system parameter drift due to the system wear can significantly influence the dynamic performance of a tidal turbine. In this context, various control techniques suitable to any particular turbine configuration can be used. In most cases, classical PI or PID control are preferred. In this section, a review of control strategies for tidal energy is presented. These strategies are mainly inspired from common practices developed for wind turbine applications.

8.2.1 General control strategy for tidal current energy extraction

Figure 8.5 shows an idealized power extracting strategy for a tidal turbine device. Such power strategy is generally used as the basis of control strategies for harnessing energy from tidal turbines.

The tidal turbine does not operate below a pre-defined cut-in tidal velocity, V_c. As the tidal velocity (V) increases above the turbine's cut-in speed, the power delivered by the generator increases proportionally to the cube of the tidal velocity by following a maximum power point tracking (MPPT) strategy. When the tidal velocity reaches the rated tidal speed V_R, the generator/converter set is delivering as

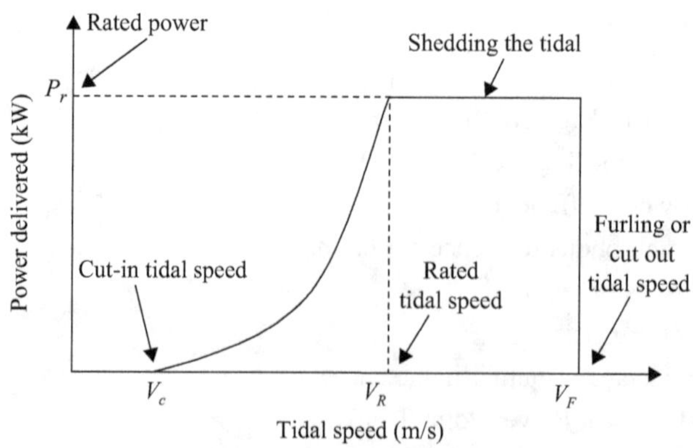

Figure 8.5 Idealized power curve

much power as it is designed for and the power has to be limited. Power output remains constant as the tidal velocity increases above the rated velocity.

The MPPT strategy and the power limitation can be achieved using (8.14) pitch control with variable pitch systems, (8.15) variable speed systems with fixed pitch systems or (8.16), a combination of these two principles.

8.2.2 Fixed-speed variable-pitch tidal turbine

In this case, optimization of the harnessing strategy involves using the $C_p(\lambda,\beta)$ characteristic ((Figure 8.6) (See §9.1.3)). If $V < V_R$, one finds the optimum of $C_p(\lambda,\beta)$ using a look-up table interpolation. Under constant-speed operation (i.e. fixed turbine rotational speed), this curve depends on the pitch angle, β, and the tidal speed V_{tides}. Optimization is achieved by changing the angle β, such that the operating point to be placed at the maximum of $C_p(\beta)$ corresponds to the tidal speed when the fluid speed is lower than V_R. If the speed is higher than V_R, the pitch angle is then chosen to limit the power or to cut out the power (for extreme tidal velocity values).

8.2.3 Variable-speed fixed-pitch turbine

Control of variable-speed fixed-pitch turbine generally aims to regulate the power harvested from the tidal currents by modifying the speed of the generator. In particular, the control goal is to capture the maximum power available from the tidal stream. For a tidal velocity lower than V_R, there is an optimum turbine rotational speed which produces a maximum power coefficient, C_p. For tidal velocities higher than V_R, the rotational speed is chosen to limit the turbine's output power to the rated power of the generator and drive set. In this case, the reference speed is chosen to be higher than the speed corresponding to the maximum value of C_p in order to limit the torque value and to operate in a stable zone of the power curve.

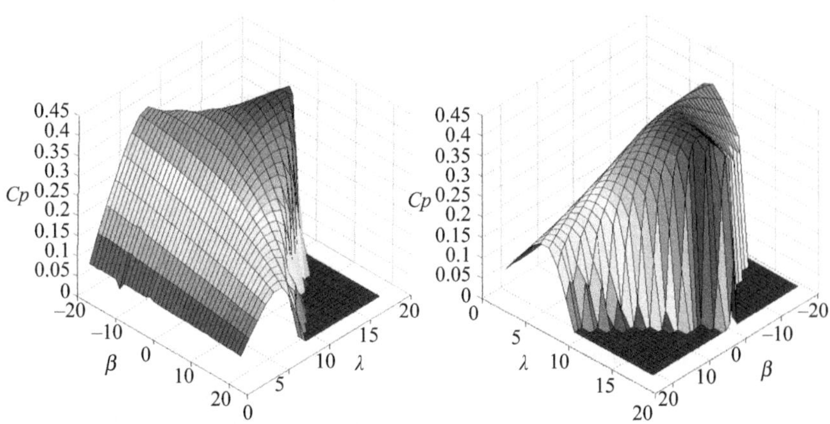

Figure 8.6 $C_p(\lambda,\beta)$ curves

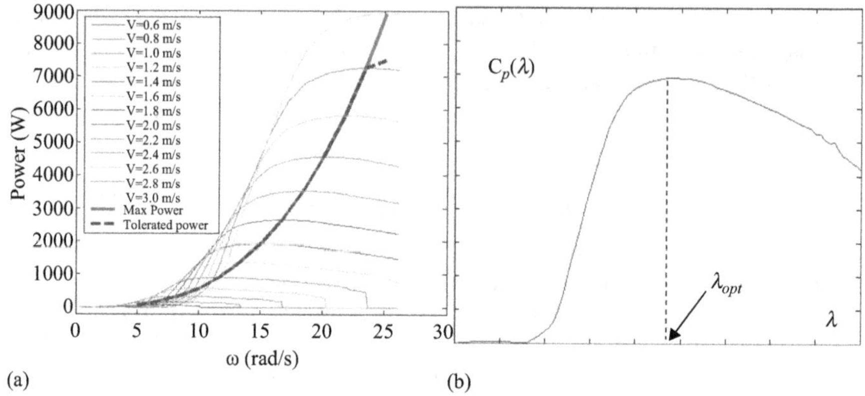

Figure 8.7 (a) Power curves for different tidal current speed; (b) power coefficient curve

All these reference speeds (as a function of the value of tidal velocity) compose what is known as the ORC (optimal regimes characteristic) in the literature (Figure 8.7(a)) [58, 59]. By keeping the static operating point of the turbine around the ORC, one ensures an optimal steady-state regime. In this case, the captured power is the maximum available from the tidal stream. This is equivalent to maintaining the TSR at its optimal value λ_{opt} for $V < V_R$ (Figure 8.7(b)). Maintaining the TSR at its optimal value can be achieved by operating the turbine at variable speed, according to the tidal speed [58]. Of course, this kind of strategy is only possible if the generator and drive set is able to control the speed in the corresponding range.

Basically, the control strategies vary in accordance with assumptions concerning the known models/parameters, the measurable variables, and the used tidal turbine model. Depending on how rich the information are about the turbine model, especially about its torque and power characteristic, the optimal control of a variable-speed fixed-pitch turbine is based upon the following approaches, when $V < V_R$.

8.2.3.1 Maximum power point tracking (MPPT) strategy

This approach is adequate when parameters λ_{opt} and $C_{pmax} = C_p(\lambda_{opt})$ are unknown. The reference of the rotational speed control loop is adjusted such that the turbine operates around the maximum power for the tidal velocity [58, 59]. To establish whether this reference must be either increased or decreased, it is necessary to estimate the current position of the operating point in relation to the maximum of the $P_{TURB}(\omega)$ curve. This can be done in two ways:

- The speed reference is modified by a speed variation, $\Delta\omega$. The corresponding change in the active power (ΔP_{TURB}) is determined in order to estimate the value $\frac{\partial P_{TURB}}{\partial \omega}$. The sign of this value indicates the position of the operating point in relation to the maximum of the $P_{TURB}(\omega)$ characteristic. If the speed reference is adjusted linearly with a slope proportional to this derivative, then the system evolves to the optimum where $\frac{\partial P_{TURB}}{\partial \omega} = 0$.

- A probing signal is added to the tidal current speed reference. This signal is a slowly varying sinusoid. Its amplitude does not significantly affect the system operation, but still produces a detectable response in the active power evolution. In order to obtain the position of the operating point in relation to the maximum, one compares the phase lag of the probing sinusoid and that of the active power sinusoidal component. If the phase lag is 0, then the operating point is placed on the ascending part of $P_{TURB}(\omega)$, while if the phase lag is π, then the operating point is placed on the descending part of $P_{TURB}(\omega)$. Therefore, the slope of the speed reference must increase/decrease. Around the maximum, the probing signal does not produce any detectable response and the speed reference does not have to change [60].

In this simplified MPPT presentation, factors like the tidal turbulence influences and system dynamics that distort information concerning the operating point position have been neglected. A more detailed description and performance analysis can be found in [58].

8.2.3.2 Shaft rotational speed optimal control

Using a set point from the turbine data

This solution can be applied if the optimal value of the TSR λ_{opt} is known. The turbine operates on the ORC if

$$\lambda(t) = \lambda_{opt} \tag{8.15}$$

which supposes that the shaft rotational speed is closed-loop controlled such that to reach its optimal value:

$$\Omega_{ref} = \frac{\lambda_{opt}}{R} v(t) \tag{8.16}$$

8.2.3.3 Active power optimal control

Using a set-point from the shaft rotational speed data

This method is used when both λ_{opt} and $C_{pmax} = C_p(\lambda_{opt})$ are known. In this case the extracted power can be written as

$$\begin{aligned} \partial P_{TURB} &= \frac{1}{2} C_P(\lambda) \rho \pi R^2 v^3 \\ &= \frac{1}{2} \frac{C_p(\lambda)}{\lambda^3} \rho \pi R^5 v^3 \end{aligned} \tag{8.17}$$

By replacing λ_{opt} and $C_{pmax} = C_p(\lambda_{opt})$, one obtains the power reference for the second region of the power–tidal speed curve.

$$P_{TURB} = P_{ref} = K\omega_{ref}^3 \tag{8.18}$$

where

$$K = \frac{1}{2} \frac{C_p(\lambda_{opt})}{\lambda_{opt}^3} \rho \pi R^5 \tag{8.19}$$

This approach supposes an active power control loop being used, whose reference is deduced from (8.19). This method is widely employed, especially for medium- and high-power wind turbine and can be exploited for marine current turbine [58, 59].

8.2.4 Tidal turbine control

In this section, we introduce some classical PI design of wind turbines that can be used for the turbine control for the three above-mentioned cases.

8.2.4.1 Torque control loop

In order to improve and maximize the captured energy, the rotor turbine must operate at the maximum possible power. Equivalently, this means imposing the electromagnetic torque (T_{ref}) which equals the tidal torque corresponding to the maximum available power. The turbine works at maximal efficiency when turning at optimal TSR λ_{opt}, so the maximum power is proportional to the cubed rotational speed eqs. (8.18) and (8.19).

$$T_{ref} = \eta K \omega_{ref}^2 \tag{8.20}$$

Of course if $V > V_R$ the reference torque must be limited to

$$T_{ref} = P_r/\omega \tag{8.21}$$

As the turbine control structure allows the tidal speed to be tracked within admissible limits of mechanical loads, this method can be used as long as it depends only on slow tidal speed variations. For high dynamics turbulent tidal current, filtering is necessary to ensure sufficiently slow closed-loop dynamics. Moreover, this method is strongly sensitive to parametric variations.

8.2.4.2 Speed control loop

The controller design is based upon the turbine linearized model. The simplified closed-loop structure is shown in Figure 8.8. A high gain, Kp, will thereby ensure better tracking performance. However, one must take account of control effort (torque) limitations, so K_p values must also be limited. The zero effect of increasing the overshoot is compensated by first-order filtering of the reference signal (Figure 8.8). Although the steady-state speed error is zero, there will always be nonzero dynamical errors due to the significantly variable reference signal ω_{ref}.

One must note that the imposed closed-loop performances are guaranteed for the chosen operating point. Both the gain and the time constant of the torque

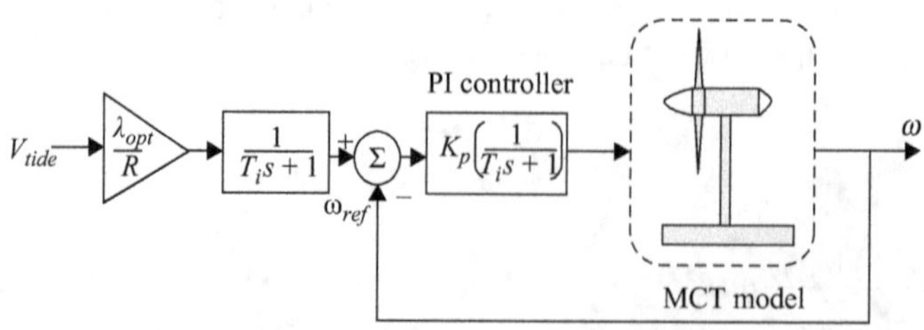

Figure 8.8 Turbine PI control structure: speed control loop case

controlled system around a certain steady-state operating point depend on that operating point (through tidal velocity and rotational speed). Therefore, the dynamic performances of the tracking system also vary upon the operating point [58, 59].

8.2.4.3 Power control loop

The input of the system is the electromagnetic torque and its output is the generated active power P. The controller design is based on the parameterization of the system response at step changes in the generator torque for a given tidal speed. The difference between the tidal torque and the electromagnetic torque leads to the variation of the rotational speed, such that the power increases according to the tidal speed dynamics until it reaches a new steady-state value [58–60].

Example: Optimal control of tidal turbine driven PMSG by PI speed control

As an example, an optimal control based on a PI speed controller has been simulated using a dedicated Matlab/Simulink® library presented in [61] for a low-power experimental variable speed fixed-pitch turbine driven PMSG. The control system is defined in the synchronous *d-q* frame. For the proposed control strategy, the permanent magnet synchronous generator dynamic model is written in the *s*-domain with the stator voltage given by:

$$\begin{cases} -(R+L_d s)i_d = V_d - \phi_q \omega_s \\ -(R+L_q s)i_q = V_q + \phi_d \omega_s \end{cases} \tag{8.22}$$

The mechanical equation is expressed by

$$T_{gen} - T_{em} = (Js+h)\Omega_{gen} \tag{8.23}$$

where $\Omega = \omega_s/p$ is the mechanical speed and T_{gen} is the mechanical torque provided by the gearbox or directly by the turbine (in direct driven system) to the generator shaft.

The electromagnetic torque T_{em} is defined by

$$T_{em} = \frac{3}{2}p(\phi_d i_q - \phi_q i_d) = \frac{3}{2}p\left[\phi_m i_q + (L_d - L_q)i_d i_q\right] \tag{8.24}$$

with $\begin{cases} \phi_d = L_d i_d + \phi_m \\ \phi_q = L_q i_q \end{cases}$

The generator chosen for simulation is a surface mounted permanent magnet synchronous generator; therefore, there are no saliency effects and $L_d = L_q$. So the electromagnetic torque T_{em} can be simplified as

$$T_{em} = \frac{3}{2}p\phi_m i_q \tag{8.25}$$

In those conditions, the PMSG Park model is illustrated by Figure 8.9.

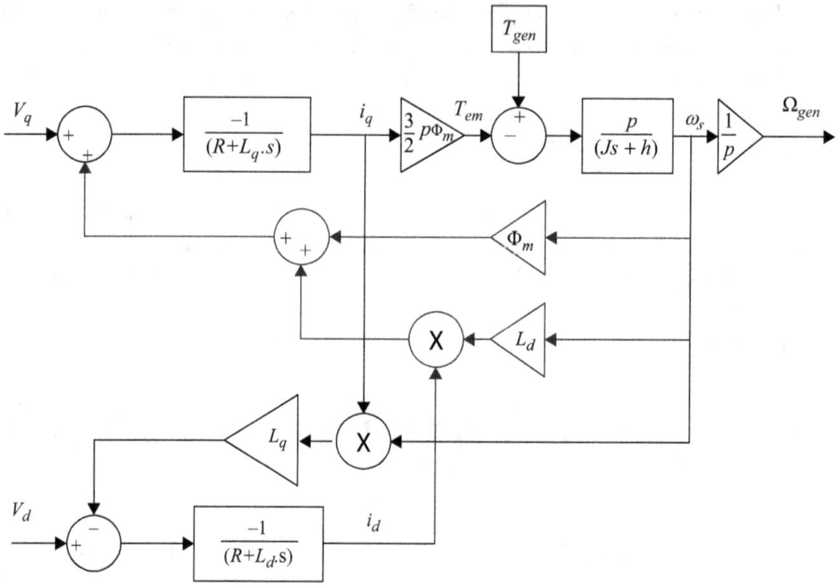

Figure 8.9 The PMSG Park model

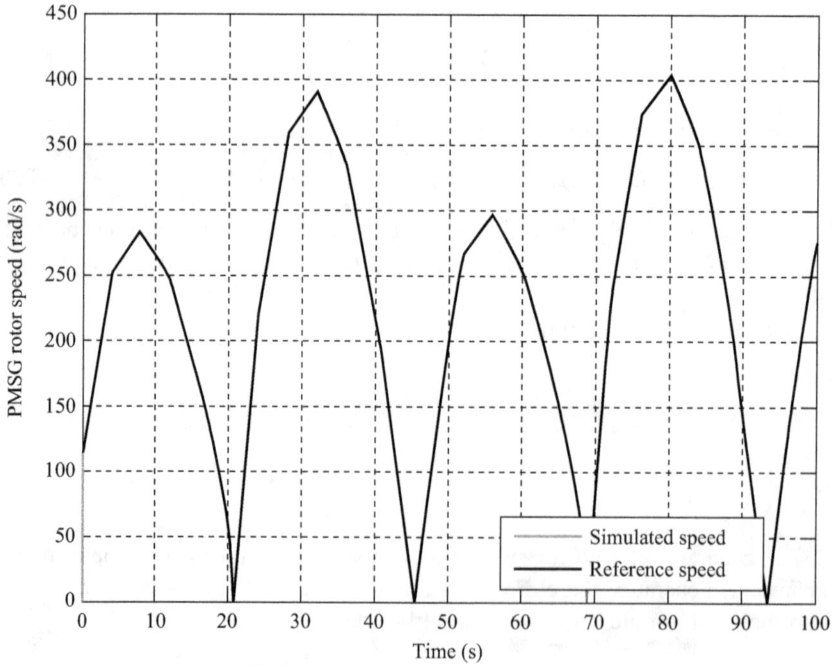

Figure 8.10 The PMSG rotor speed and its reference

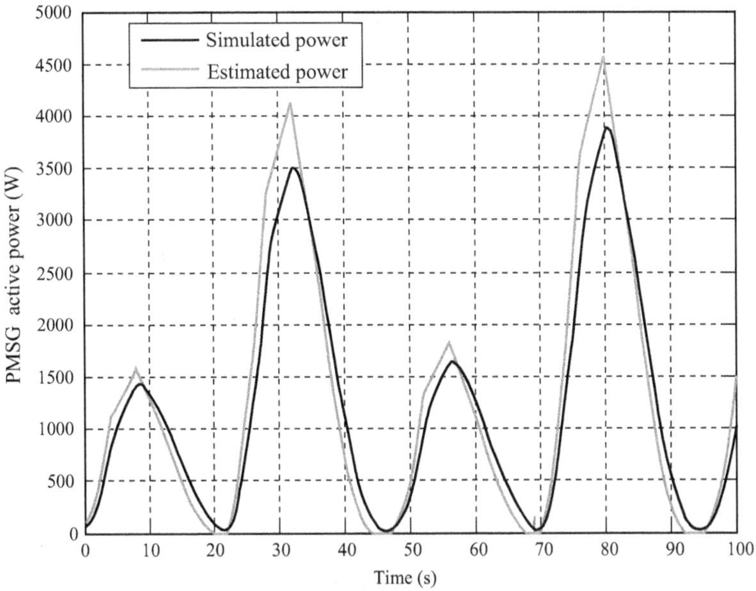

Figure 8.11 The PMSG active power

An inner loop is needed for the current (torque) control. Then an outer loop can be used to control the turbine shaft speed. Indeed, as shown in Figure 8.9 the PMSG rotor speed can be controlled through the rotor current I_q which is proportional to the torque. It can be seen that in the case of a fixed-pitch turbine, power limitation can be achieved using overspeed operations (the generator/converter set must be designed to be controlled in the corresponding speed range).

To illustrate the control behaviour, a PMSG-based turbine is simulated for a varying tidal speed. The PI control strategy is tested for a low power experimental marine current turbine of 1.44 m diameter and 7.5 kW PMSG [62–63].

For speed references given by the MPPT strategy ($V < V_R$) , the PMSG-based turbine control performances are shown in Figure 8.10 and 8.11 respectively illustrating the rotor speed tracking performance and the generated active power.

The estimated power presents the maximum power that can be extracted. It depends on the tidal speed variation and is deduced from the tidal turbine hydrodynamic model. The simulated power presents the measured power generated by the turbine simulator. The obtained results show good tracking performances of the PMSG rotor speed. However, the tidal current turbulences and the system parameter drift due to the system wear out can significantly influence the turbine dynamic performance. In this context, various control techniques, suitable to any particular configuration, can be used. In most cases, classical PI or PID control are preferred. However, advanced control techniques can be used in order to ensure better performances, especially for guaranteeing robustness to modelling uncertainties [62–65].

8.3 Implications of control schemes for electrical system design in wave energy converters

M. Molinas and *E. Tedeschi*

8.3.1 Introduction

In the most general case, the electrical system for a WEC is composed of an electrical machine controlled by a power electronics unit and a grid interface power electronics unit. The selection of these components is strongly affected by the control strategy to be applied regarding both the topology and the rating. For instance, the goal of maximizing the average power extraction from the waves generally requires variable speed electrical machines. Moreover, control strategies implying bidirectional power flow call for motor/generator operation of the electrical machine and a fully bidirectional power electronics converter. Such aspects will be treated in the next paragraphs following a twofold approach:

(a) Design of the PTO system based on control (i.e. impact of control on the rating of the PTO when there is not yet a given PTO rating).
(b) Tuning of the control based on a given PTO (when the PTO rating has been pre-defined beforehand).

Illustrative test cases will be provided, referring to a spherical point absorber in heave coupled to an all-electric PTO system, as this is found to be the WEC concept where the most critical impact of the intermittency of the sea waves is observed.

8.3.2 Relationship between control schemes and the WEC electrical system

A WEC with fixed mass and geometry is characterized by a specific natural (or eigen) period. The maximum power extraction is obtained when incident waves match the WEC natural period, so that the WEC operates in resonance conditions. In practical applications, the increase of the power extraction by achieving such resonance condition will entirely rely on the quality of the WEC control. The main goal of WEC control is therefore to actuate the PTO to modify the amplitude and/or the phase of the motion to make it close to the optimum condition. It is then clear that identifying the most suitable control strategy is a challenging issue for all the types of WEC, since it strongly affects power absorption and, ultimately, the profitability. It is worth noting, however, that active control is even more crucial for devices that are small compared to the waves' wavelength (i.e. point absorbers), since their resonance bandwidth is inherently very small.

Up to now, the most commonly adopted control techniques for point absorbers are:

- *Complex-conjugate control* – in which case the WEC is controlled to operate in a resonant condition, by actively modifying both the amplitude and the phase of its motion. For this purpose, the PTO exerts on the prime mover a force having one component proportional to the WEC velocity and another component proportional to the WEC position or acceleration.
- *Passive loading* – in which case the PTO applies a force that is proportional to the buoy velocity, which modifies its dynamic resistance (or damping). In this case, only the amplitude of motion is changed to achieve a relative maximum of the extracted power.
- *Latching and clutching control* – in which case the point absorber is held in the upper and/or lower position in due intervals, so that its velocity stays in phase with the excitation force. When the device is released its damping is set as in passive loading. Clutching differs from latching only in that the resistance is switched between zero and a (different) constant value.
- *Model predictive control* – in which a constrained optimization of the expected response of the system is performed over a short future horizon. The implementation of this control requires a discrete model of the WEC system and a prediction of the excitation force.

8.3.2.1 Design of the PTO system based on control

When the WEC PTO rating is not a-priori decided, the control strategy selected for the WEC will have a direct impact on the rating of the PTO system components. This is because each control scheme of the WEC will result in a distinct peak-to-average power extraction ratio.

The average power extraction is the main reference parameter when considering the PTO design and consequent performance evaluation of a WEC. To achieve a fully feasible system design, it is important to consider the electromechanical constraints imposed by the PTO at the initial stages of the design process. Another important parameter is the peak power produced, which impacts on the sizing of both the electrical machine and the power converters. The peak power will determine the choice of a suitable electrical generator, and the maximum current limit for the associated power electronics interface. The relevance of the value of peak power is clearly explained in [66]. Power electronics sizing is especially critical due mainly to the small time constant of the component, which is typically in the order of hundreds of milliseconds [67]. Electrical machines, and other electrical components such as transformers and cables, have time constants in the order of minutes, thus they can be operated transiently at higher peak power than their mean rating. Despite being less demanding from the thermal perspective, the rating of the electrical generator is strongly affected by its functionality within the context of WEC system operation. This is because in most WECs, the electrical machine is not only used for the electro-mechanical conversion, but also it is actively involved in the optimization of the prime mover efficiency. In fact, several WECs, such as overtopping devices or OWCs with variable speed air turbines, require a wide speed and torque control range to ensure optimized performance. This can be even more crucial for point absorbers with and without hydraulic PTO, where the generator is controlled to directly influence the prime mover motion [68]. It is worth underlining that WEC performances have been traditionally evaluated through the peak-to-average power ratio, which is a relevant metric to cope with different issues (from the PTO design to the WEC-grid connection [69, 70]). It is suitable for several types of systems [68, 71], but it is especially used for point-absorber direct-driven applications [66, 72, 73].

For a specific point absorber application including stroke limitations, one comparative study has shown [74]that traditional passive loading can lead to a peak-to-average power ratio of 12 in irregular waves. In the same sea conditions all reactive control strategies actually increase the average power production but at the expense of very high peak-to-average power ratios (17 for complex-conjugate control and more than 35 for model predictive control). Latching control shows comparably high peak-to-average power ratios (11–18), which can instead be reduced to 6–9 by using clutching control. Both latching and clutching, however, perform poorly in small waves.

8.3.2.2 Tuning of the control based on a given PTO

When the PTO for the WEC has been defined beforehand, specific constraints on the maximum torque/force and speed that the PTO can stand must be respected in addition to the peak power limit. Table 8.1 shows relevant ranges of electrical and mechanical parameters for four different WEC types with their corresponding peak-to-average power ratios. Thus, the control to be implemented on the WEC calls for the integration of the physical system constraints into the control algorithm itself to enable a realistic solution that can potentially reach the market. This confines the WEC power extraction problem into a constrained optimization problem with a critical compromise between the:

1. Rating of the PTO system and
2. Optimal power extraction within the given rating constraints.

Several works [75, 76] found that in order to optimize power extraction with a constrained PTO, different real-time control strategies should selectively be applied following the changes in wave's frequency and amplitude. In [77], a wave–to–wave adaptive control strategy based on a simplified monochromatic approach and selective inclusion of a reactive control component depending on the incident wave showed a potential for increased power extraction compared to controls with constant parameters.

Table 8.1 Ranges for electrical and mechanical parameters for four different WEC types and PTOs (adapted from [68])

WEC Type	Prime mover speed range	Generator peak torque range	Generator type	Example of reported peak/ average power
OWC	600–1500 rpm	2–4 pu; reduces with added inertia	Variable Speed: SCIG/SG/PMG	10 [3]
Point absorber with hydraulic PTO	1000–3000 rpm	Close to 1 pu with high accumulator storage, up to 4 pu as storage reduces	Fixed Speed: SG/SCIG-SVC for high-storage designs Variable Speed: SG/PMG/SCIG for low storage designs	N/A
Overtopping/ pump devices with hydro PTO	100–250 rpm	Close to 1 pu	GBC: FS SG/SCIGSVC or VS PMG/SCIG Low speed VS PMG	N/A
Point absorber with direct PTO	0–3 m/s 0–400 rpm	2–5 pu; reduces with added inertia	Linear: PMG (custom) Rotary: GBC VS SG/PMG	7.7–17.1 (passive loading) or higher for other control strategies[9,20]

Case study

For the study reported here, reference is made to a direct coupled spherical point absorber (radius = 5 m) with an all-electric PTO system composed of a squirrel-cage induction machine and a back-to-back (BTB) power electronics converter as the interface to the grid. The induction machine rated power is here assumed to be 850 kW, and its rated speed is 2985 r/min.

If a reference velocity of 3 m/s is considered for the point absorber, with a pinion radius of 0.1 m, a gear ratio of 10 is required to couple the point absorber and the PTO. The considered electrical machine has a rated torque of 2716 Nm and a peak torque of 7333 Nm. They correspond to a rated force of 271.6 kN and a peak force of 733.3 kN respectively.

Referring, for the sake of explanation, to the simplified case of sinusoidal incident waves having $T = 9$ s, such adaptive control strategy proves to be an intermediate solution that ensures the (relative) maximization of the average extracted power while reducing the peak-to-average power ratio with respect to complex-conjugate control with constant parameters (as can be seen in Figure 8.12). Such an approach, which can be easily extended to include all the main constraints in the system, leads to a map of the advisable control strategies

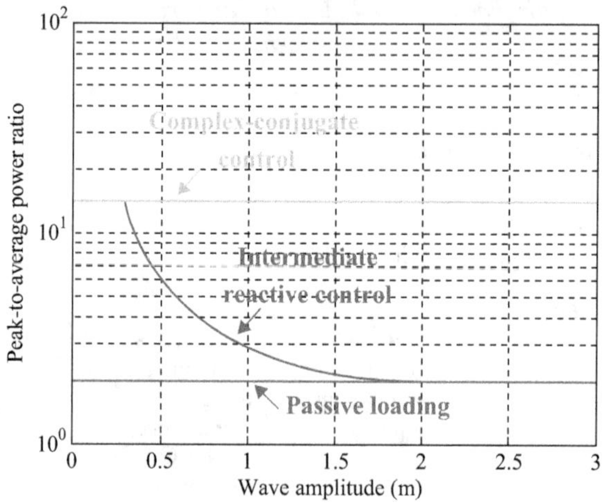

Figure 8.12 Peak-to-average power ratio with three different control strategies for sinusoidal incident waves having $T = 9$ s

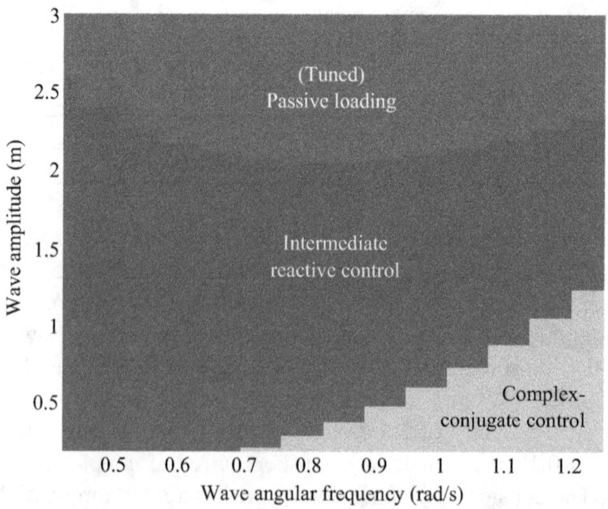

Figure 8.13 Map of the advisable control strategies according to the changes in the sinusoidal incident wave amplitude

according to changes in the incident wave amplitude and frequency for each specific WEC (such map is shown in Figure 8.13 for the presented test case).

Performance in irregular waves

A 20 min irregular wave profile was generated from a Bretschneider spectrum [78, 79] having significant wave height $H_s = 2.12$ m and energy period $T_e = 9$ s.

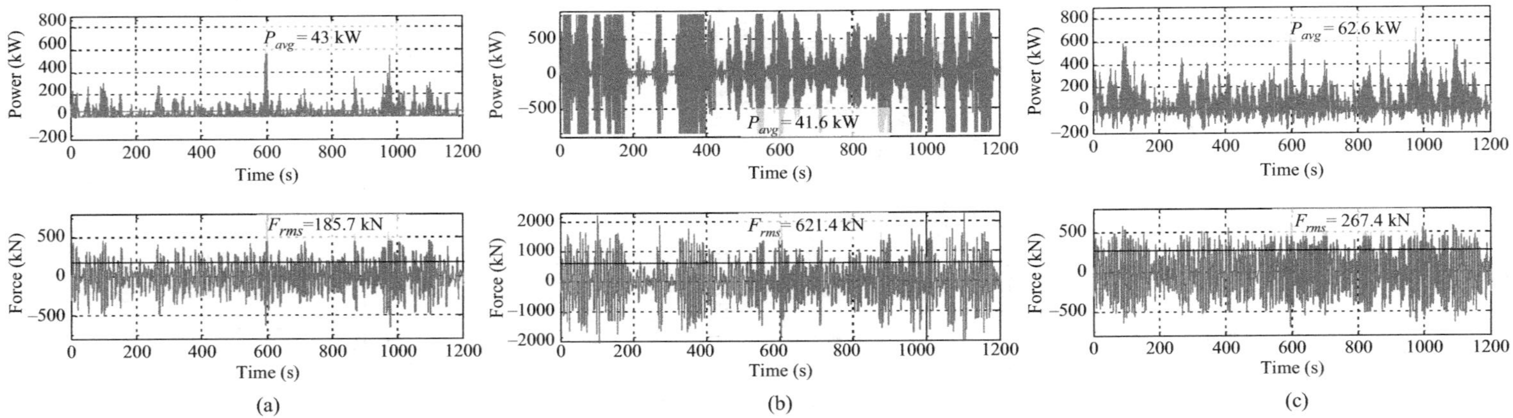

Figure 8.14 Instantaneous and average power (top) and instantaneous and rms PTO force (bottom) in the cases of constant passive loading (a), constant complex-conjugate control (b) and proposed adaptive control (c)

Applying traditional passive loading with constant damping value being optimal for a sinusoidal wave having $T = 9$ s, the average power extraction over the 1200 s simulation is $P_{avg} = 43$ kW, and the maximum power kept for a few seconds is $P_{max} = 602$ kW (Figure 8.14(a)). In this case, the rms value of the PTO force ($F_{rms} = 185.7$ kN) stays below the corresponding rated force. The peak value of the force is $F_{max} = +696.5$ kN, which is also below the peak force limit. When complex-conjugate control with constant parameters is used in the same sea state and with the same PTO rating of 840 kW, the peak power limit is repeatedly reached for tens of seconds every time. The average power absorption is, however, lowered to 41.6 kW, due to the reverse power flow required during large parts of the operation (Figure 8.14(b)). From the mechanical perspective, complex-conjugate control would be unfeasible with the selected machine as it requires the electrical generator to apply an rms force of 621.4 kN, with a maximum value over 2000 kN. With the adaptive control strategy (intermediate reactive control) a peak power $P_{max} = 719.17$ kW is reached, with an average extracted power of $P_{avg} = 62.6$ kW (Figure 8.14(c)). The 46% increase in the average power extraction indicates the clear advantage of intermediate reactive control compared to traditional passive loading and the better exploitation of the PTO equipment. Figure 8.14(c) shows that, in small waves, when a reactive control component is applied, the power flow is bidirectional. On the other hand, in high waves the power is unidirectional for the use of a tuned passive loading. The mechanical force required to the PTO, ($F_{rms} = 267.4$ kN), does not violate the machine constraints, improving the use of the PTO compared to passive loading. The force peak value, $F_{max} = -689.8$ kN, is still within the physical constraint of the machine. Tests under irregular waves provide the order of magnitude of the required oversizing of the real PTO when compared to the ideal case of sinusoidal waves. This does not undermine the role of preliminary analyses developed under sinusoidal operations [71, 80–82] but stresses the need to consider them as preliminary guidelines, to be followed by more realistic tests (Table 8.2).

Table 8.2 Summary of results

	Passive loading	Complex conjugate	Adaptive
P_{avg} (kW)	43	41.6	62.6
P_{max} (kW)	602	840	719.2
F_{rms} (kN)	185.7	621.4	267.4
F_{max} (kN)	696.5	≥ 2000	–689.8

8.3.3 Impact of efficiency on power extraction

It is of paramount importance to consider the efficiency of the overall power conversion when analysing the power extraction capabilities of a WEC system [69, 83, 84]. Electric machines and power electronic interfaces have efficiency properties that depend on many factors (such as the rotational speed and the loading conditions) and which decrease rapidly when the extracted power is a small percentage of the rated power. This property will have a different impact depending on the selected control strategy. The most important effect to be taken into account is that under comparable conditions, complex-conjugate control requires much higher peak power circulation to achieve the same average power extraction as passive loading. As a consequence, higher losses will be observed in complex-conjugate control due to non–unity efficiency of the PTO. The losses are observed alternatively in the two directions of power (from the WEC to the grid and from the grid to the WEC) whenever complex-conjugate (or intermediate reactive) control is considered. Such losses can be extremely severe in the complex-conjugate control case, this being a clear disadvantage with respect to passive loading.

In the simpler case of sinusoidal waves, the different effect of non-unity efficiency on passive loading and complex-conjugate control is exemplified in Figure 8.15(b) and (c). The incident wave amplitude has been adjusted to ensure the same average power for both cases when considering ideal efficiency. Due to the difference in the system operating conditions and in the maximum power value, a 12% average power reduction due to non-ideal efficiency is experienced with complex-conjugate control, while only 5% is observed in the case of passive loading.

Case study

The effect of a non-ideal efficiency of the electric PTO (Figure 8.15(a)) on the power conversion and its specific impact on the different control strategies in irregular waves is analysed in detail for a point absorber directly connected to the electric PTO operating in a sea state with energy period $T_e = 6$ s and significant

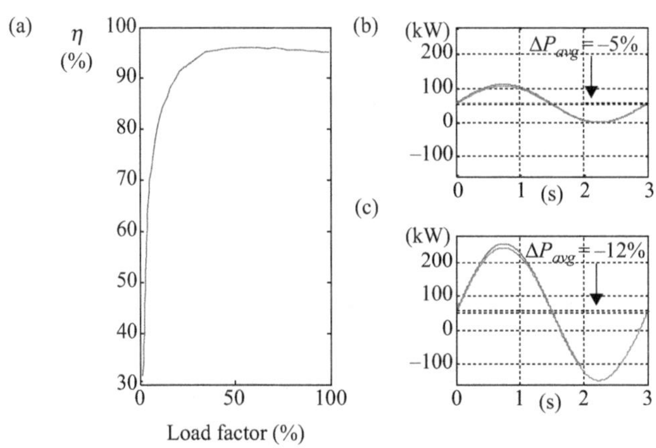

Figure 8.15 Example of efficiency of the PTO equipment (a) and its effects on power extraction with passive loading (b) and complex-conjugate control (c)

wave height $H_s = 1.41$ m [85]. In Figure 8.16, the average power extraction as a function of the PTO power rating is shown for both unity and non-unity conversion efficiency, for both passive loading and complex-conjugate control.

It can be noticed that for both the control strategies, the non-ideal efficiency property determines a maximum point in the average power extraction, corresponding to a specific rating of the PTO. This means that even a higher rating of the PTO would bring no increase in the final power extraction. The reason for this is that when the PTO is highly oversized to stand the variability of irregular waves, it mostly works in low load conditions and this makes the entire power conversion more and more inefficient.

In case of very low PTO ratings, passive loading gives higher average power extraction than complex-conjugate control, even more in the case of non-unity efficiency. From the non-ideal curves of Figure 8.16, it can be clearly noticed that the maximum power extracted with passive loading ($P_{avg} = 16.7$ kW) is higher than the maximum power extracted with complex-conjugate control, even in its most favourable condition ($P_{avg} = 15$ kW).

Moreover, in the passive loading case, the highest power extraction is obtained with lower rated PTO equipment ($P_{rat} = 75$ kW), compared to the complex-conjugate control case (having its maximum at $P_{rat} = 500$ kW). The proven superiority

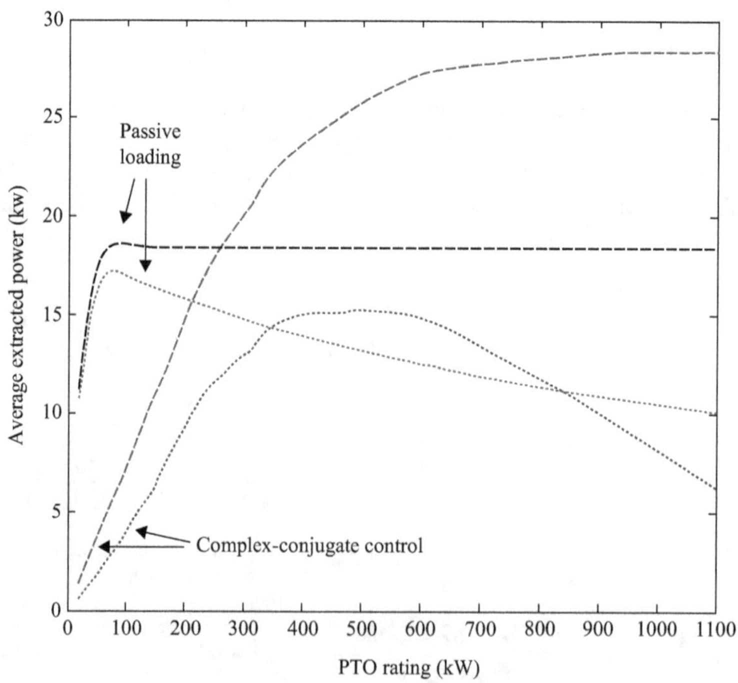

Figure 8.16 Average power extraction as a function of the PTO rating in presence of unity (dashed line) and non-unity (continuous line) efficiency of the power conversion for passive loading

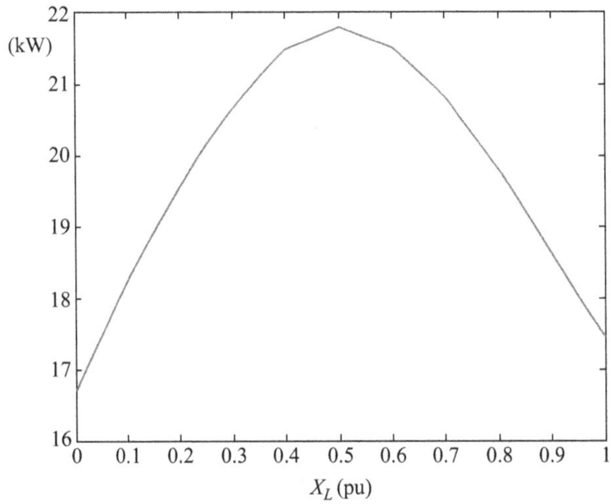

Figure 8.17 Average power extracted with non-unity efficiency as a function of the reactive component, evaluated as a fraction of the value used in complex-conjugate control. Resistive component is the one used in passive loading

of passive loading over complex-conjugate control with non-ideal efficiency has been shown. As a further step it can be also proved that an intermediate control strategy employing a suitably adjusted reactive component is advisable even under non-unity efficiency conditions. Indeed it should be carefully optimized to maximize the final power extraction as exemplified in the following: System average extracted power was evaluated assuming a PTO rating of 110 kW and considering a resistive control component fixed as in the passive loading case plus a reactive component being an increasing fraction of the one corresponding to complex-conjugate control. As shown in Figure 8.17, when applying a reactive component that is 50% of the value previously adopted in complex-conjugate control, an average extracted power of 21.8 kW is achieved, representing an almost 30% increase compared to pure passive loading. Similar power extraction improvements were confirmed by full-system simulations including the detailed (switching) model of the PTO system [85].

8.3.4 Discussion

This section has discussed the impact that WEC control strategies will have on the power extraction capabilities of a given point absorber type WEC and thus on the sizing of the all-electric PTO components. Two ways in which the sizing of the WEC PTO components can be treated are presented, one in which the rating is pre-defined beforehand (constrained control problem) and the other in which the sole control strategy will determine the ratings by careful selection of the most suitable peak-to-average ratio of power extraction (unconstrained control problem).

In a more rigorous manner, a preliminary sizing of the PTO should be considered by careful observation of the impact of unconstrained control strategies on energy extraction over an entire year or a few years. With this information and by selection of the most suitable peak-to-average ratio, a constrained optimized extraction problem should be solved to decide the control strategies to be implemented for maximizing the power extraction of the WEC. It has also been clear from the case study presented that power constraint alone will not ensure proper functionality of the WEC device if physical limitation such as stroke, speed and force are not taken into consideration when deciding the control strategy to be applied. This implies that a careful selection of constraints should be first considered in the process of defining/formulating the optimization problem, making sure that a feasible solution that respects the physical limits of the WEC device is obtained and implemented.

A further reduction effect of power extraction is observed when introducing a non-ideal efficiency into the calculation of the impact of control strategies, depending on the control strategy applied. This can also determine the advantage of one control strategy over the others. By these results it becomes clear that several parameters (electrical and mechanical) should be taken into consideration when formulating the optimization problem for wave energy extraction.

8.4 References

[1] J.-C. Gilloteaux and J. Ringwood, 'Control-informed geometric optimisation of wave energy converters', *Proceedings of the IFAC Conference on Control Applications in Marine Systems (CAMS)*. Rostock, Germany, 2010, pp. 399–404.

[2] L. Pao and K. Johnson, 'Control of wind turbines: Approaches, challenges and recent developments', *IEEE Control Sytems Magazine*, vol. 31, no. 2, pp. 44–62, 2011.

[3] A. Babarit and A. Clement, 'Optimal latching control of a wave energy device in regular and irregular waves', *Applied Ocean Research*, vol. 28, no. 2, pp. 77–91, 2006.

[4] R. Costello, B. Teillant and J. Ringwood, 'Techno-economic optimisation for wave energy converters', *Proceedings of the International Conference on Ocean Energy (ICOE*). Dublin, 2012.

[5] G. Goodwin, M. Seron and J. de Dona. *Constrained Control and Estimation*. Springer-Verlag, London, 2005.

[6] F. Lewis and V. Syrmos, *Optimal Control*. Wiley-Interscience. John Wiley and Sons, New Jersey, 1995.

[7] H. Noura, D. Theillol, J.-C. Ponsart, and A. Chamseddine, *Fault-Tolerant Control Systems*. Springer-Verlag, London, 2009.

[8] W. Levine. *The Control Handbook*. The Electrical Engineering Handbook Series. CRC Press, Boca Raton, Florida, 1996.

[9] Y. Kim and L. Keel and S. Bhattacharyya, 'Transient response control via characteristic ratio assignment', *IEEE Transactions on Automatic Control*, vol. 48, no. 12, pp. 2238–2244, 2003.

[10] B. Quo, *Automatic Control Systems*, 8th edn. Wiley, New Jersey, 2007.

[11] F. J. Brasch and J. Pearson, 'Pole placement using dynamic compensators', *IEEE Transactions on Automatic Control*, vol. 15, no. 1, pp. 34–43, February 1970.

[12] B. Anderson and J. B. Moore, *Linear Optimal Control*. Prentice-Hall, Englewood Cliffs, New Jersey, 1971.

[13] S. Skogestad and I. Postlethwaite, *Multivariable Feedback Control Analysis and Design*. Wiley, Chichester, England, 2007.

[14] L. Ljung, *System Identification: Theory for the User*. Pearson, New Jersey, 1998.

[15] K. Narendra and A. Annaswamy, *Stable Adaptive Systems*. Dover Publications, Mineola, New York, 2005.

[16] W. Leithead and B. Connor, 'Control of variable speed wind turbines: Design task', *International Journal of Control*, vol. 73, no. 13, pp. 1189–1212, 2000.

[17] F. Bianchi, H. De Battista and R. Mantz, *Wind Turbine Control Systems: Principles, Modelling and Gain Scheduling Design*. Springer-Verlag, London, 2007.

[18] W. Cummins, 'The impulse response function and ship motions', *Schiffstechnik*, vol. 9, pp. 101–109, 1962.

[19] M. Eriksson, R. Waters, O. Svensson, J. Isberg and M. Leijon, 'Wave power absorption: Experiments in open sea and simulation', *Journal of Applied Physic*, vol. 102, no. 8, 084910–084910–5, 2007.

[20] A. Merigaud, J.-C. Gilloteaux and J. Ringwood, 'A nonlinear extension for linear boundary element methods in wave energy device modelling', *Proceedings of the Conference on Offshore Mechanics and Arctic Engineering (OMAE)*. Rio de Janeiro, Brazil, 2012.

[21] M. Bhinder, A. Babarit, L. Gentaz and P. Ferrant, 'Effect of viscous forces on the performance of a surging wave energy converter', *Proceedings of the Conference of the Intenational Society of Offshore and Polar Engineers (ISOPE)*. Rhodes, Greece, 2012, pp. 545–550.

[22] A. O'Dwyer, *Handbook of PI and PID: Controller Tuning Rules*. Imperial College Press, London, 2003.

[23] F. J. Doyle III., H. S. Kwatra and J. S. Schwaber, 'Dynamic gain scheduled process control', *Chemical Engineering Science*, vol. 53, no. 15, pp. 2675–2690, 1998.

[24] N. Hori, R. J. Cormier and K. Kanai, 'Matched pole-zero discrete-time models', *IEE Proceedings on the Control Theory and Applications D*, vol. 139, no. 3, pp. 273–278, May 1992.

[25] K. Astrom and B. Wittenmark, *Computer-controlled Systems: Theory and Design*. Prentice-Hall, New Jersey, 2011.

[26] K. Furuta, A. Sano and D. Atherton, *State Variable Methods in Automatic Control*. Wiley, New York, 1988.

[27] M. Grimble and M. Johnson, *Optimal Control and Stochastic Estimation: Theory and Applications*. Wiley, New York, 1986.

[28] R. Soeterboek, *Predictive Control: A Unified Approach*. Prentice-Hall, New Jersey, 1992.

[29] J. Marshall, *Control of Time-Delay Systems*. Peter Peregrinus, London, 1979.

[30] J. Maciejowski, *Predictive Control: With Constraints*. Pearson Education, Prentice-Hall, New Jersey, 2002.

[31] K. Hunt, D. Sbarbaro, R. Zbikowski and P.J. Gawthrop, 'Neural networks for control systems: A survey', *Automatica*, vol. 28, no. 6, pp. 1083–1112, 1992.

[32] C. Rao, S. Wright and J. Rawlings, 'Application of interior-point methods to model predictive control', *Journal of Optimization Theory and Applications*, vol. 99, no. 3, pp. 723–757, 1998.

[33] A. Packard and J. Doyle, 'The complex structured singular value', *Automatica*, vol. 29, no. 1, pp. 71–109, 1993.

[34] I. Petersen, V. Ugrinovskii and A. Savkin, *Robust Control Design using H-Infinity Methods*. Springer-Verlag, London, 2000.

[35] D. McFarlane and K. Glover, 'A loop-shaping design procedure using H∞ synthesis', *IEEE Transactions on Automatic Control*, vol. 37, no. 6, pp. 759–769, 1992.

[36] M. Kothare, V. Balakrishnan and M. Morari, 'Robust constrained model predictive control using linear matrix inequalities', *Automatica*, vol. 32, no. 10, pp. 1361–1379, 1996.

[37] J. Helton and M. James, *Extending H-Infinity Control to Nonlinear Systems: Control of Nonlinear Systems to Achieve Performance Objectives*, vol. 1. SIAM, Philadehphia, 1987.

[38] L. Ghaoui and S. Niculescu, *Advances in Linear Matrix Inequality Methods in Control*. SIAM, Philadehphia, 1987.

[39] K. Narendra and K. Parthasarathy, 'Identification and control of dynamical systems using neural networks', *IEEE Transactions on Neural Networks*, vol. 1, no. 1, pp. 4–27, 1991.

[40] H. Khalil and J. Grizzle, *Nonlinear Systems*, vol. 3. New Jersey: Prentice-Hall, 1996.

[41] C. Kwan and F. Lewis, 'Robust backstepping control of nonlinear systems using neural networks', *IEEE Transactions on Systems, Man and Cybernetics, Part A: Systems and Humans*, vol. 30, no. 6, pp. 753–766, 2000.

[42] V. Utkin, 'Sliding mode control design principles and applications to electric drives', *IEEE Transactions on Industrial Electronics*, vol. 40, no. 1, pp. 23–36, 1993.

[43] K. Ariyur and M. Krstic, *Real-Time Optimisation by Extremum-seeking Control*. Wiley, New York, 2003.

[44] S. Draper, *Tidal Stream Energy Extraction in Coastal Basins*. PhD thesis, University of Oxford, 2011.

[45] J. Falnes, *Ocean Waves and Oscillating Systems: Linear Interactions Including Wave-Energy Extraction*. Cambridge, University Press, Cambridge, 2002.

[46] D. Evans, 'Maximum wave-power absorption under motion constraints', *Applied Ocean Research*, vol. 3, no. 4, pp. 200–203, 1981.

[47] J. Hals, J. Falnes and T. Moan, 'Constrained optimal control of a heaving buoy wave-energy converter', *Journal of Offshore Mechanics and Arctic Engineering*, vol. 133, no. 1, 2011.

[48] J. Cretel, *et al.*, 'Maximisation of energy capture by a wave-energy point absorber using model predictive control', *Proceedings of the IFAC World Congress*, Milan, 2011, pp. 3714–3721.

[49] G. Bacelli, J. Ringwood and J. Gilloteaux, 'A control system for a self-reacting point absorber wave energy converter subject to constraints', *Proceedings of the 18th IFAC World Congress* (2011), International Federation of Automatic Control (IFAC), pp. 11387–11392.

[50] F. Fusco and J. V. Ringwood, 'A simple and effective real-time controller for wave energy converters', *IEEE Transactions on Sustainable Energy*, vol. 4, no. 1, pp. 21–30, 1981.

[51] K. Budal and J. Falnes, 'Interacting point absorbers with controlled motion', In: *Power from Sea Waves*. B. Count (Ed.), Academic Press, London, 1980.

[52] A. Babarit, M. Guglielmi and A. Clément, 'Declutching control of a wave energy converter', *Ocean Engineering*, vol. 36, no. 12, pp. 1015–1024, 2009.

[53] M. Lopes, *et al.*, 'Experimental and numerical investigation of non-predictive phase-control strategies for a point-absorbing wave energy converter', *Ocean Engineering*, vol. 36, no. 5, pp. 386–402, 2009.

[54] J. Scruggs and S. Lattanzio, 'Optimal causal control of an ocean wave energy converter in stochastic waves', *Proceedings of the European Wave and Tidal Energy Conference*. Southampton, UK, 2011.

[55] F. Fusco and J. Ringwood, 'A study of the prediction requirements in real-time control of wave energy converters', *IEEE Transactions on Sustainable Energy*, vol. 3, no. 1, pp. 176–184, 2012.

[56] F. Fusco and J. Ringwood, 'Short-term wave forecasting for real-time control of wave energy converters', *IEEE Transactions on Sustainable Energy*, vol. 1, no. 2, pp. 99–106, 2010.

[57] P. Mhaskar, J. Liu and P. Christofides. *Fault-Tolerant Process Control*. Springer-Verlag, London, 2013.

[58] I. Boldea, *et al.*, *The Induction Machine Handbook*. CRC Press, 2001.

[59] I. Munteanu, *Contribution to the Optimal Control of Wind Energy Conversion Systems*, PhD Thesis, University of Galaty, 2006.

[60] I. Munteanu, *et al.*, *Optimal Control of Wind Energy Systems: Towards a Global Approach*. Springer, 2008.

[61] S. E. Benelghali, *On Multiphysics Modeling and Control of Marine Current Turbine Systems*, PhD thesis, University of Brest, France, 2009 (free link: http://tel.archivesouvertes.fr/docs/00/52/16/15/PDF/THESE_BENELGHALI_LBMS_UBO_2009.pdf, last consultation September 2012).

[62] S.E. Ben Elghali, *et al.*, 'High-order sliding mode control of a marine current turbine driven permanent magnet synchronous generator', *Proceedings of the IEEE IEMDC'09*, Miami, FL, pp. 1541–1546, May 2009.

[63] S.E. Ben Elghali, *et al.*, 'Experimental validation of a marine current turbine simulator: Application to a permanent magnet synchronous generator-based system second-order sliding mode control', *IEEE Transactions on Industrial Electronics*, vol. 58, no. 1, pp. 119–126, 2011.

[64] S.E. Ben Elghali, *et al.*, 'High-order sliding mode control of a marine current turbine driven doubly-fed induction generator', *IEEE Journal on Oceanographic Engineering*, vol. 35, no. 2, pp. 402–411, 2010.

[65] S.E. Ben Elghali, *et al.*, 'Performance comparison of three- and five-phase permanent magnet generators for marine current turbine applications under open-circuit faults', *IEEE Powering Proceedings* – 2011 IEEE Powereng, Espagne, 2011.

[66] M. Stalberg, R. Waters, O. Danielsson and M. Leijon, 'Influence of generator damping on peak power and variance of power for a direct drive wave energy converter', *Journal Of Offshore Mechanical and Arctic Engineering*, vol. 130, pp. 1–4, August 2008.

[67] D. O'Sullivan, D. Murray, J. Hayes, M. G. Egan and A. W. Lewis (2011). The benefits of device level short term energy storage in ocean wave energy converters, In , R. Carbone (Ed.), *Energy Storage in the Emerging Era of Smart Grids* ISBN: 978-953-307-269-2, InTech. Available at: http://www.intechopen.com/books/energy-storage-in-the-emerging-era-of-smart-grids/the-benefits-of-device-level-short-term-energy-storage-in-ocean-wave-energy-converters

[68] D.L. O'Sullivan and A.W. Lewis, 'Generator requirements and functionality for ocean energy converters', 2010 XIX International Conference on Electrical Machines (ICEM), pp.1–7, 6–8 September 2010.

[69] D. Ravi Kiran, A. Palani, S. Muthukumar and V. Jayashankar, 'Steady grid power from wave energy', *IEEE Transactions on Energy Conversion*, vol. 22, no. 22, pp. 539–540, June 2007.

[70] S. Muthukumar, S. Kakumanu, S. Sriram, R. Desai, A. A. S. Babar and V. Jayashankar, 'On minimizing the fluctuations in the power generated from a wave energy plant', *Proceedings on the IEEE International Conference on Electric Machines and Drives*, San Antonio, TX, pp. 178–185, June 2005.

[71] J. Hals, T. Bjarte-Larsson and J. Falnes, 'Optimum reactive control and control by latching of a wave-absorbing semisubmerged heaving sphere', *Proceedings of the 21st International Conference on Offshore Mechanics and Artic Engineering*, Oslo, Norway, pp. 1–9, June 23–28, 2002.

[72] M. Ruellen, H. BenAhmed, B. Multon, C. Josset, A. Babarit and A. Clement, 'Design methodology for a SEAREV wave energy converter', *IEEE Transactions on Energy Conversion*, vol. 25, no. 3, pp. 760–767, September 2010.

[73] M. Molinas, O. Skjervheim, P. Andreasen, T. Undeland, J. Hals, T. Moan, and B. Sorby, 'Power electronics as grid interface for actively controlled energy converters', *Proceedings of the International Conference on ICCEP*, Capri, Italy, May 21–23, 2007, pp. 188–195, 2007.

[74] J. Hals, J. Falnes and T. Moan, 'A comparison of selected strategies for adaptive control of wave energy converters', *Journal of Offshore Mechanics and Arctic Engineering*, vol. 133, no. 3, August 2011, pp. 1–32.

[75] E. Tedeschi and M. Molinas, 'Control strategy of wave energy converters optimized under power electronics rating constraints', *Proceedings of the 3rd International Conference on Ocean Energy*, Bilbao, Spain, 2010.

[76] M. Schoen, J. Hals and T. Moan, 'Wave prediction and fuzzy logic control of wave energy converters in irregular waves', *Proceedings of the Mediterranean Conference on Control and Automation*, Ajaccio-Corsica, France, pp. 767–772, 2008.

[77] E. Tedeschi and M. Molinas, 'Tunable control strategy for wave energy converters with limited power take-off rating', *IEEE Transactions on Industrial Electronics*, vol. 59, no. 10, pp. 3838–3846, 2012.

[78] W. H. Michel, 'Sea spectra revisited', *Marine Technology*, vol. 36, no. 4, pp. 211–227, 1999.

[79] G. P. Harrison and A. R. Wallace, 'Sensitivity of wave energy to climate change', *IEEE Transactions on Energy Conversion*, vol. 20, no. 4, pp. 870–877, December 2005.

[80] P. R. M. Brooking and M. A. Mueller, 'Power conditioning of the output from a linear vernier hybrid permanent magnet generator for use in direct drive wave energy converters', *Proceedings of the Institute of Electrical Engineers — Generation, Transmission and Distribution*, vol. 152, no. 5, pp. 673–681, September 2005.

[81] J. K. H. Shek, D. E. Macpherson, M. A. Mueller and J. Xiang, 'Reaction force control of a linear electrical generator for direct drive wave energy conversion', *IET Renewable Power Generation*, vol. 1, no. 1, pp. 17–24, March 2007.

[82] V. Delli Colli, P. Cancelliere, F. Marignetti, R. Di Stefano and M. Scarano, 'A tubular-generator drive for wave energy conversion', *IEEE Transactions on Industrial Electronics*, vol. 53, no. 4, pp. 1152–1159, August 2006.

[83] T. K. A. Brekken, H. Hapke and J. Prudell, 'Drives comparison for reciprocating and renewable energy applications', *Proceedings of the 24th Applied Power Electronics Conference and Exposition*, APEC 2009, Washington, DC, pp. 732–738, 15–19 February 2009.

[84] H. Lendenmann, K.-C. Stromsem, M. Dai Pre, W. Arshad, A. Leirbukt, G. Tjensvoll and T. Gulli, 'Direct generation wave energy converters for optimized electrical power production', *Proceedings of the 7th EWTEC*, Porto, Portugal, September 11–13, 2007, pp. 1–10, 2007.

[85] E. Tedeschi, M. Carraro, M. Molinas and P. Mattavelli, 'Effect of control strategies and power take-off efficiency on the power capture from sea waves', *IEEE Transactions on Energy Conversion*, vol. 26, no. 4, pp. 1088–1098, December 2011.

Chapter 9
Modelling and simulation techniques

9.1 Resource to wire modelling for tidal turbines

S. Benelghali, J.F. Charpentier and *M.E.H. Benbouzid*

Nomenclature

ρ	Fluid density
A	Cross-sectional area of the marine turbine
V_{tide}	Fluid speed
C_{p}	Power coefficient
β	Pitch angle
C	Tide coefficient
V_{st} (V_{nt})	Spring (neap) tide current speed
λ	Tip speed ratio
s, (r)	Stator (rotor) index (superscripts)
d, q	Synchronous reference frame index
V (I)	Voltage (current)
P (Q)	Active (reactive) power
ϕ	Flux
ϕ_m	Permanent magnet flux
T_{em} (T_m)	Electromagnetic torque (mechanical torque)
R	Resistance
L (M)	Inductance (mutual inductance)
σ	Total leakage coefficient, $\sigma = 1 - M^2/L_sL_r$
θ_r	Rotor position
ω (ω_s)	Rotor electrical speed (electrical synchronous speed)
ω_r	Rotor current frequency ($\omega_r = \omega_s - \omega$)
Ω	Mechanical speed ($\Omega = \omega$/p)
f	Viscosity coefficient
J	Rotor inertia

p	Pole pair number
L_{ss}	Stator inductance
M_s	Stator mutual magnetizing inductance
M_{jr}	Rotor mutual magnetizing inductance in phase j, $j = a$, b or c
λ_j	PMSG rotor flux in phase j, $j = a$, b or c

9.1.1 Modelling requirements

Tidal turbine dynamic performance analysis requires the use of computational models representing the nonlinear differential-algebraic equations of the various system components such as tidal resource, turbine, generator, converter, control system and grid connection, as shown in Figure 9.1. However, the main difficulty is to include a variety of sub-models with different timescales for hydrodynamic loads (turbine, mechanical systems, generators, power electronics and other components). The user is therefore concerned with selecting the appropriate models for the problem at hand and determining the data to represent the specific turbine equipment.

Appropriate model choice depends mainly on the timescale of the simulation. Figure 9.2 shows the principal tidal turbine dynamic performance areas displayed on a logarithmic timescale ranging from microseconds to days. The lower end of the band for a particular item indicates the smallest time constants that need to be included for a precise modelling of each element.

The upper end indicates the approximate length of time that must be analysed. It is possible to build a turbine simulation model that includes all dynamic effects from very fast dynamics to very slow hydrodynamic loads; however, this solution can be highly time consuming to develop and execute. For efficiency and ease of analysis, normal engineering practice dictates that only models incorporating dynamic effects relevant to the particular performance of the concerned area to be used. Some basic models of each subsystem will be presented in the next section: resource, turbine, generator and converter.

9.1.2 Resource modelling

First of all, a basic model of the tidal resource is presented. The proposed method is illustrated by the calculation of the extractable power from the Raz de Sein (Brittany, France) as it is one of the more characterized sites in terms of tidal

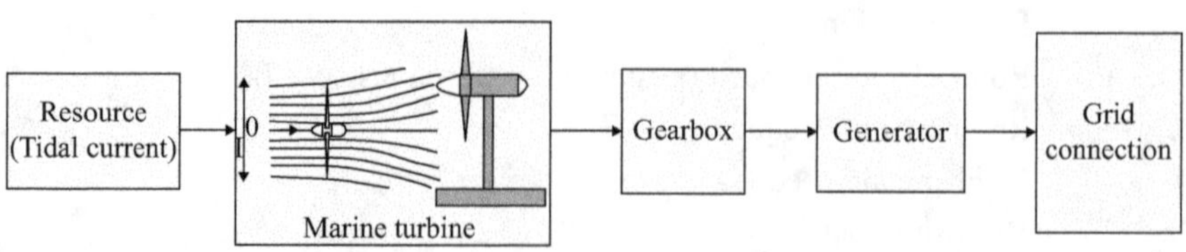

Figure 9.1 Tidal turbine global scheme

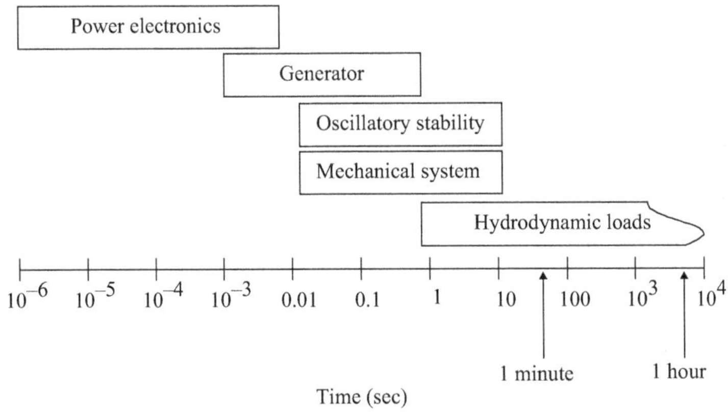

Figure 9.2 Timescale in a tidal turbine

current. The following modelling approach can be extended to any other sites. This site was chosen above several others listed in the European Commission report EUR16683 due to the presence of high velocity current coupled with appropriate depths suitable for tidal turbine [1]. Moreover, the marine current velocity distribution for most of the time is greater than the minimum velocity required for economic deployment of marine turbines, estimated to be 1 m/s [1, 2]. It should be noted that tidal current data are provided by the SHOM (French Navy Hydrographic and Oceanographic Service) and it is available for various locations in chart form [3].

Oceanographic services (such as the French SHOM for the illustration site) produce charts giving the current velocities for spring and neap tides at a specific site. These values are given at hourly intervals starting at 6 tidal hours before high waters and ending 6 hours after (tide hours). Therefore, knowing tide coefficients, it is easy to derive a simple and practical model to calculate the tidal current velocity vector for each tidal hour for a given coefficient $V_{tide}(Hm, C)$, where Hm is the tidal hour (i.e. the time defined by the semidiurnal tide period divided by 12) [10].

$$\overrightarrow{V_{tide}}(Hm, C) = \overrightarrow{V_{nt}(Hm)} + \frac{(C-45)(\overrightarrow{V_{st}} - \overrightarrow{V_{nt}})}{95-45} \tag{9.1}$$

C is the coefficient that characterises each tidal cycle (95 and 45 are, respectively, the spring and neap tide medium coefficient). This coefficient is determined by the astronomic calculation of earth and moon positions. V_{st} and V_{nt} are, respectively, the spring and neap tide current velocities for hourly intervals starting at 6 hours before high waters and ending 6 hours after.

For example, consider the case of a semidiurnal cycle that is characterised by a tide coefficient of $C = 80$ in a given geographical point. At this point, 3 hours after high tide, the data charts give $V_{st} = 1.8$ knots and $V_{nt} = 0.9$ knots, thus the speed calculated using (9.1) is $V_{tide} = 1.53$ knots. This first-order model is then used to calculate the tidal velocity vector for each hour. The implemented model will allow

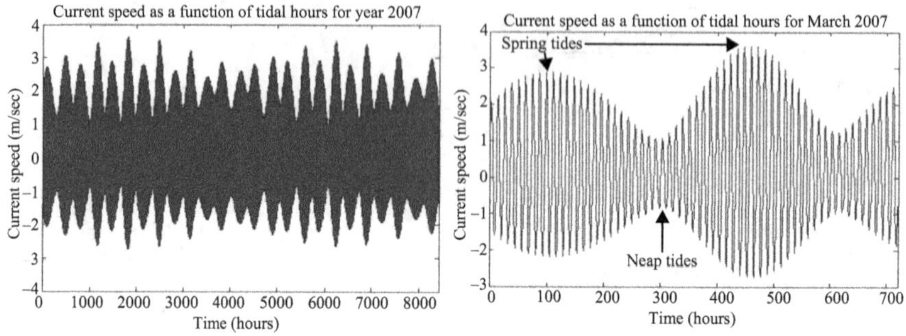

Figure 9.3 Tidal velocity in the Raz de Sein for the year 2007 and March 2007

the user to compute tidal velocities in a predefined time range. Figure 9.3 shows the model output for a month (March 2007) and for a year (2007).

The first-order model which has been adopted for the resource has several advantages including its modularity and its simplicity. Indeed, the marine turbine site can be changed, the useful current velocity can be varied, and the time range taken into account can also be altered from one month to one year. For a simulation of short periods (less than a few minutes), the tidal velocity can be considered as constant. However, some faster disturbances such as swell effects and turbulence can also be taken into account for a greater level of accuracy.

9.1.3 Hydrodynamic modelling of an horizontal axis turbine

Several technologies have been proposed to convert tidal power into electrical power. Most of them are based upon the use of horizontal axis turbines [4], which have been successfully utilized to harness wind energy [3]. Therefore, many techniques can be transferred from the design and operation of wind turbines [5]. There are, however, a number of fundamental differences in the design and operation of marine turbines. Particular differences entail changes in force loadings, immersion depth, different stall characteristics, and the possible occurrence of a cavitation phenomenon in the blades. Much information is however available on the cavitations and stall characteristics of marine propellers [6], which can provide useful information for marine turbines [7–8].

A first theoretical analysis of the hydrodynamic behaviour of a turbine can be done using the Rankine–Froude actuator disk model. This model consists of replacing the rotor with a theoretical actuator disk, which is a circular surface of zero thickness that can support a pressure difference, and thus decelerate the tidal current through the disk. The principal use of the actuator disk model is to obtain a first estimate of the wake-induced flow, and hence the total induced power loss. Note that the actual induced power loss will be larger than the actuator disk result because of the non-uniform and unsteady induced velocity. The assumptions on which the Rankine–Froude actuator disk theory is based are well-detailed in [5]. Among these assumptions, one requires that the disk slows the tidal current equally at each radius, which is equivalent to assuming uniform thrust loading at the disk.

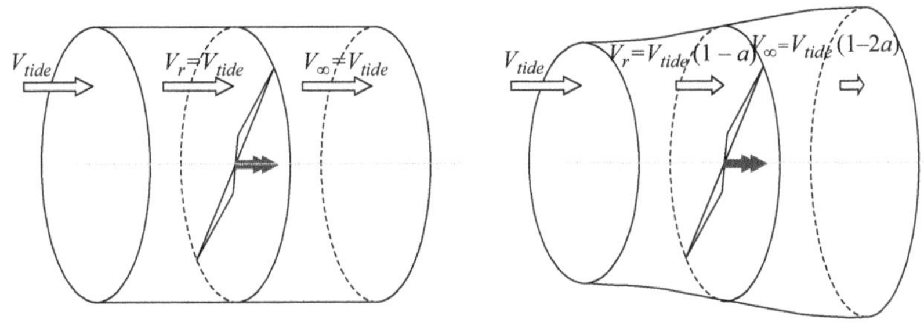

Tidal turbine with (right) and without (left) loading

Figure 9.4 The actuator disk model

Uniform thrust loading is, in turn, equivalent to considering an infinite number of rotor blades.

Figure 9.4 illustrates the one-dimensional flow through the actuator disk plane for a non-loaded and loaded machine. For instance, for a turbine with zero loading, the current velocity in the rotor plane (V_r) is equal to the undisturbed tidal current velocity (V_{tide}); while for a loaded turbine, the rotor current velocity is reduced. If the decreased velocity induced by the rotor is V, then the velocity at the disk is $V_{tide} - V = V_r$, and far downstream, at section 1, the current has been slowed further to velocity V_∞. The difference between the axial component of the current velocity and the axial flow velocity in the rotor plane is usually called *induced velocity*. Thus, the velocity at the disk is the average of the upstream and downstream velocities. Defining an axial induction factor, a, as the fractional decrease in current velocity between the free stream and the rotor plane represented by (9.2)

$$a = \frac{V}{V_{tide}} \tag{9.2}$$

The results in (9.3):

$$\begin{cases} V_r = V_{tide}(1-a) \\ V_\infty = V_{tide}(1-2a) \end{cases} \tag{9.3}$$

For $a = 0$, the current is not decelerated and no power is extracted, whereas for $a = 0.5$, the far wake velocity vanishes, and, without presence of flow behind the turbine, no power is generated. The power extracted from the tidal current by the rotor is given by (9.4)

$$P = \frac{1}{2}\rho A V_r (V - V_\infty)(V + V_\infty) \tag{9.4}$$

Substituting V_r and V_∞ from (9.3), we find that

$$P = \frac{1}{2}\rho A V_{tide}^3 4a(1-a)^2 \tag{9.5}$$

A power coefficient C_p is then defined as

$$C_p = \frac{P}{\frac{1}{2}\rho A V_{tide}^3} = 4a(1-a)^2 \tag{9.6}$$

where the denominator represents the global kinetic energy of the free-stream current contained in a stream tube with an area equal to the disk area. The extracted power is expressed by (9.7)

$$P = \frac{1}{2}\rho C_p A V_{tide}^3 \tag{9.7}$$

The theoretical maximum value of the power coefficient C_p occurs when $a = 1/3$. Hence, $C_{p,max} = 16/27 \approx 0.59259$, $V_r = \frac{2}{3} V_{tide}$ and $V_\infty = \frac{1}{3} V$. Thus, the theoretical maximum amount of energy extraction equals the 16/27th part of the kinetic energy in the current. This limit is often referred to as the *Betz limit*, or more accurately the *Lanchester–Betz limit*. In practice this limit cannot be reached and the maximal values of the C_p of real turbines are often in the 0.4–0.5 range.

C_p varies according to the pitch angle of the blades (β) and the tip-speed ratio (λ), where λ is the ratio between the tip-speed of the rotor and the tidal velocity. A $C_p(\lambda, \beta)$ curve, such as the one shown in Figure 9.5, is typically developed by manufacturers as a means of characterizing the performance of the turbine.

This very simple model gives an indication of the turbine behaviour but is not sufficient to calculate the real performance of a given turbine for a working point (given blade shape and geometry and given flow and rotation speeds).

One of the more easily used models that can be used for this purpose is the blade element momentum (BEM) theory, which is one of the oldest and most commonly used methods for calculating induced velocities on wind turbine blades. This theory is an extension of the actuator disk theory [5, 9] extended to each blade sections. This theory is fully described in [11].

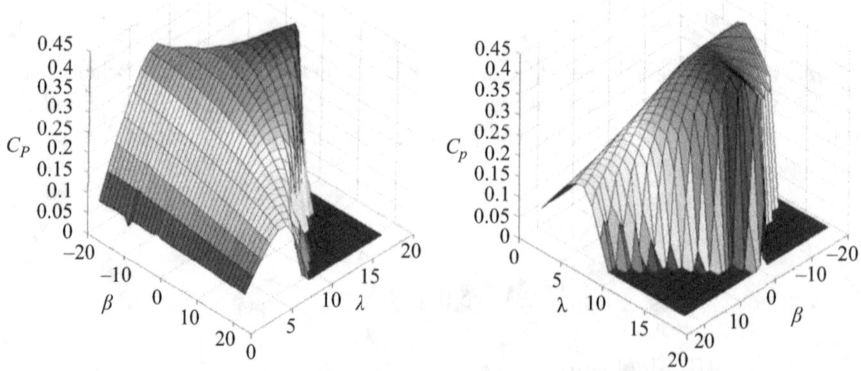

Figure 9.5 $C_p(\lambda,\beta)$ curves

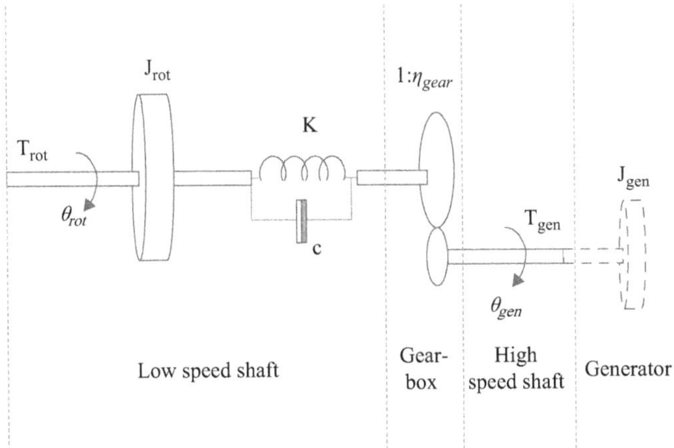

Figure 9.6 Drive train model

An example of simulation results using BEM theory, for a given three-blade turbine, is given by Figure 9.5, which presents the rotor power coefficient (Cp) according to the pitch angle variation of the blades β and the speed ratio λ.

9.1.4 Drive train modelling

For mechanical modelling, emphasis is put only on those parts of the dynamic structure of the marine current turbine that contributes to the interaction with the grid. Therefore, only the drive train is considered because this part of the marine turbine has the most significant influence on the power fluctuations. The structural components of the tidal turbines are not considered in this first order approach.

The mechanical drive train can be considered as a two-mass model, namely a large mass representing the rotor, and a smaller mass representing the generator (Figure 9.6). These rotating masses have inertia J_{rot} and J_{gen}, respectively. The (low-speed) rotor shaft is connected to the (high-speed) generator shaft via a 1:N gearbox. The low-speed shaft is modelled by a stiffness k and a damping coefficient c, while the high-speed shaft is assumed stiff.

The drive train transfers the hydrodynamic rotor torque, T_{rot}, to the low speed shaft torque T_{gen}. The mechanical model dynamical description consists of the following equations in (9.8)

$$\begin{cases} \dfrac{d\theta_{rot}}{dt} = \omega_{rot} \\ \dfrac{d\theta_k}{dt} = \omega_{rot} - \dfrac{\omega_{gen}}{\eta_{gear}} \\ \dfrac{d\omega_{rot}}{dt} = \dfrac{(T_{rot} - T_{gen})}{J_{rot}} \end{cases} \tag{9.8}$$

where θ_k is the angular difference between the two ends of the flexible shaft. For simplicity, θ_k can be considered as a constant and the drive train model can be simplified as shown in Figure 9.7.

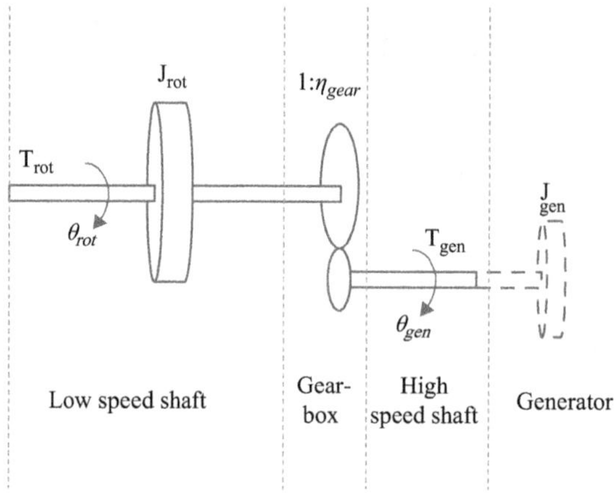

Figure 9.7 Simplified drive train model

9.1.5 Generator modelling

Much of the technology that has been suggested for tidal current energy extraction is comparable to that used in wind energy applications. It is then obvious that some wind generator topologies could be used for tidal turbines. This section will focus on modelling a PMSG associated with a back-to-back IGBT converter; other solutions based on Induction Generators or Double Fed Induction Generators (DFIG) can be modelled using similar methods.

For a PMSG, the stator voltage equations in stationary reference frame are given by (9.9)

$$[V_{abc}] = [R][i_{abc}] + [L]\frac{d[i_{abc}]}{dt} + \frac{d\lambda_{abc}}{dt} \tag{9.9}$$

where $[L] = \begin{bmatrix} L_{ss} & M_s & M_s \\ M_s & L_{ss} & M_s \\ M_s & M_s & L_{ss} \end{bmatrix}$

The PMSG is modelled under the following simplifying assumptions:

- sinusoidal distribution of stator winding,
- electric and magnetic symmetry,
- negligible iron losses and unsaturated magnetic circuit.

Under these assumptions, the generator model in the so-called steady-state (or stator) coordinates is first obtained.

From (9.9), a simple model suitable for simulation and control can be obtained in *d-q* rotor coordinates. Conversion between (*a*, *b*, *c*) and *d-q* coordinates can be realized by means of the classical Park transform. Using this transform, the *d-q* PMSG voltages and fluxes become

$$\begin{cases} V_d = Ri_d + L_d\dfrac{di_d}{dt} - L_q i_q \omega_s \\ V_q = Ri_q + L_q\dfrac{di_q}{dt} + (L_d i_d + \phi_m)\omega_s \end{cases} \tag{9.10}$$

The electromagnetic torque is obtained as in (9.11)

$$T_{em} = \frac{3}{2}p(\phi_d.i_q - \phi_q.i_d) = \frac{3}{2}p\left[\phi_m.i_q + (L_d - L_q)i_d.i_q\right] \tag{9.11}$$

If the permanent magnets are mounted on the rotor surface, then $L_d = L_q$ and the electromagnetic torque becomes (9.12)

$$T_{em} = \frac{3}{2}p.\phi_m.i_q \tag{9.12}$$

This equation can be linked with the mechanical equation of the rotating parts (generator, rotor, drive train, turbine). As an example for a direct drive system the mechanical equation is

$$T_{gen} - T_{em} - T_l = J\frac{d\Omega_{gen}}{dt} \tag{9.13}$$

where T_{gen} is the mechanical torque which comes from the gearbox or directly from the turbine (in case of direct driven systems) and is an input for the generator. This torque can be deduced from the gearbox characteristics, the power curves of the turbines, the flow speed and the turbine rotational speed. T_l is a torque which represents the mechanical and iron losses.

Knowing the main electrical parameters of the generator (L_d, L_q, ϕ_m and generator mechanical parameters), (9.10), (9.11) and (9.12) can be used to simulate the electromechanical behaviour of the generator. Of course, similar electromechanical models can be found and used for the other possible kind of generator (induction, double fed induction generator).

9.1.6 Global model of the system

It can be noted that for each model, input or output corresponds to output or input of others models. For example the input of the model of the turbine is the rotating speed related to the mechanical equation and current flow speed related to resource model. So the presented basic models can be associated in a modular environment dedicated for system simulation as for example the Matlab/Simulink environment. In this kind of environment, each of the previously described equation sets (resource, turbine, drive train, converter, or generator models) can be implemented in simulation blocks and connected to form a global simulation model. This allows for the calculation of the electrical and mechanical behaviour of the global system

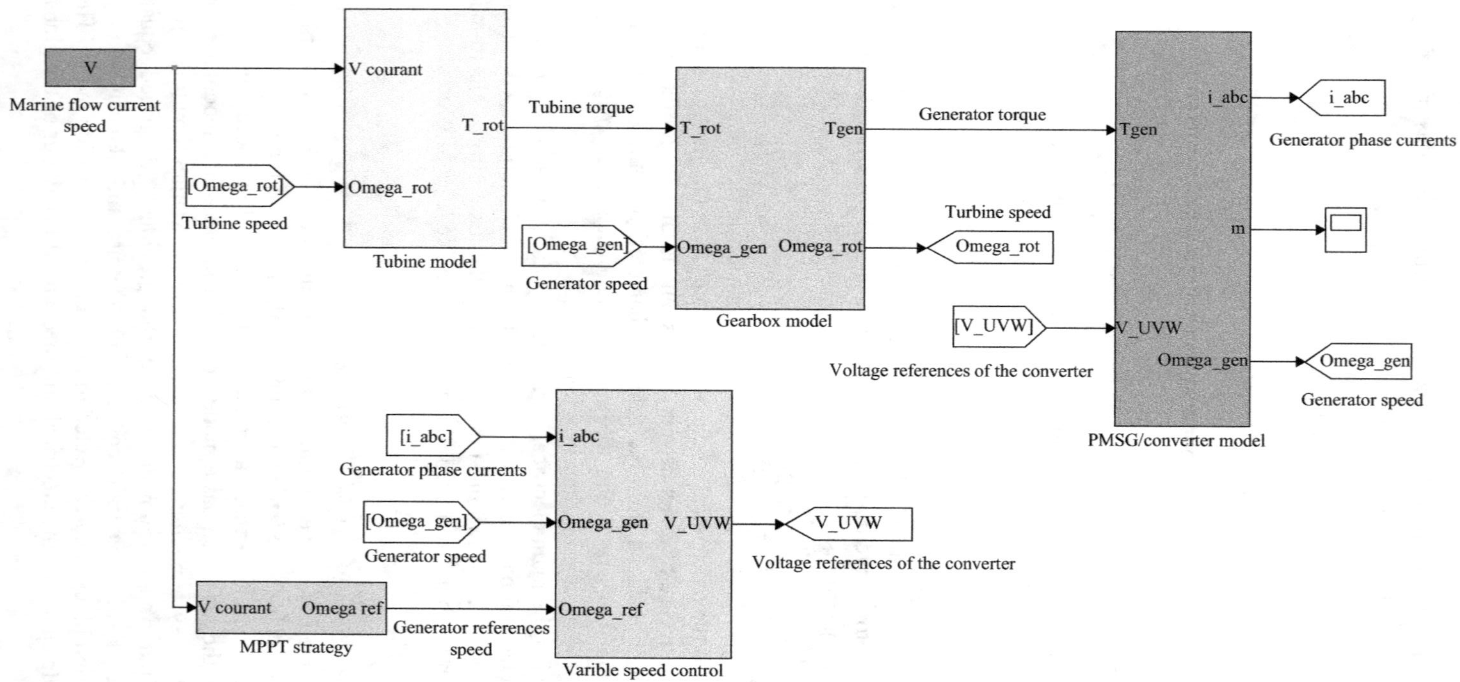

Figure 9.8 General view of a turbine simulation model in Matlab/Simulink environment

for a given tidal resource and a given energy harnessing control strategy. In this global simulation scheme, a control block which models the control laws (as described in Chapter 8) has also to be connected to the converter/generator block to take into account the variable speed control strategy as illustrated by Figure 9.8.

9.2 Resource to wire modelling techniques for wave energy converters

M. Santos-Mugica, J. L. Mendia and *F. S. Fernandez*

Nomenclature

A	Wave amplitude
$A_{jk\infty}$	Added mass at infinite frequency
K_{jk}	Radiation impulse response function for an oscillation in the degree of freedom j due to a unit amplitude oscillation in the degree of freedom k
B_{jk}	Radiation damping coefficient
τ	Time-lag
ρ	Water density
g	Gravity acceleration
f_e	Non-dimensional excitation force
x	Vertical direction coordinate
Γ	Excitation force coefficient
t	Time
S_B	Wetted surface of a body immersed in water
A	Wave amplitude
W	Wave frequency
Σ	Solution of a numerical problem

There are a multitude of designs for wave energy devices, at various stages of development, employing many means of energy conversion. Each device has its own particular advantages and disadvantages, but there is as of yet no clear indication which technology type or group of technologies will emerge as viable from an engineering and economic point of view. A typical resource-to-wire model of a wave energy device, generally consists of five subsystems related to the conversion process stages (Figure 9.9) [12].

Most of the studies carried out for wave energy converters (WECs) are focused on assessing either the performance of the device or grid integration issues, such as voltage stability, frequency variations, flicker and voltage ride through capability. For each type of study there is a suitable model. Consequently, depending on the purpose of the analysis, the level of detail in the representation of the components of the system will differ. The type of study not only determines

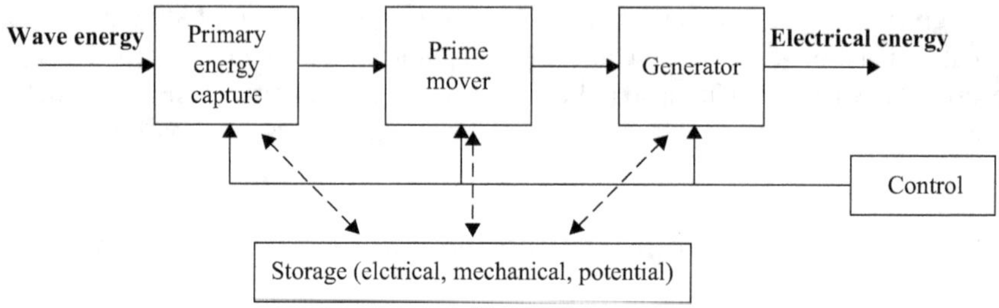

Figure 9.9 Typical wave energy conversion process

the modelling approach, but also the simulation software and necessary simplifications [13, 15].

9.2.1 Performance analyses

The objective of the performance analyses is to have a correct estimation of the expected power output of the devices. To have a preliminary idea of the performance of a device frequency domain analyses are made. However, to evaluate the performance of a WEC with realistic PTO configurations, moorings, control systems and other contributions, time-domain models are required to reflect the non-linearity arising from the different elements of the model.

9.2.1.1 Frequency domain analyses

Hydrodynamic optimization of WEC concepts is usually carried out by means of a frequency-domain analysis. From the mathematical point of view, the modelling and analysis of wave energy converters consists primarily in finding the solution for its dynamic equation. Equation (9.14) shows an example of the dynamic behaviour of a heave oscillating wave energy converter:

$$M\ddot{x}_3(t) = F_{e3}(t) - A_{33}\ddot{x}_3(t) - B_{33}\dot{x}_3(t) - \rho g S_w x_3(t) - C_{PTO}\dot{x}_3(t) - K_{PTO}x_3(t) \tag{9.14}$$

The sum of terms that are directly proportional to the velocity and acceleration of the heave oscillating WEC represent the radiation forces only by virtue of the hypothesis of monochromatic excitation and linear system [otherwise the radiation impulse response function (RIRF) formulation would be required].

If the system is linear and the excitation sinusoidal, then the motion response will be sinusoidal as well and can be represented as (9.15)

$$x_i(t) = \mathrm{Re}\{\hat{X}_i e^{i\omega t}\} \tag{9.15}$$

Equation (9.14) above can now be rewritten

$$-\omega^2 M\hat{X}_3 = \hat{F}_{e3} + \omega^2 A_{33}\hat{X}_3 - i\omega B_{33}\hat{X}_3 - \rho g S_w\hat{X}_3 - i\omega C_{PTO}\hat{X}_3 - K_{PTO}\hat{X}_3$$

And this leads to the straightforward solution for the amplitude of motion

$$\hat{X}_3 = \frac{\hat{F}_{e3}}{-\omega^2(M + A_{33}) + i\omega(B_{33} + C_{PTO}) + \rho g S_w + K_{PTO}}$$

The former is a well-known expression of the so-called Response Amplitude Operator (RAO) in heave. The main distinction with the usual expression applied in offshore engineering consists in the introduction of the simplified power take-off (PTO) coefficients that will clearly affect the device dynamics.

The instantaneous power absorbed by the PTO system is given by (9.16)

$$P(t) = F_{PTO}\dot{x}_3(t) = C_{PTO}\dot{x}_3^2(t) + K_{PTO}x_3(t)\dot{x}_3(t) \tag{9.16}$$

By considering sinusoidal motion, the average absorbed power in regular waves can be written as

$$\bar{P} = \frac{1}{2}C_{PTO}\left|\hat{V}_3\right|^2 = \frac{1}{2}C_{PTO}\omega^2\left|\hat{X}_3\right|^2$$

By using the previous equations, it can make explicit the expression of the power:

$$\bar{P} = \frac{1}{2}\frac{C_{PTO}\omega^2\left|\hat{F}_{e3}\right|^2}{\left(-\omega^2(M + A_{33}) + \rho g S_w + K_{PTO}\right)^2 + \omega^2(B_{33} + C_{PTO})^2} \tag{9.17}$$

9.2.1.2 Time domain analyses

To prove the validity of a PTO design or a control strategy, a more realistic approach is often needed, particularly when real sea-state performance is the determining criterion to validate a solution. In such cases, a time-domain model is generally unavoidable.

When nonlinear effects such as viscous forces or other interactions are considered, the linearity assumption is no longer valid. The most general and fundamental approach to deal with these cases is a full nonlinear time-domain analysis that requires the solution of the flow equation in the time domain, possibly with complex and time-consuming computational fluid dynamic (CFD) codes that, at the current stage, are really difficult to implement and not always accurate enough.

Another way to take into account nonlinearities, particularly when they can be modelled as time-varying coefficients of a system of Ordinary Differential Equations (ODEs), is to apply the linear time-domain model based on the Cummins equation [16], whose use is widespread in seakeeping applications. This is based on a vector integro-differential equation which involves convolution terms to account for the radiation forces. The following equation is an example of the formulation of an oscillating point absorber in one degree of freedom.

$$(M + A_{33\infty})\ddot{x}_3(t) + \int_{-\infty}^{t} K_{33}(t-\tau)\dot{x}_3(\tau)d\tau + \rho g S x_3(t) + F_{ext}(\dot{x}_3, x_3, t) = F_e(t) \tag{9.18}$$

In this equation, all the possible nonlinearities are included in the term F_{ext}, which represents the external forces that are applied to the system, such as PTO or mooring forces, and that could be possibly linked to other independent variables that form a set of ODEs.

Notice that the excitation force in the time domain is formally related to its frequency-domain expression by an inverse Fourier transform [17] which is, in general, not causal. For regular monochromatic waves, the excitation force can be expressed as (9.19)

$$F_e(t) = |\Gamma(\omega)| A\cos(\omega t + \phi) \tag{9.19}$$

where an excitation force coefficient, $\Gamma(\omega)$, proportional to the wave amplitude, has been introduced. In irregular waves, the excitation force can be simply modelled as a linear superposition of N independent sinusoidal components (ideally, N $\rightarrow +\infty$) such as

$$F_e(t) = \sum_{i=1}^{N} |\Gamma(\omega_i)| A_i \cos(\omega_i t + \phi_i) \tag{9.20}$$

The amplitudes A_i of each frequency component are defined from the energy spectral density $S(w)$ as Rayleigh distributed random values with mean square $2S(\omega)\Delta\omega$. Phases, ϕ_i are randomly selected assuming a uniform distribution within $[0; 2\pi]$. With this assumption, the randomness of the elevation process is properly reproduced and its statistical properties are correctly modelled.

The time-varying Radiation Impulse Response Function (RIRF) can be derived by a number of methods, including linear time-domain Boundary Element Methods (BEM) codes like TiMIT [18] or ACHIL3D [19] or indirectly by firstly solving the linear problem in the frequency domain and then using the computed frequency-dependent hydrodynamic coefficients in (9.21):

$$K_{33}(t) = \frac{2}{\pi} \int_0^{\infty} B_{33}(\omega) \cos(\omega t) d\omega \tag{9.21}$$

The integral in this equation has to be evaluated after a truncation at a properly defined frequency. Since the radiation damping coefficient $B(\omega)$ tends asymptotically to zero as the frequency tends to infinity, it is sufficient to introduce an upper limit for this coefficient to be negligible. For instance, taking as the truncation frequency as the frequency above which $B(w)$ is less than one thousandth of its maximum value produces satisfactory accuracy.

However, this is not always possible with commercial BEM codes since the largest frequency at which the coefficients are evaluated is limited by the number of panels (or, more correctly, the average panel size in comparison with the wave length) and the radiation damping coefficient might be still much larger than zero at the largest frequency assumed in the computation.

A preliminary simple model, which permits also to check the quality of the time-domain results comparing them with the ones given by the frequency-domain, consists in assuming just a damper as Power Take-Off, capable of opposing a force proportional to the velocity of the buoy. Additionally, one can assume the presence of a spring term (i.e. proportional to the displacement), that is required to guarantee the stability of the system if no restoring force is present.

9.2.2 Grid integration analysis

Studies carried out on the power system are based on the simulation of real phenomena that represent the same behaviour as the real system. Before developing a model for grid integration analyses, it is mandatory to understand and quantify as accurately as possible the behaviour and the parameters that define any of the elements of the system as well as the different phenomena that concern the power system [15]. These phenomena cover a wide range of time intervals, from micro seconds to several hours.

Due to the complexity of the power system and as wave energy penetration level increases it would be unfeasible to simulate wave energy devices behaviour with a high level of detail. For this reason, it is necessary to develop simplified models. In these models is very important to take into account the influence of the control system when analysing a phenomenon. The control system does not influence the case of fast phenomena (milliseconds to a few seconds). However, for slow phenomena (minutes to hours), the control system must be considered. So, the kind of analysis and the model of the device and the power system are related to the duration of the phenomena under study. An example of the type of model required depending on the type of analysis desired is shown in Table 9.1:

Table 9.1 Model types versus analysis type [13, 14]

Type of analysis	Model
Voltage variation **Load flow** **Short-circuits**	**Static model**
Transient stability **Small-signal stability** **Transient response** **Steady-state waveforms** **Synthesis of control** **Optimization**	**Dynamic models** **Functional models**
Start-up transient effects **Load transient effects** **Fault operation** **Harmonics and sub harmonics** **Detailed synthesis of control** **Detailed optimization**	**Dynamic models** **Mathematical physical models (power electronic)**

As can be seen in Table 9.1, two different models can be used to carry out grid integration studies. On one hand static models, simple and ease to create, and on the other hand dynamic models, both functional and mathematical. The mathematical physical model includes detailed power electronics model [13, 14].

9.2.2.1 Static model

In these models just the electrical variables at fundamental frequency are considered and the study can be limited to an equivalent single phase analysis in balanced systems or to sequence components in unbalanced systems. When the steady-state voltage level is of interest, the device active and reactive power can simply be represented as sources of active and reactive power. Active power is given by the static power curve of the wave energy converter, the relation between the sea state and the produced active power, so the hydrodynamic model is not necessary, while the amount of reactive power exchanged with the grid is decided according to the control objectives of the device.

9.2.2.2 Dynamic model

Dynamic models provide a grid operator with a means of assessing the impact of renewable energy generators on the local and national grid from the point of view of system stability, dynamic voltage variation, and fault performance and ratings [12]. Power system dynamic models will be discussed in more detail in the following section.

9.3 Power system dynamic models

D. Mollaghan

9.3.1 Dynamic models for power systems

Dynamic models for power system studies are used primarily for analysis of power system stability. Power system stability refers to the ability of a power system to remain in a steady-state under normal operating conditions and to regain an acceptable state of equilibrium after being subjected to a disturbance [20]. This type of analysis is critical for ensuring a reliable electricity supply, and would not be possible without dynamic models. Standard models have been developed for conventional power plant equipment, but as electricity generation from renewable sources has become more prominent in recent years, the range of generation equipment in use has become more diverse and dynamic models are perhaps more important now than ever before. Renewables such as wind and ocean energy have a varying power source and implement a wide variety of power system configurations such as fixed and variable speed induction generators, doubly-fed induction generators, and synchronous generators with back-to-back convertors. It is necessary for grid operators to study the behaviour of this new generation on the grid, thus emphasising the need for accurate and reliable dynamic models.

Many grid codes throughout the world specify a requirement for dynamic models to be supplied before a plant can procure a grid connection [21]. The generating plant operators must provide a dynamic model of their power system that is compatible with the network operator's software. This model must be capable of simulating the plant's output electrical power and other dynamic characteristics, during both normal and faulted operation. These models are a requirement as they are used by the TSO to ensure that any new generation on their system:

- Is stable under normal and fault conditions
- Does not affect other users/generators
- Produces acceptable power quality
- Does not overload existing protection circuitry and power lines
- Allows load flow analysis to be performed

9.3.2 Model development and analysis

A power system dynamic model consists of a number of sub-models for each of the components used in the power system, e.g. generator, transformers, reactive compensation devices, power electronics etc. Many dynamic simulation software packages, such as Siemens PSS/e and DIgSILENT PowerFactory, contain a library of models of standard equipment. A network model is used to represent the power system's electrical equipment configuration from the generator(s) to the point of common coupling (PCC) with the national grid. The model developer must decide which equipment models are required, how the network is to be laid out, and must also parameterise each of the models according to the equipment specifications. For a full dynamic study, further models such as power take-off models, control algorithms and protection mechanisms would also be required. Figure 9.10 shows an example of a network model (left) and an electrical control topology for a wind farm (right).

There is a distinction between a power system model and a design level model for a plant. Design level models would often include a detailed physical model of the mechanical power train from the energy source to the generator. These models would typically be used for performance assessment or system optimisation. However, this level of detail is not necessary for power system studies as many of the variables used in a design-level model would not influence the electrical power system over the relatively short timescale used for power system simulations. Furthermore, the computational effort for these design-level models would be too demanding for a large-scale grid study. For these reasons, a power system dynamic model has a simplified power take-off model in order to reduce the computational effort. The model developer must find the right balance between obtaining reasonable simulation run-times and maintaining accuracy in the model.

There are a number of stages required when performing a dynamic analysis. A typical analysis process is outlined below:

1. Load flow: A load-flow (or power-flow) analysis is the calculation of power flows and voltages at each of the nodes in a network model. The calculations are solved iteratively using techniques such as the Newton–Raphson method.
2. Initialisation: Model initialization is required to ensure the simulation begins in a steady state, thus avoiding any unwanted transients. The initialization stage uses results from the load flow to configure internal state variables within the model.

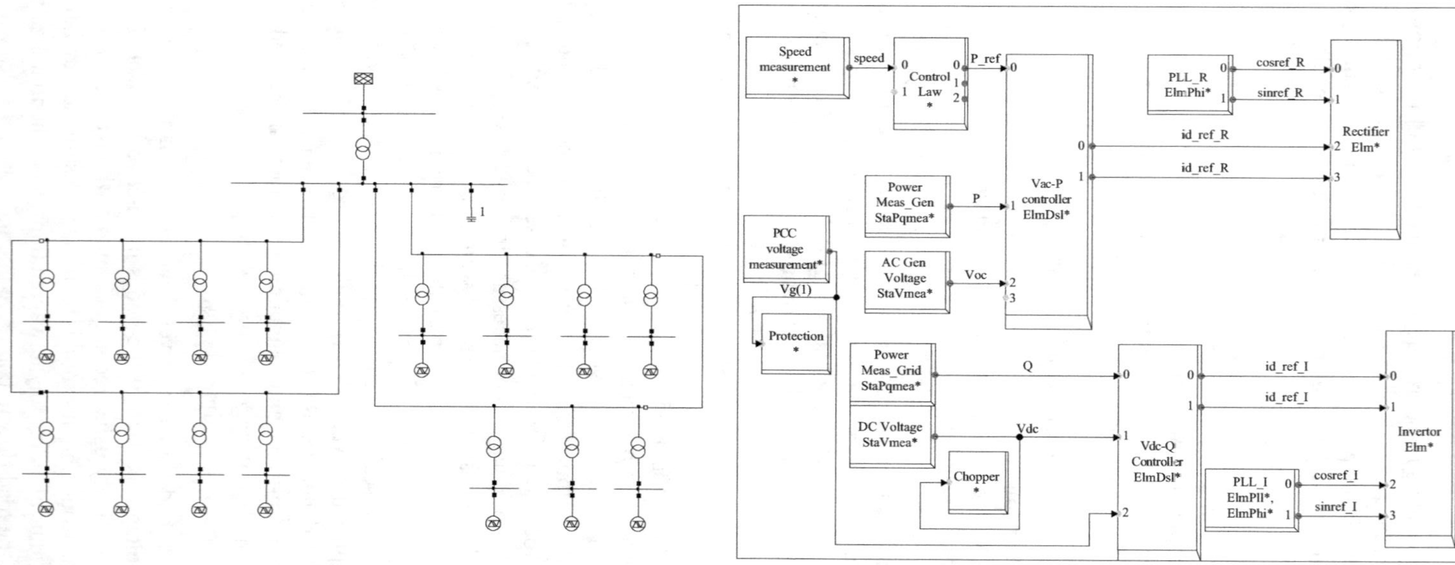

Figure 9.10 Wind farm network model (left) and electrical control and protection model overview (right), developed using DIgSILENT PowerFactory

3. Steady-state analysis: A simulation is performed in steady state to assess the plant's performance.Simulations are typically carried out over a period less than 30 seconds but can vary depending on the type of analysis.
4. Disturbance analysis: A disturbance is applied to a point in the network for a suitable time. The plant's ability to recover from a disturbance is assessed.

9.3.3 Experience from the wind industry

Wind turbine (WTG) dynamic models have been in development for a number of years, many have been made publically available by manufacturers and research institutions [22–25]. In addition to the electrical system models (e.g. generator, power electronics, protection etc.), WTG models typically consist of a mechanical power-train model, which incorporates pitch control, turbine power and drive shaft dynamics. Wind power is calculated according to (9.22):

$$P_{wind} = \frac{1}{2}\rho A v^3 C_p(\lambda, \beta) \tag{9.22}$$

where ρ is air density, A is rotor blade swept area, v is wind speed and C_p is the performance coefficient, which is a function of the tip-speed ratio and the pitch angle. A typical WTG power-train model is outlined in Figure 9.11:

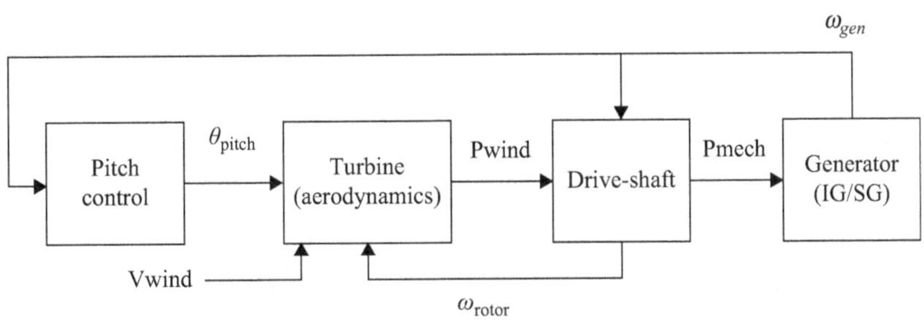

Figure 9.11 WTG power-train model

9.3.3.1 Model aggregation

The above method of modelling is suitable for single turbines or small wind farms, however the simulation run-times become longer and more complex as the size of the wind farm gets larger. There is currently a lot of focus on model aggregation, which aims to reduce the computational demands of large wind farm modelling by creating one or more equivalent models to represent an entire wind farm. Conventional power plants are generally modelled individually as there are a relatively small amount of such plants in a typical grid network. However, countries with high levels of wind power on their networks would have multiple wind farms containing a large amount of wind turbines. In such cases, modelling each turbine individually would be unfeasible for a grid study, mainly due to the lengthy simulation times for this approach.

The simplest method of model aggregation is to use one generator model to represent all of the wind farm's generators, where the rated power of the equivalent generator is equal to the rated power of each individual generator times the number of generators. The electrical system is also reduced to one equivalent circuit. This method is suitable for fixed-speed WTGs, where there is only a small generator speed deviation across the wind farm. Furthermore, if the wind speed is assumed to be constant across the farm, one equivalent drive-train model (scaled appropriately) can be used to represent all of the wind turbines in the farm. This method is the most simplistic representation of a wind-farm; the computational effort required would be dramatically reduced compared to detailed models of individual turbines. However, it is limited by the many assumptions used in this approach. For instance, inter-array effects such as wake effects and wind speed deviations are neglected, as are any generator speed deviations across the farm.

If wind speed across the farm cannot be assumed constant, or if the wind turbines have variable speed generators, then a compromise can be made by only aggregating some parts of the model. For example, one approach to model a variable speed WTG farm [26] is to only aggregate the electrical system. One equivalent model is used for the power electronic convertors and controls, and also the electrical part of the generators. The generator inertia, aerodynamics and pitch controllers are not part of the aggregation. Overall there are a number of methods which can be used for aggregate modelling; the most adequate method will depend on the area of interest for the study.

9.3.3.2 Generic modelling

The wind energy industry is currently at a more advanced stage than the ocean energy industry, so there is a good opportunity to learn from the issues encountered by the wind energy sector, from a dynamic modelling perspective. The modelling issues encountered when numerous wind turbine dynamic models were being tested by the Irish grid operators, Eirgrid, were described in [27]. Some of the issues encountered included:

1. Numerical instabilities occurring in the simulations due to the use of different combinations of integration algorithms and dynamic systems.
2. Inadequate model initialisation, particularly at part load. Thought to be due to inadequate model development for part load simulations.
3. To protect intellectual property, WTG manufacturers supplied their dynamic models in a proprietary 'black-box' format, without giving access to the underlying source code. This led to difficulties diagnosing problems encountered in the simulation.

In an attempt to reduce to dynamic modelling effort for a number of interested parties (e.g. grid operators, device manufacturers, project developers), a considerable effort is being made towards the creation generic standard models for wind turbines. Groups such as WECC Wind Generation Modelling Group and the IEEE Working Group on Dynamic Performance of Wind Power Generation have been engaged in research in this area for the past number of years. Furthermore, a new

IEC standard (IEC 61400-27) for generic simulation models for wind power generation is currently in development. The working group for this standard consists of a range of parties involved in the wind industry, including manufacturers, grid operators, project developers, software developers and research institutions.

> *The purpose of IEC 61400-27 is to define standard, public dynamic simulation models for wind turbines and wind power plants, which are intended for use in large power system and grid stability analyses, and should be applicable for dynamic simulations of power system events such as short circuits (low voltage ride through), loss of generation or loads, and typical switching events (e.g. line switching).* [28]

Four types of wind turbine technology have been identified for generic standardised modelling:

- Type 1: Induction generator;
- Type 2: Induction generator with variable rotor resistance;
- Type 3: Doubly-fed induction generator;
- Type 4: Asynchronous or synchronous generator with full convertor interface.

These model types represent the most common generator types used by the wind industry. In addition to a generator/convertor module, the models will also include an active and reactive power control module, and a mechanical drive-train module. A future phase of the standard definition will specify standard models for aggregated wind farms. This standardisation is an important step forward for the industry and will aid grid operators, project developers and device developers in areas such as performance assessment, design, planning studies and comparative studies.

9.3.3.3 Validation

Model validation is an important part of the modelling process. Validation is required to ensure the model accurately reflects the dynamic performance of the physical device(s) for a power system stability study. This can be achieved by comparing model simulation results with measured data from physical devices. Physical testing can be performed in the lab or in the field, where an event, such as a voltage dip, can be implemented and the results recorded. The measured results are then compared with the model's simulation results. However, this method is not always feasible, particularly when dealing with fault analysis.

In the early stages of wind turbine development, there was not a lot of field data available, one validation approach used was to compare the power system models with manufacturer's detailed design-level models. The main disadvantage with this approach is obviously the lack of measured data, although it could be a useful first approximation when no field data is available. Another approach would be to monitor an existing device, measuring various disturbances over time and comparing these measurements to the simulation model. However, this approach could take a considerable amount of time to gather enough information to validate a model [29].

9.3.4 Requirements for OE industry

The technology for ocean energy conversion is still in its relative infancy. At the time of writing, the installed capacity of ocean energy devices supplying to national grids worldwide is less than 10 MW. There are a multitude of designs for wave and tidal energy devices, at various stages of development, employing many means of energy conversion. Each device has its own particular advantages and disadvantages, but there is as of yet no clear indication which technology type or group of technologies will emerge as viable from an engineering and economic point of view. Globally there are only a few devices that have exported power to national grids. Some examples of these are:

- Marine Current Turbines (tidal) in Strangford Lough, Northern Ireland
- Pelamis Wave Power (wave) in Aguçadoura, Portugal
- Oceanlinx (wave) in Port Kembla, Australia
- Pico Power Plant (wave) in the Azores islands
- Wavegen Power Plant (wave) in Islay, Scotland

The diversity of the ocean energy technologies clearly presents a problem for grid operators from a power system perspective, thus emphasising the need for power system dynamic models for ocean energy systems. Modelling techniques employed by developers tend to focus on a high level of accuracy and are targeted at device and performance optimisation. These models would typically involve some form of hydrodynamic study of the device. A fully descriptive and exact dynamic model would require much of the following information, and more:

- Hydrodynamic coefficients in all degrees of freedom
- Hydrodynamic and thermodynamic models of the primary power capture stage
- Full geometric device information
- Prime mover dynamic pressure and flow characteristics
- Full knowledge of control strategy

This type of model strives to be highly accurate in its representation of device dynamics. Constructing a model in this manner is clearly quite a complex and time consuming process. Also as previously discussed, the level of detail in this type of model is not necessary for the purpose of dynamic modelling for grid connectivity. A successful power system dynamic model should be able to capture the main

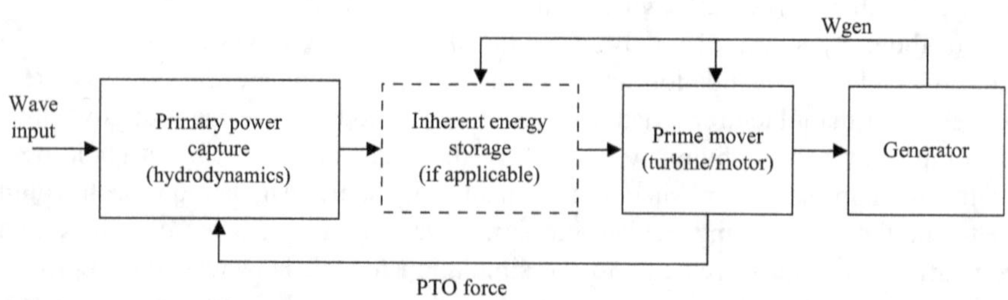

Figure 9.12 Typical wave energy device model overview

dynamics of an ocean energy device's electrical system, initially for a single device, but ultimately for an aggregated farm of devices.

Under normal conditions, the dynamics of the power transfer from the ocean to the prime mover will be dominant over the electrical power dynamics. However, under fault conditions, the electrical power dynamics change rapidly and the system must respond accordingly. The fast electrical power dynamics means that the input power from the ocean has less of an effect on the overall system dynamics during the fault period. Also, grid codes specify that faults must be dealt with within a specified time (typically three seconds) [30], either by implementation of a fault ride-through mechanism or else disconnect from the grid. Thus to model a system under fault conditions, a higher order model is required, but over a much shorter timescale.

A distinction needs to be drawn between a power system dynamic model and a design-level model. In a power system dynamic model, power output variation, protection mechanisms, and generator and prime mover dynamics need to be modelled to a reasonable level of accuracy, whereas hydrodynamic modelling is not so important due to the timescale of the modelling. The use of inherent energy storage methods in an ocean energy device, such as reservoirs or hydraulic accumulators, could also have an effect on the electrical dynamics in the event of a fault, so would need to be included in the power system model.

Tidal turbine dynamic models share a lot of similarities with wind turbine dynamic models, so the tidal sector can benefit greatly from the advances made in the wind power dynamic models. Wave energy dynamic modelling is perhaps a more complex issue, largely due to the diversity of the technology. Despite this diversity, the overall power conversion process is similar amongst many devices, as shown in Figure 9.12. Some work has been carried out to investigate the generic modelling approach for ocean energy systems, such as an outline proposed by Khan [31], an IEA-OES report [32], and SEAGRID, a dynamic modelling tool for ocean energy systems, which is currently under development at the HMRC [33]. This is an ongoing topic and is sure to become more prevalent as the technologies become closer to commercial readiness.

This is an important area for grid integration of ocean energy. Important lessons were learned through the integration of wind energy and it is hoped the ocean energy sector won't encounter the same pitfalls seen by the wind sector. To aid future power system analysis and grid planning studies for ocean energy, the ocean energy sector should strive to have good quality, validated, dynamic models prepared in anticipation of large-scale grid integration.

9.4 References

[1] EU Commission, 'The exploitation of tidal marine currents', *Report EUR16683EN*, 1996.

[2] L. Myers *et al.*, 'Simulated electrical power potential harnessed by marine current turbine arrays in the Alderney Race', *Renewable Energy*, vol. 30, pp. 1713–1731, 2005.

[3] I. G. Bryden *et al.*, 'ME1 – Marine energy extraction: Tidal resource analysis', *Renewable Energy*, vol. 31, pp. 133–139, 2006.

[4] S. E. Ben Elghali *et al.*, 'Marine tidal current electric power generation technology: State of the art and current status', in *Proceedings of IEEE IEMDC'07*, Antalya (Turkey), vol. 2, pp. 1407–1412, May 2007.

[5] E. Bossanyi, *Wind Energy Handbook*. New York: Wiley, 2000.

[6] J. S. Carlton, *Marine Propellers and Propulsion*, 2nd edn. Oxford: Butterworth-Heinemann, 2007.

[7] W. M. J. Batten *et al.*, 'Experimentally validated numerical method for the hydrodynamic design of horizontal axis tidal turbines', *Ocean Engineering* (2006), doi:10.1016/j.oceaneng.2006.04.008.

[8] W. M. J. Batten *et al.*, 'Hydrodynamics of marine current turbines', *Renewable Energy*, vol. 31, pp. 249–256, 2006.

[9] H. Glauert, *The Elements of Airfoil and Airscrew Theory*, 2nd edn. Cambridge University Press, Cambridge, England, 1959.

[10] Notice Courants: Les courants de marée des côtes de France (Manche/Atlantique)-Edition 2005-Service Hydrographique et Océanographique de la Marine (SHOM) France.

[11] H. Glauert, *The Elements of Airfoil and Airscrew Theory*, 2nd edn. Cambridge University Press, Cambridge, England, 1959.

[12] A. Blavette, D. O'Sullivan, D. Mollaghan and R. Alcorn, 'Dynamic characteristics of wave and tidal energy converters & a recommended structure for development of a generic model for grid connection', *OES-IA Document T0321*, HMRC-UCC, 2010.

[13] A. Sudrià, M. Chindris, A. Sumper, G. Gross and F. Ferrer, 'Wind turbine operation in power systems and grid connection requirements', *International Conference on Renewable Energies and Power Quality ICREPQ'05*, 16–18 March 2005, Zaragoza, Spain.

[14] Z. Lubosny, *Wind Turbine Operation in Electric Power Systems: Advanced Modeling*. Springer-Verlag New York, LLC, September 2003, ISBN: 354040340X.

[15] M.ª Inmaculada Zamora, *Simulación de sistemas eléctricos*. Pearson Prentice Hall, Madrid, Spain, 2005, ISBN: 84-205-4808-1.

[16] W. E. Cummins, The Impulse Response Function and Ship Motions. *Schiffstechnik*, vol. 9, no. 1661, pp. 101–109, 1962.

[17] J. Falnes, *Ocean Waves and Oscillating Systems: Linear Interactions Including Wave-Energy Extraction*. Cambridge University Press, Cambridge, United Kingdom, 2002.

[18] F. T. Korsmeyer, H. B. Bingham and J. N. Newman, 'TiMIT—A panel method for transient wave–body interactions', USA: Res. Lab. of Electronics, M.I.T, 1999.

[19] A. H. Clément, 'Using differential properties of the green function in seakeeping computational codes', *Proceedings of 7th International Conference on Numerical Ship Hydrodynamics*, pp. 1–15, Paris, France, 1999.

[20] P. Kundur, *Power System Stability and Control*. McGraw-Hill, United States, 1993.

[21] ECAR, 'Wind integration: International experience WP2: Review of grid codes', 2011.

[22] M. Singh, 'Dynamic models for wind turbines and wind power plants', NREL Report SR-5500-52780, October 2011.

[23] A. I. Estanqueiro, 'A dynamic wind generation model for power systems studies', *IEEE Transactions on Power Systems*, vol. 22, no. 3, pp. 920–928, 2007.

[24] J. O. G. Tande, J. Eek, E. Muljadi and O. Carlson, 'Final technical report dynamic models of wind farms for power system studies', IEA WIND Annex XXI, 2007.

[25] K. Clark, N. W. Miller and J. J. Sanchez-gasca, "Modeling of GE wind turbine-generators for grid studies," version 4.5, April 16, 2010.

[26] M. Poller and S. Achilles, 'Aggregated wind park models for analyzing power system dynamics', in *4th International Workshop on Large-Scale Integration of Wind Power and Transmission Networks for Offshore Wind Farms*, 2003, pp. 1–10.

[27] Y. Coughlan, P. Smith, A. Mullane, M. O. Malley and S. Member, 'Wind turbine modelling for power system stability analysis—a system operator perspective', *IEEE Transactions on Power Systems*, vol. 22, no. 3, pp. 929–936, 2007.

[28] P. Sørensen, B. Andresen, J. Fortmann, K. Johansen and P. Pourbeik, 'Overview, status and outline of the new IEC 61400 -27–Electrical simulation models for wind power generation', in *The 10th International Workshop on Large-Scale Integration of Wind Power into Power Systems*, 2011.

[29] M. Asmine *et al.*, 'Model validation for wind turbine generator models', *IEEE Transactions on Power Systems*, vol. 26, no. 3, pp. 1769–1782, Aug. 2011.

[30] Eirgrid, 'EirGrid Grid Code Version 3.4.' p. PCA.4.10.1, 2009.

[31] J. Khan, A. Moshref and G. Bhuyan, "A generic outline for dynamic modeling of ocean wave and tidal current energy conversion systems", Power & Energy Society General Meeting, 2009. PES '09. IEEE, vol., no., pp.1,6, 26–30 July 2009.

[32] D. Mollaghan, S. Armstrong, D. O. Sullivan, A. Blavette and R. Alcorn, 'SEAGRID: A new dynamic modelling tool for power system analysis of ocean energy devices', in *4th International Conference on Ocean Energy*, 2012.

[33] D. O' Sullivan, D. Mollaghan, A. Blavette and R. Alcorn, 'Dynamic characteristics of wave and tidal energy converters and a recommended structure for development of a generic model for grid connection', a report prepared by HMRC-UCC for the OES-IA Annex III. Available: www.iea-oceans.org [online].

Chapter 10

Economics of ocean energy electrical systems

10.1 Economic challenges and optimisation of ocean energy electrical systems

F. Sharkey

10.1.1 Introduction and components of ocean energy electrical systems

Ocean energy systems are relatively immature and there is limited experience of the costs associated with connecting large-scale arrays at present. There is therefore some uncertainty over the overall Capex of such projects. However there is a credible ambition to bring the costs of ocean energy systems in line with those of offshore wind. There are some similarities and some key differences between offshore wind farms electrical systems and those of offshore wave and tidal farms [1], and with much higher installed capacity the offshore wind industry can be used to inform the ocean energy industry. Electrical systems for offshore wind farms typically cost 20–25% of the overall system Capex [2] and the same is expected for ocean energy farms [3], perhaps a higher proportion for wave farms and lower for tidal farms. Note that this assumes that the overall costs for ocean energy reach similar levels to offshore wind. For early-stage arrays the percentage of Capex for electrical systems will be lower as the cost of the actual converters will be much higher.

This chapter aims to present the expected costs for ocean energy electrical systems and some of the major challenges faced by ocean energy in this area. The chapter also looks at techno-economic optimisation of array layouts and goes on to explore some potential strategies to reduce the cost of ocean energy electrical systems. The focus is mainly on wave energy electrical systems. However, there are several commonalities to tidal energy electrical systems and the challenges are broadly the same.

Although there are numerous wave energy converter (WEC) and tidal energy converter (TEC) types and there is some variation in the electrical collection and export concepts, marine energy converter (MEC) arrays will typically have the following components that are explained in more detail in other chapters of this book:

- Generators and balance of onboard electrical plant (power electronic converters, transformers, switchgear, etc.).
- Dynamic power cables (floating MEC only).
- Submarine connectors and other submarine electrical systems.

- Submarine power cables.
- Offshore substations.
- Onshore substations and grid connections.

There are, of course, some exceptions (such as nearshore WECs with hydraulic transmission) but the components listed above are considered 'typical' of an MEC array.

10.1.2 Expected costs for electrical system components

So if the target cost of ocean energy systems is that of offshore wind systems [4] and the proportion of that cost for electrical systems is 20–25% of the overall cost then we would expect the following costs for ocean energy electrical systems.

Ocean energy target installed costs: €4 m/MW [4]
Ocean energy electrical systems target costs: €1 m/MW (25% of the above)

Therefore all of the electrical components in the ocean energy system must cost less than €1 m/MW to be comparable to offshore wind. Although the cost of the MEC is expected to come down dramatically as the industry reaches maturity, the cost of the electrical system is predominantly mature at present as it uses mostly mature technologies. There are, however, some design criteria which will increase the electrical system costs and also some potential strategies for reducing costs which are discussed in later sections.

It is difficult to give actual costs for electrical system components as these are volatile over time and can tend to be project specific rather than generic. However, some euro figures are given in Table 10.1 for the major components which may be

Table 10.1 Typical component cost ranges for MEC arrays

Component	**Suggested cost**
Generators*	€40–70/kW [4–7]
Power converters*	€90–110/kW [6–7]
Power transformers*	€40–60/kW [6–7]
Switchgear*	€60–100/kW [6–7]
Submarine connectors	Splice housing: €60–100k per connector (est.) Dry mate connector: €100–200k per connector (est.) Wet mate connector: €150–250k per connector (est.)
Dynamic power cables – medium voltage (MV)	€300–800/m (est.)
Submarine array cables (MV)	€300–800/m [4, 8–14]
Submarine export cables – high voltage (HV)	€1000–2000/m [4, 8–14]
Offshore substation	Foundation and structure: unknown Topside: €120–150k/MW (electrical plant only)
Onshore substation and grid connection	Suggested costs: €100–250k/MW [15]

*These components are part of the MEC itself (in most cases) therefore they would be included in the MEC cost and not in the 'electrical system' costs. They are included for reference and completeness.

suitable for preliminary assessments. However, they are strictly not suitable for budgeting purpose. They are based on what could be considered 'off the shelf' components and do not cover bespoke or specialist installations.

10.1.2.1 Submarine cable cost model

The cost of submarine power cables is extremely volatile in that there are numerous factors that can affect the overall cost of the cable and its installation; namely materials cost (particularly copper and steel), mobilisation costs (significant for remote sites), seabed conditions (affecting installation method), downtime (determined by prevalent weather) and availability of equipment (determined by market demand). Therefore, it is difficult to put a euro price on cables that will remain relevant across all projects which can be seen by the range shown above in Table 10.1. Another approach is to look at the factors that make up the installed price of a cable and develop a normalised cost model that will be valid with all else being equal in the cost of cables and installation methods across a particular project. This method disregards contract strategies such as bulk purchasing or multi-project, which are not possible to model.

By looking at the elements of each factor of the cable cost a normalised cost model can be established. The main factors affecting the cable cost are:

1. The voltage rating of the cable (i.e. the insulation rating/thickness).
2. The cross-sectional area (CSA) of the conductor.
3. The installation costs.

For simplicity we will assume three core cross-linked polyethylene (XLPE) cables with copper conductors and a single layer of armouring for all cases as these are common cables in the offshore wind industry.

As this is a normalised cost model a base case is required to normalise against. The base case will be a 10 kV, $1 \times 3 \times 95\,\text{mm}^2$ cable. This cable will have an installed normalised cost of 1.0 and all other cables will be represented as a multiple of this. The cost model was developed primarily using the formulae given by Lundberg [8] and also verified by comparing against numerous sources such as [9–14]. The developed normalised costs are shown in Table 10.2.

For example, a 33 kV, $240\,\text{mm}^2$ cable is 58% (1.58/1.0) more expensive than the base 10 kV, $95\,\text{mm}^2$ cable. Also a 20 kV, $500\,\text{mm}^2$ cable is 165% (2.25/0.85) more expensive than a 20 kV, $50\,\text{mm}^2$ cable.

The cost model presented in Table 10.2 will be used for analysis of electrical network options throughout this chapter.

10.1.3 Economic challenges for ocean energy electrical systems

There are a number of economic challenges for ocean energy electrical systems, which are discussed and analysed in more detail below. Building a business case for early stage arrays will be challenging and it is likely that there will be pressure to reduce the costs of elements such as the electrical system. Therefore, there will be a drive towards the least cost option. However designers must be wary to not compromise critical safety and functionality issues with this pressure for minimising cost.

Table 10.2 Installed cable normalised costs

Cable CSA (mm^2)	Voltage			
	10 kV	**20 kV**	**33 kV**	**132 kV**
35	0.79	0.82	0.85	–
50	0.81	0.85	0.88	–
70	0.85	0.89	0.94	–
95	1.00	1.05	1.11	–
120	1.05	1.11	1.18	–
150	1.10	1.17	1.25	–
185	1.25	1.34	1.43	–
240	1.35	1.46	1.58	–
300	1.65	1.80	1.97	–
400	1.80	1.99	2.21	2.79
500	2.00	2.25	2.53	3.25
630	2.25	2.55	2.89	3.75

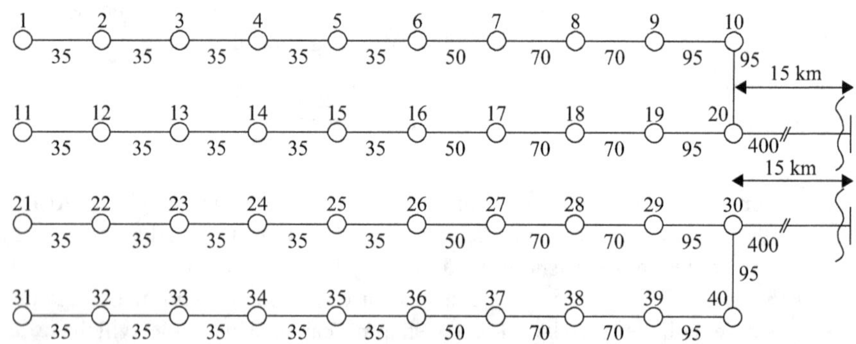

Figure 10.1 Candidate, 40 MW, array

A 'medium' size, 40 MW, WEC array is taken from [1] as the candidate array (Figure 10.1). This candidate array has the following, base case, assumptions:

- Each MEC (node) is rated at 1 MW with unity power factor.
- Each MEC has a 30% capacity factor.
- The inter-MEC spacing is 400 m (inter-MEC cables are 400 m + twice the depth).
- The water depth is 100 m.
- The export distance is 15 km.

This will be used in conjunction with the normalised cable cost model given in the last section in order for an economic analysis to be undertaken.

10.1.3.1 Individual MEC ratings

At the current stage of the industry's maturity there is a trend, both in wave and tidal sectors, towards devices with ratings of ~1 MW. Individual device ratings of 1 MW are, therefore, used as the base case in any analysis done. There are of course a number of exceptions to this trend. Offshore wind turbines are mostly rated

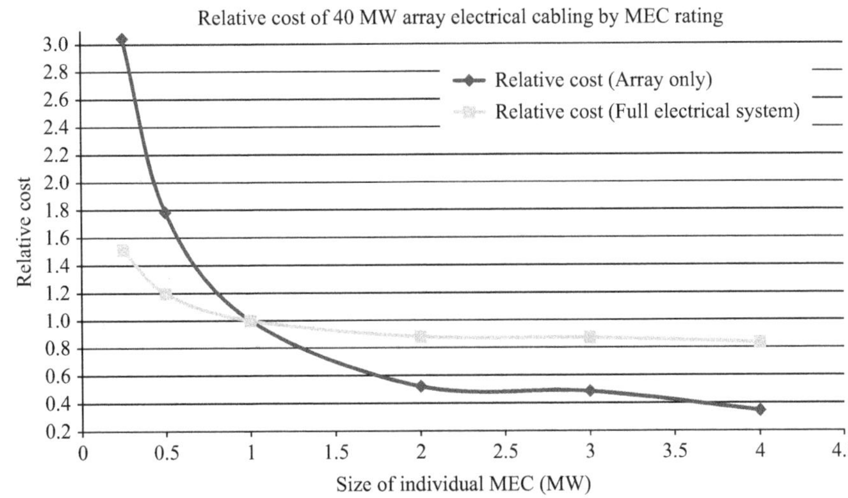

Figure 10.2 Relative cost of 40 MW array electrical cabling based on device rating

around 3–4 MW with a trend towards higher power units (5 MW+). These small MEC unit sizes will present a challenge to the economics of ocean energy electrical systems as each device in an array will require dynamic cables (floating MEC), submarine connectors, and a cable connection to the next device in the array. Naturally the more devices in the array will mean additional cost for the array, certainly on a per MW level.

Just the cost of the dynamic and static submarine cables will be evaluated here. The relative cost of the array (versus the base case) is established for a 40 MW array with 250 kW, 500 kW, 1 MW (base case), 2 MW and 4 MW individual MEC ratings. The total rating of the array remains at 40 MW in all cases, i.e. the quantity of MECs changes depending on the MEC rating. The array and export voltage is also 20 kV in all cases.

The relative cost as a percentage of the base case is shown in Figure 10.2. The relative cost is shown for the array only and the full electrical system (i.e. array and export cable). This shows that as expected the relative cost is higher for smaller devices and lower for larger devices. The increase can be as much as 3 times for the array cable costs. It should be noted that the costs do not decrease as much or as exponentially for larger individual devices with decreases to as low as 0.4 times possible for the array cable costs.

The focus here is on the electrical system only; however, it is worth noting that lower MEC ratings will increase other elements of Capex such as installation, moorings, etc.

10.1.3.2 Device capacity factor

The capacity factor of offshore wind turbines is typically in the region of 30–40% [16] depending on turbine type, location, yearly wind speed, etc. Given the variety of wave and tidal energy devices available it is unclear what capacity factors these

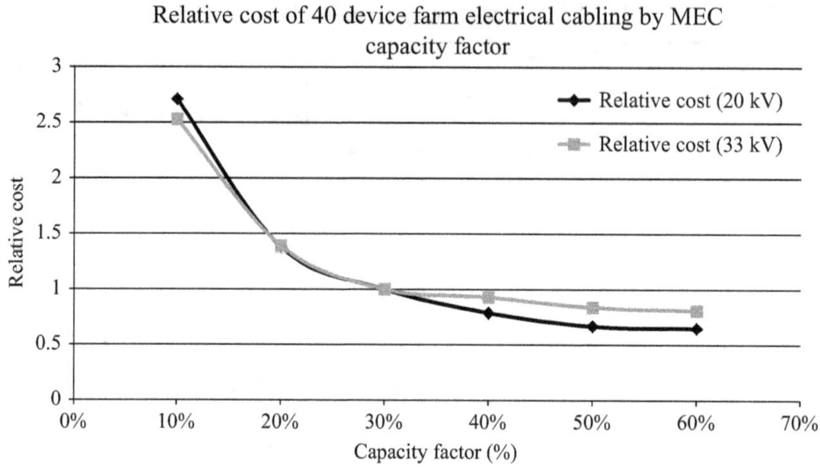

Figure 10.3 Relative cost of 40 device array electrical cabling based on device capacity factor

devices will have. For 'direct drive' wave energy converters the capacity factor could be very low, <20%, due to a high peak to average output ratio. Conversely some tidal turbines may achieve capacity factors of over 60% at high energy sites.

The relative cost of the array electrical network (versus the base case) is established for the candidate array with capacity factors of 10%, 20%, 30% (base case), 40%, 50% and 60%. The overall average output of the array remains at 12 MW (base case 40 MW × 30%) in all cases but the peak power output changes with the capacity factor.

The relative cost as a percentage of the base case is shown in Figure 10.3. The relative cost is shown for the full electrical system only (i.e. array and export cable). This is because capacity factor effects both array and export systems. The relative cost is assessed at two voltage levels (20 kV and 33 kV). This shows that as expected the relative cost of the electrical network is higher for devices with lower capacity factor and lower for device with higher capacity factor. Halving the capacity factor from 30% to 15% would almost double the cost of the electrical network. Doubling the capacity factor form 30% to 60% would decrease the costs by up to 40%.

10.1.3.3 Submarine connectors and other submarine electrical systems

In offshore wind farms the cables are routed, through J-tubes, straight into the turbine tower. This is not the case with ocean energy arrays as the devices are required to be removed for maintenance on a regular basis. This presents a number of issues, including redundancy in the electrical network, which is discussed in the next section. For floating MECs there will be a connection required between the dynamic cable and the static cable. In some cases there will be a requirement for the device to be quickly and repeatedly connected and disconnected from the electrical network, although more so at prototype stage. Therefore some type of connector is required. These connectors are discussed in Chapter 3.

However, as these connectors are a requirement for ocean energy electrical systems, which does not exist in offshore wind, they will naturally add to the overall cost of the system. In some cases, where a radial circuit it used (see next section), there will be a requirement of two connectors per device. As mentioned in section 10.1.2 the electrical system will need to cost less than €1 m/MW. Also in section 10.1.2 it is shown that electrical connectors could cost anywhere from €60k to 250k per installed connector. Naturally if we do not want to exceed the threshold of €1 m/MW then 2 × €250k per connector is not feasible, i.e. €0.5 m/device on connectors alone. So, although wet-mate connectors may increase the functionality of the device, they may be unfeasibly due to cost in the medium term. Either way submarine connectors will be required and it is a matter of trying to balance the cost with the functionality of the connector. This is explored further in the next section.

There are other potential submarine electrical systems which could be utilised in MEC arrays. These could be simple junction boxes (such as 'Wavehub'), submarine switchgear modules (such as OPT's undersea substation pod), or more complicated 'submarine substations' in place of platform based substations. It is not clearly understood what the potential costs of these components would be; however, these would on the whole represent additional costs over the traditional offshore wind farm electrical system. It is highly likely that such systems would be expensive to build, install and maintain and costs would be in the millions of euro. There are also practical functional and safety concerns with such systems. Submarine electrical systems have been successfully installed in deepwater oil and gas fields; however, the economics is not comparable to ocean energy, and so these systems may not be suitable.

10.1.3.4 Cable installation and protection

The cable costs given in section 10.1.2.1 are based on the assumption that a standard installation method can be used, ploughing or jetting the cable into the seabed sediment. In truth every site is different but the seabed is predominantly rocky along the western seaboard of the United Kingdom and Ireland, and areas of high tidal flows are likely to be swept clear of most sediment. These conditions present extremely challenging cable laying conditions and expensive installation and protection methods must be used such as rock trenching, rock dumping, armour casings, concrete mattresses or horizontal directional drilling [17]. These methods could more than double the cost of the cable installation and so are huge challenges to the sector. A high number of installed wave and tidal facilities have required these measures such as:

- EMEC (armour casings and concrete mattresses).
- Wavehub (rock dumping).
- MCT SeaGen (horizontal directional drilling).

The careful selection of sites with sufficient sediment may allow the avoidance of expensive cable installation methods and this may go hand in hand with mooring requirements for wave energy arrays.

There is also a challenge in the protection of dynamic power cables as this will require numerous additional components such as bend restrictors, stress relievers, floatation module and scour protection. Again this will add to the cost of the electrical system; however, this is expected to be relatively modest.

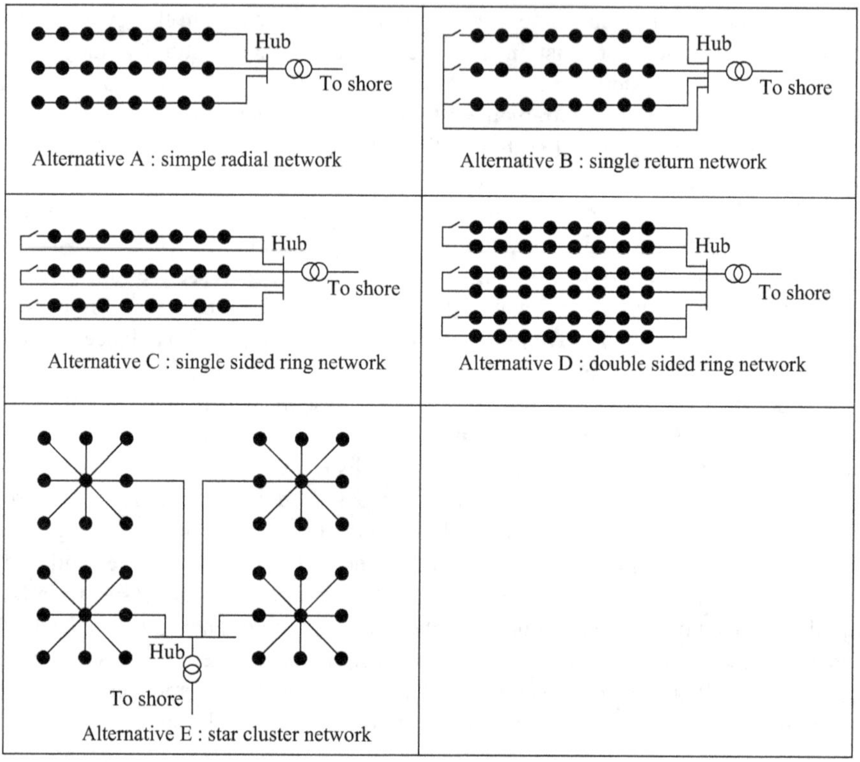

Figure 10.4 Alternative network configurations

10.1.4 Techno-economic optimisation of ocean energy electrical systems

This section examines at some of the issues above and considers the optimal electrical network configuration for an MEC array based on the technical and economic aspects.

10.1.4.1 Optimal array electrical configuration

One major factor in the cost and functionality of the electrical system is the configuration of the MEC array electrical network. There are a variety of alternative configurations as shown in Figure 10.4. For MEC arrays some proposals have been made for submarine 'hubs', which could act as an aggregation point in a star network. These are discussed further in section 10.1.4.3.

We can evaluate the candidate wave farm using the alternative configurations as shown in Figure 10.4 under a number of criteria.

The following assumptions are made in addition to those shown in section 10.1.3:

- The physical grid layout of the devices is assumed to be maintained at all times, for all configurations.

Table 10.3 Cost of alternative array network configurations

Network configuration	**Relative cost (array only)**	**Relative cost (array and export)**
Radial network (A)	1.0	1.0
Single return ring network (B)	2.58	1.39
Single sided ring network (C)	1.8	1.2
Double sided ring network (D)	1.69	1.17
Star cluster network (E)	1.54	1.13

- Redundant circuits are assumed to be rated for worst case full load, i.e. they are 100% redundant.
- No bespoke equipment such as submarine switchgear is considered at this stage and all switching operations are assumed to be contained within the MEC or in the onshore substation.

Cost (Relative to (A))
Table 10.3 shows the relative cost of the array only, and the array and export cabling for the various alternative configurations detailed in Figure 10.4. This shows that the radial network is the least cost solution from an array configuration perspective. This is primarily due to additional cabling required for the proposed alternatives. Also, in order to allow redundancy and bidirectionality in some of the circuits, the cross-sectional area (CSA) of some of the cables must be increased, thus increasing cost. The Star Cluster Network (E) shows an relative cost of 1.54 with the existing physical grid layout. It is expected that with optimisation of the Star Cluster Network this could be on par with the radial network for cost; however, the electrical cost may only be one optimisation factor for the selection of physical layout.

Installation
The radial network would allow the simplest installation with multiple short cable runs. The installation process for the alternative array configurations may be more complex involving additional and longer cable runs and possible cable crossings.

Operation
The radial circuit has no redundancy in the array network meaning that in the event of a fault during normal operations all upstream MECs in the circuit will be disconnected from the system. All of the alternatives offer some level of redundancy in the circuit which has been shown to increase availability of the overall array [18].

Maintenance
A unique characteristic of deepwater wave farms is that individual WECs will require removal for routine and non-routine maintenance. Similar to the comments in 'Operation' above a radial circuit would have no redundant circuit. The alternative configurations would be more suitable to overcome this but there are solutions to overcome the lack of redundancy in radial circuits. These solutions are discussed below.

Isolation and Protection

How the individual MECs and array cables are isolated is an important consideration for safe operation of a MEC array. The operation of a radial circuit is well understood where any MEC or cable can be simply isolated by switching out the connection at either side. More complicated switchgear and isolation systems may be required for the alternative networks.

What can be concluded from the above discussion is that the simple radial network appears to be the most advantageous in terms of cost; however, the radial network is less suitable where redundancy is required. In reality, as shown in section 10.1.2, the cost of the electrical system would need to be kept as low as possible, therefore any other technical or functional considerations may not be valid. Thus radial networks are selected here as the most suitable array network configuration for MEC arrays.

This has proven the case with offshore wind farms, with radial networks being used in all offshore wind farm array configurations and few wind farms having any redundancy in the electrical system. However, with offshore MEC arrays we have the issue of removal of MECs in the circuit which needs to be resolved. This can be done with a number of options including:

1. 'Standby' or 'dummy' MECs to 'slot' into place.
2. A system for temporarily 'bridging' the gap left by the MEC in the electrical circuit.
3. Submarine switchgear allowing continued operation of the infield circuit (see next section).

It is likely that that option 2 here would be the least cost solution to this issue.

10.1.4.2 Key electrical interfaces

If the MEC array network configuration is to be a radial network then the key interfaces between the MEC and the radial network need to be optimised. This means achieving a balance between the functionality of these interfaces and cost.

These key interfaces are categorised as:

1. Dynamic cable to MEC interface.
2. Dynamic cable to static cable interface.
3. MEC MV switchgear interface.
4. Offshore substation.

There is certain functionality required at the key interfaces between the electrical system and the MECs including the following:

- Multiple connection/disconnection of the MEC.
- Initial cable installation.
- Electrical protection.
- Electrical isolation (and earthing).
- Cable hull penetration.
- Circuit continuity (i.e. redundancy).

Although the maximum functionality in the key electrical interfaces would be desirable, the cost of the key interfaces must also be minimised. The expected costs may limit the functionality that can be viably achieved in the key interfaces. The key interfaces are not outlined in any more detail here but a techno-economic optimisation can be found in [19].

10.1.4.3 Other bespoke solutions

The focus here has been on offshore MEC arrays with radial array networks. Other bespoke solutions have been proposed which all fall into a general category of submarine 'hubs' utilising star cluster type network configurations.

These hubs in general collect the generated power from several MECs and condition it for transmission to shore. These hubs can contain one or all of the below equipment:

- Power electronic converters.
- Low voltage (LV) & MV switchgear.
- Power transformers.
- Energy storage solutions.
- Battery chargers and auxiliary systems.

Although these are not explored in detail here there are several major challenges that must be overcome in order to make these types of solutions viable. They are the same challenges that apply to larger submarine offshore substations. These challenges are outlined here for information only:

- Access to complicated equipment such as power electronic converters, digital protection relays, battery chargers etc. would be required in the event of even a simple fault. This operation alone would be a huge cost.
- There are safety implications with having a point of isolation and earthing in a location where it cannot be verified or locked out.
- The practicalities of connecting multiple LV and MV cables to a submarine hub are onerous. This would require multiple expensive mate-able connector and/or remotely operated vehicles (ROV) operations.
- The potential construction and installation costs of a submarine hub are very large and there is little experience here apart from the oil and gas industry.
- There are other, less technically and economically challenging options for electrical connection schemes which should be explored first.

10.1.5 Cost reduction of ocean energy electrical systems

The purpose of this section is to explore strategies to reduce the Capex of the electrical network of MEC arrays, i.e. to maximise the value of the electrical network asset with particular emphasis on the cabling system. This in turn will reduce the overall Capex of MEC arrays. As shown in section 10.1.3 increasing MEC capacity factor or increasing the unit rating will reduce the electrical system Capex.

There are a number of other strategies which are explored in this section in order to achieve this increase in the value from the MEC array electrical network.

1. Less than 100% rating based on statistical data.
   - Based on the idea that a MEC arrays rarely output 100%.
2. Dynamic rating based on environmental data.
   - Based on the idea that cable rating change depending on the environmental conditions.
3. Dynamic rating based on real-time measurement.
   - Based on the idea that actual cable ratings can be measured in real time.

These strategies are outlined briefly below and more detailed analysis can be found in [20].

10.1.5.1 Less than 100% rating based on statistical data

It could be assumed that an array of MECs would rarely reach 100% output based on resource availability and MEC reliability. This leads to the possibility that the electrical export system could be rated at less than 100% of 'nameplate' rating. In this case the rating will mean that the cable is under-rated when the MECs do reach maximum output simultaneously, leading to either output curtailment or a combination of one of the other techniques described in this section. However, any loss in energy may be offset by the savings gained from using a lower rated cable.

A small MEC array is modelled so that the effect of <100% rating of the cabling can be evaluated. The proportion of time (and generated energy) when the cable limits are exceeded is calculated. The effect on the annual energy yield of the array can be established and it can be seen whether this is offset by the savings in the Capex of the electrical network.

A small array of devices is examined to assess the possibility of lowering the rating of some of the cables thus realising cost savings. For simplicity a 5 WEC array is considered here. It should be noted that, unlike the candidate wave farm (see Figure 10.1), the physical spatial arrangement of the devices is considered here. All WECs are considered identical and interference between WECs, either destructive or constructive, is not taken into account. Interference is an area of significant interest to the wave energy industry; however, it is not considered to be sufficiently developed to be included in this study.

In order to avoid simultaneous operation the array layout is staggered so that some devices will be out of phase with others regardless of the angle of incidence. This means that the 5 WECs may not react simultaneously to the oncoming wavefront, although there may be a combination of wave period and approach angle that allows this to occur. This array is shown in Figure 10.5.

Focusing on the export cable only (5-Grid), reducing the cable CSA from 95 to 70 mm^2 would reduce the export capacity from 5 to 4.15 MVA or 83% of the rated array output. From the normalised cost model in section 10.1.2.1 this will give a saving of 15% for the export cable. The time series output from the five devices is assessed to see if or when the overall output exceeds 4.15 MVA. This will allow a cost benefit analysis to be carried out to see if the potential savings outweigh the possible loss of annual energy from the array.

This showed that the 5-WEC array reached 100% output (5 MVA) for 3.2% of the year. The output was >83% (i.e. >4.15 MVA) for 6.2% of the year and this

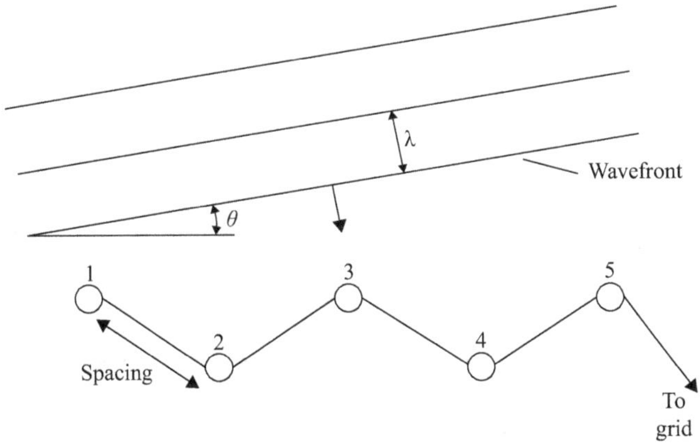

Figure 10.5 Concept of array for analysis (θ = angle of incidence, λ = wavelength)

contributed to 2.98% of the overall annual energy production. So if the export cable capacity was reduced from 5 to 4.15 MVA a saving of 15% of the cable cost would be made; however, this would result in a 2.98% energy curtailment per annum. A breakeven analysis will show if this curtailment can be justified based on the proposed saving.

10.1.5.2 Dynamic rating based on environmental data

The current carrying capacity (ampacity) of power cables is calculated according to IEC60287 [21]. The maximum permissible continuous current is based on the maximum conductor operating temperature as defined by the cable manufacturer. For XLPE insulated cables this temperature is typically 90 °C. The cable must dissipate heat during normal operation so the maximum permissible current is calculated based on the thermal properties of all of the components of the cable (insulation, screens, sheaths, filler, armour, and serving), the cable geometry and the thermal properties of the surroundings.

The current ratings given in submarine cable specifications use assumed values for the ambient conditions and surroundings such as those given below:

- Ambient temperature of 20 °C.
- Sheaths bonded at both ends and earthed.
- Burial depth of 1 m.
- Thermal resistivity of surroundings of 1 Km/W.

The ambient temperature, burial depth and thermal resistivity of the surroundings are somewhat within the control of the designer. These vary over time and over the length of the cable route. Therefore the maximum permissible current will vary also.

By focusing on our candidate wave farm (see Figure 10.1) and in particular the export cables which would be 400 mm^2 for 20 kV and 150 mm^2 for 33 kV, we can evaluate the effect of lowering the cable CSA.

Focusing on the west coast of Ireland, Figure 10.6 shows that the seawater temperature varies seasonally from approximately 6–15 °C. Also the air temperature for the land based portion of the cable is important and this is shown in Figure 10.7 and varies seasonally from approximately 3–17 °C although with some extremes. This implies that the cable ampacity will vary throughout the year due to ambient temperatures.

It is assumed for this analysis that the worst thermal resistivity along the route is 1.0 Km/W and that the burial depth is 1.0 m along the entire cable route. From this information we can show the available and required ampacity across the year for the selected cable and the next lowest cable CSA. The air temperature is used for the calculation as it has higher extremes than the seawater temperature and the land section of the submarine cable would be expected to be a 'bottleneck' as a result.

Figure 10.8 shows the results of the seasonal adjustment for a 20 kV system. Based on the adjustment of the seasonal temperatures alone we can show that a 300 mm^2 cable is more suitable for this application. The output of the array almost reaches the ampacity limit in the summer months; however, this is only when the output of the array is 100%. Thus by understanding the environmental data the cable CSA may be decreased from that using the assumed values.

For the 20 kV array the reduction in cost of the export cable by reducing the cable from 400 to 300 mm^2 would be approximately 10%. This saving only considers the export cable. Further savings to the overall electrical system costs could

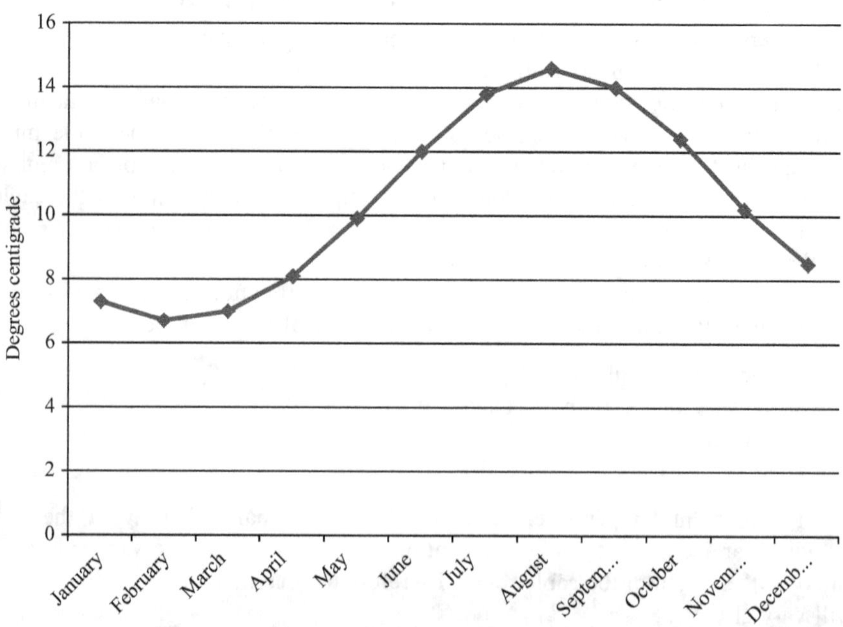

Figure 10.6 Average monthly seawater temperature at Malin Head 1961–1990 [Source: Met Eireann]

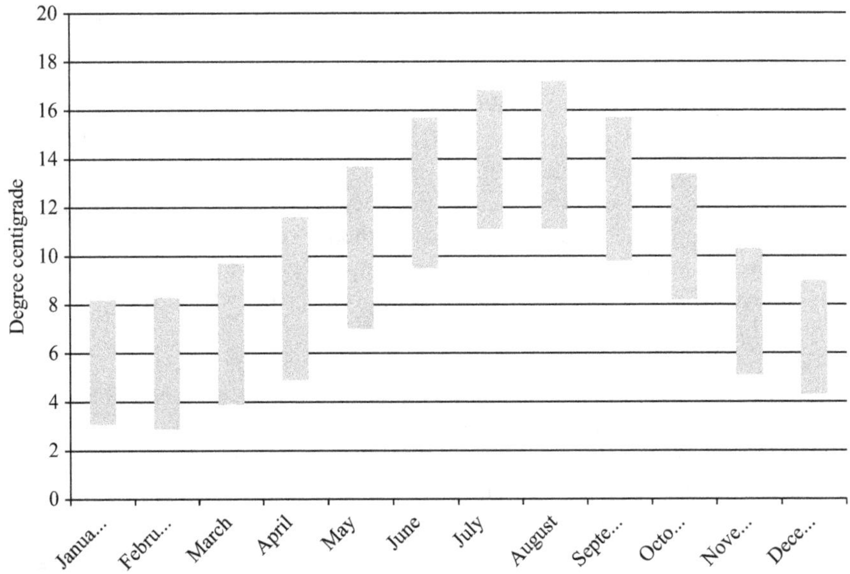

Figure 10.7 Average monthly air temperature range at Belmullet 1961–1990 [Source: Met Eireann]

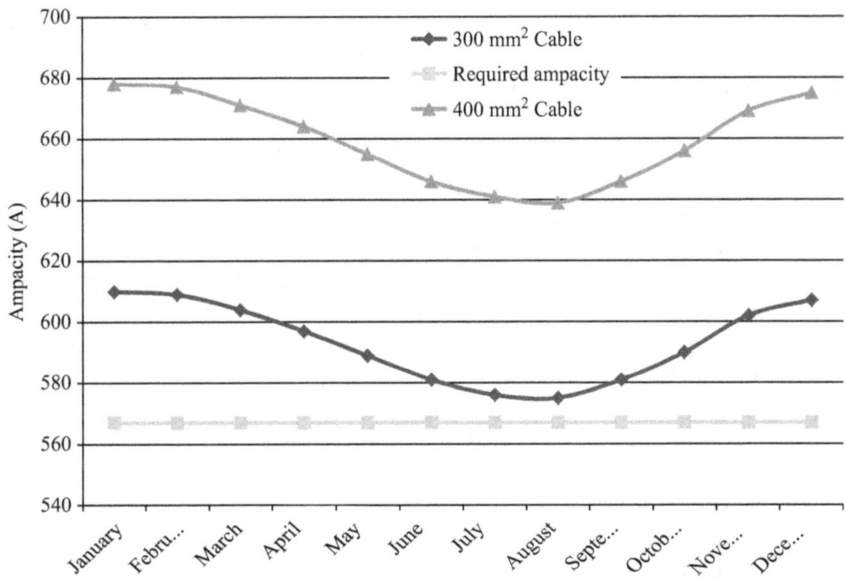

Figure 10.8 Seasonal ampacity of 20 kV cables

be made by reducing the inter-array cables CSA, particularly those nearest the export side, using the same method.

10.1.5.3 Dynamic rating based on real-time measurement

Dynamic or real-time thermal rating (RTTR) systems have been developed in order to utilise the 'headroom' available in transmission assets to increase the capacity at a given location. These systems monitor the environmental conditions (such as temperature, humidity, etc.) and/or measure/model the temperature of the conductors themselves to allow dynamic constraints to be set on the system. This has been shown to allow 10–30% increased capacity over the static thermal rating of overhead lines [22].

To date this has been utilised successfully, with varying levels of complexity, on transmission systems in a number of countries. It has also been utilised for offshore wind farm export cables [23].

These measurement technologies ensure that an accurate figure of the cable ampacity is maintained at all times thus allowing the cable asset to be utilised to its actual full permissible rating when required. Similar to the above methodology in the previous section, this would give greater accuracy and confidence regarding the actual maximum current rating at any given time.

The methodology in the previous section above carries a certain amount of risk as there may be times when the air temperature is significantly higher than the average for a given month. Therefore, the system is normally designed for extremes to introduce a factor of safety.

In order to remove this risk real-time measurement may be utilised to ensure that the ampacity of the cable is calculated in real time and the cable is never at risk of becoming overloaded. This can be done by simply measuring the ambient temperatures at several locations along the route and using a model of the cable to calculate ampacity. However, this does not give actual real-time data about the conductor temperature and simply gives a calculated ampacity at a given time. More complex distributed temperature sensing (DTS) systems which measure the actual temperature of the conductor across the entire cable route will allow a very high degree of certainty in the loading at a given time.

DTS systems can use fibre optic technology which through a combination of back scattered light intensity and time domain reflectometry can measure the temperature to one metre resolutions in cables up to 30 km in length [23–24]. This can give a temperature profile of the entire length of the cable thus allowing accurate loading of the cable, i.e. accurate dynamic ampacity ratings, and identification of hotspots along the route. While the DTS fibre optic cable can be installed after cable manufacture, it is preferable to install the sensing cable during manufacture as this will improve response time and make the system integral to the power cable.

Such a real-time system would allow the operator to use the strategies given in this chapter with full confidence that the power cable asset will be maintained within safe limits. It also means that any output curtailment will be kept to an absolute minimum. Naturally such a system will increase the costs of the installation but this would be expected to be a marginal increase.

10.2 Ocean energy system economics and cost of electricity

G. Dalton

10.2.1 Introduction

Energy economics is an established theory for conventional fossil fuels, and more recently for onshore wind. The wave and tidal sector is a recent emerging sector. There are no commercial units to date; therefore, solid economic and commercial figures are not yet available. Therefore, much of the current economic analysis and data used for feasibility analysis for these sectors are bases on similar experiences from offshore wind as well as extrapolating from early pre-commercial to commercial costs.

One of the largest gaps in knowledge is therefore conjecturing what will be the Capex of wave energy, as well as the costs per MW, as cost of electricity (€/kWh). This chapter will assess the cost drivers for various components of the wave energy sector, and provide referenced articles on current research into what will be the cost breakdown of these devices.

Commercial viability will depend on commercial scale farms. The theory of learning and bulk discount rates will be discussed as well as supply and demand bottle necks that might arise. Economic feasibility will also depend on viable O/M rates as well as revenue support mechanisms. These mechanisms will be discussed as well as the problem of their changing dynamics, the fact of which substantially adds risk to the sector.

10.2.2 Capex

10.2.2.1 Capex cost breakdown

The percentage cost of components and procedures is an important baseline that must be established at the beginning of a project. As wave energy is a new and diverse industry field, percentage cost breakdowns vary substantially. An example of two breakdowns is presented in Figures 10.9 and 10.10.

Price of steel

Steel is currently the main material constituent of a WEC, and thus has the largest influence on initial cost (IC). Steel has had major price fluctuations over the past few years (Figure 10.11). Recent factors influencing steel price fluctuation were increasing demand from China for raw materials, which led to a price escalation [27]. The price peaked in 2008, and has since stabilised back to what prices were in 2007. It is anticipated that the prices will start to rise again once the recent recession of 2008–2013/2014 is over. The final cost of manufactured steel, typically grade 50 (S355), painted with corrosive protection, can cost anywhere from €5000 to 7000/ton.[1] The example of this analysis can be performed for any material e.g. concrete. Most current generations of devices, however, have been of steel fabrication.

[1]Personal communication Paul Collins, Malacky Walsh Engineers, Cork.

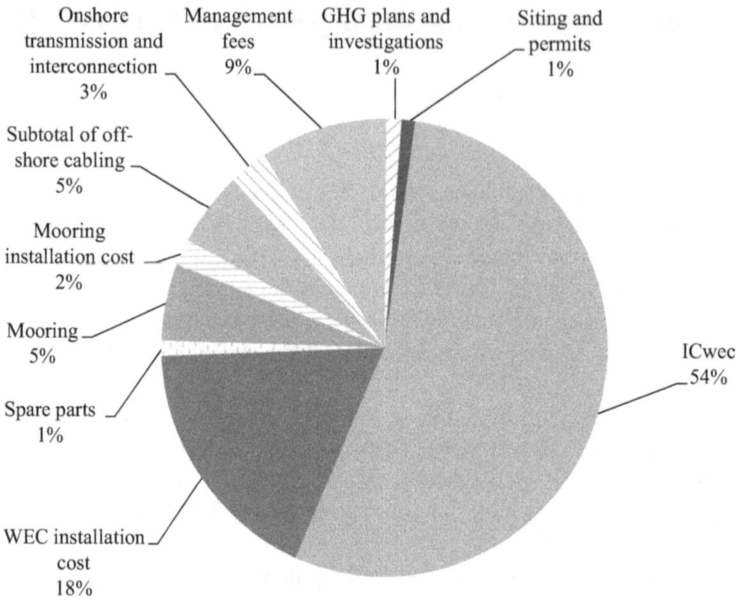

Figure 10.9 Capex breakdown of a sample wave energy farm project – taken from a Pelamis study by O'Connor and Dalton [25]

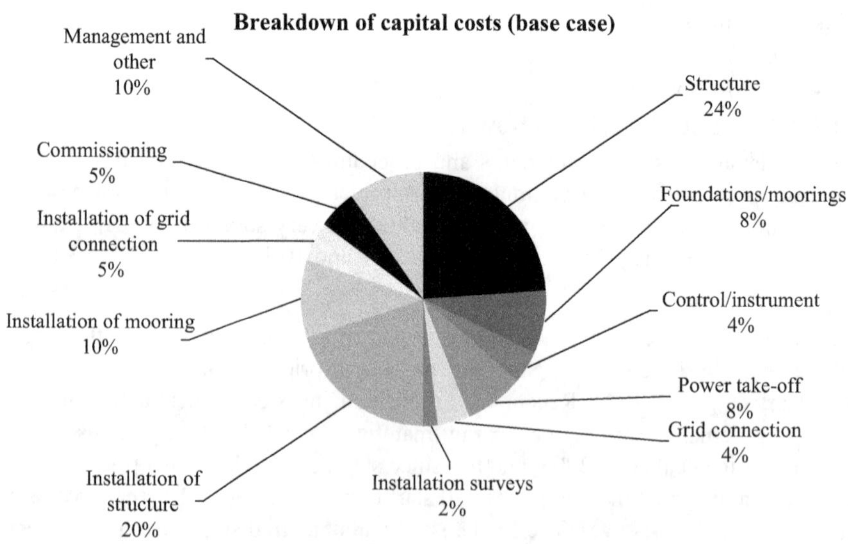

Figure 10.10 Capital cost breakdown [26]

Cost per kW or Capex per kW

The cost per kW is an easy costing method to compare prices of different technologies. A review of cost/kW for offshore wind and wave is presented in Table 10.4. Cost/kW usually decreases due to economies of scale as discussed later.

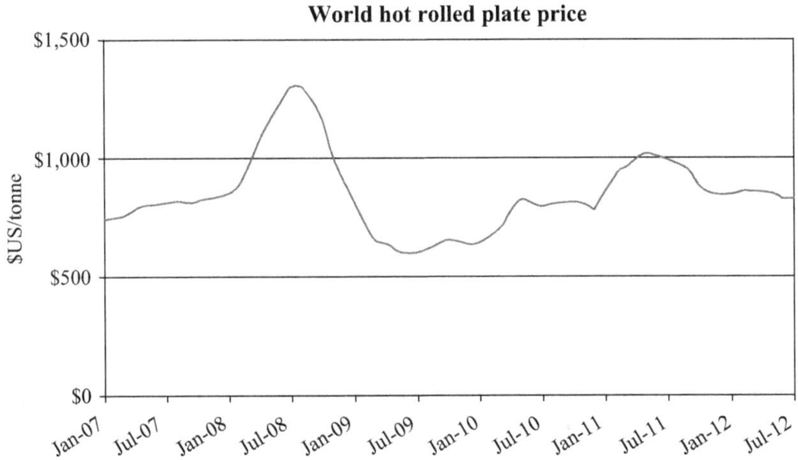

Figure 10.11 World hot rolled plate price 2007–2012 (supplied by MEPS steel www.meps.co.uk/)

Table 10.4 Cost per kW for various offshore wind and wave energy studies. (€1 = $1.4US, €1 = £0.82)

Technology	Author	Reference	Turbine or farm size	Euros/kW
Offshore wind	Snyder and Kaiser	[28]	1–2 MW turbines	€1500– €3000
			2–5 MW turbines	€2000–€3000
			1–50 MW farms	€1500–€3000
			50–200 MW farms	€2000–€3000
	Fingersh *et al.*	[31]	3 MW	€1500 ($2100)
	DETI (quoted in SEI)	[32]	Not quoted	€1400–€2000
	Horns Reef (quoted in SEI)		160 MW	€1700
	Barthelmie *et al.*	[33]	Not quoted	€1650
	Luypaert *et al.*	[34]	Not quoted	€2500
Wave energy				
	Previsic (calculated from report by the author)	[29]	1 MW	€5350 ($7500)
			105 MW farm	€1900 ($2600)
	Carbon Trust	[35]	Commercial	€1400–€3500
	Weiss *et al.*	[36]	90 MW	€1800 ($2600)
	Dunnett and Wallace	[37]	Not quoted	€2500 ($3500)
	ARUP	[38]	Not quoted	£7,300–£10,300

However, this is not always the case, as Snyder and Kaiser [28] observe where sometimes costs are greater for larger farms than for smaller ones despite bulk discount rates. Explanation of this trend is discussed later.

Costs reported for wave energy farms vary substantially with size, ranging from a high of ~€5300/kW [29] for small wave farm developments to €1400 for larger ones [30]. It must be noted that the €5400/kW cost is from 2004, and costs may have increased substantially since that time.

Quotes for cost per kW or MW can be often misrepresented and used without qualification or quantification. The following are important factors need specification:

- Year of the quote, as inflation, cost on materials and supply/demand have major influence on annual fluctuations in prices.
- Definition of objects quoted for i.e. device only, device plus installation or full project Capex including installation.
- Number of object quoted i.e. size of the farm, which is influenced by bulk discount rates.

Table 10.5 displays an example of the variation in €/kW costs dependent on the factors just discussed. In this example, €/kW results are much higher for 2010 than for 2004, due mostly to increase in the cost of materials. There is almost double the €/kW cost for device-only as compared to a quotation for a whole project cost, emphasising the necessity for clarity in definition of objects quoted. There is a 30% drop in €/MW costs between 1 and 20 MW farm, demonstrating the bulk purchase impact on Capex.

10.2.2.2 Installation costs

There are four common cited methods for reporting the costs as part of a project:

1. As a percentage of the total project cost (or total project Capex).
2. As a percentage of the device initial cost.
3. €/MW.
4. €/MWh.

Costing methods 3 and 4 are the most useful from a project cost analysis point of view, as these figures can be used independently of the total project. However, costing methods 1 and 2 are the most common and rely on the total project cost or the device cost being available for the relevant figure to be calculated. Table 10.6 provides a review of the literature of costing method 1, showing a wide variation on percentages, varying from approximately 20% to 35% of the total project costs.

Table 10.5 The initial cost (IC) of the device only and the total project cost or Capex for a 1 MW project and 20 MW project [39]. 2010 costs are based on a factor of 2, based on the doubling of the cost of steel in that time period

		2004	**2010**
1 MW farm	WEC only IC	€2100/kW[a]	€4300/kW
	Total project IC or Capex	€4000/kW	€8000/kW
20 MW farm	20 WEC only IC	€1500/kW	€3100/kW
	Total Project IC or Capex	€3000/kW	€5800/kW

[a]Based on €1,600,000 for 0.75 MW device [29].

Table 10.6 Installation costs as a percentage of total Capex

	Reference	Installation as a % of total project Capex
Garrad Hassan	[40]	19% (5% WEC, 7.5% foundation, 6.5% electrical)
Carbon Trust	[26]	35% (WEC 20%, mooring 10%, cable 5%)
Walker *et al.*	[41]	30%
Beyene and MacPhee	[42]	25%
O'Connor *et al.*	[43]	2%

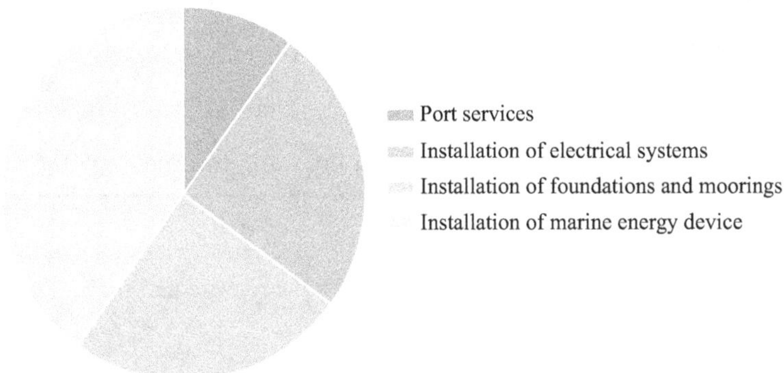

Figure 10.12 Types of installation as a % of total installation procedures [44]

Installation can be further broken down into the installation costs for various components as well as the port services. An example of the split is demonstrated in Figure 10.12, demonstrating an even split between the subgroups.

10.2.2.3 Learning rates

Learning rates refer to the projected cost reduction of a product based on experience over time. For each doubling of cumulative installed capacity, costs fall to a percentage of those in the reference year by a factor defined as the progress ratio. In general, progress ratios in the range 85–90% have been applied to the cost of energy from wave energy systems. Learning rates are not used where there is a project which has one-off Capex purchase and installation based in the first year. The term 'learning curve' has been defined by Neij [45] as the cost reduction of a standardised product within a single firm, while an 'experience curve' may also describe cost reductions of non-standardised products on a national or a global level. He observed a number of problems with using experience curves for all sectors of an industry e.g. for wind turbines, wind farms, wind electricity, which cannot be compared directly.

Table 10.7 presents a summary of the publications in this area. Both Batten and Bahaj [46] and Gross *et al.* [47] concluded that 'learning curves cannot be used to assess wave and tidal energy as there is little or no market experience and hence no data' and 'to apply such an approach for the immature ocean energy industry with varied technologies would be little more than guess work'.

Table 10.7 Literature review of learning curve rates

Author	Reference	Technology	Learning curves rate
IEA	[48]	PV/wind	82%/96%
McDonald and Schrattenholzer	[49]	Wind	68–96%
Junginger *et al.*	[50]	Offshore wind	23%[a]
Junginger *et al.*	[51]	Wind (review of 20 studies)	88–90%
Batten and Bahaj	[46]	Wave	85–90%
Carbon Trust	[35]	Wave	Scenario A 90%, Scenario B 85%
Hoffmann	[52]	PV	80–82%
Pelamis and Carcas	[53]	Wave	85%
Allan *et al.*	[54]	Wave	82%
Bedard	[55]	wave	82%
RenewableUK	[56]	wave	85–90%
Zwaan *et al.*	[57]	Offshore wind	95–97%

[a]Referred to time rather than cost reduction.

Learning curve formula derived by Hau [58] is represented in

$$P = N^{(\ln(lc)/\ln(2))} \tag{10.1}$$

where N is the number of WEC components and lc is the 'learning curve' factor.

10.2.2.4 Bulk discount costs

Purchasing multiple devices is cheaper than buying a single device, due to discounts provided by any manufacturer to encourage multiple purchases. The discount is based on a cumulative sum of the factorial reduction in price. The cumulative total of n number of devices is derived in

$$\text{Total WECIC} = \sum\nolimits_1^n P_n \times \text{WECIC} \tag{10.2}$$

where WECIC is the WEC initial cost, and P is the percent reduction [58] used in IC costing for WEC.

10.2.3 OPEX costs

10.2.3.1 Operation and maintenance

Metrics relating to O/M expense are defined in the literature by either of four following metrics and statistics, which are also summarised in Table 10.8:

1. *€/MWh*: This is the most commonly used metric, which provides a cost based on the relationship between the total initial cost (TIC) of the project and the annual energy output, and is the most commonly quoted metric. Its main advantage is that it can be used as a performance indicator, as the result is directly proportional to the device performance at the location. It can be used

Table 10.8 Literature review of operation and maintenance cost for onshore and offshore wind, and wave energy studies. Four metrics are presented: €/MWh, % of TIC, % of OPEX and % of cost of electricity (COE)

Wind/ wave	Location	Author	Reference	€/MWh	% of total TIC[a]	% of OPEX	% of COE
Wave	USA	Bedard *et al.*/ Siddiqui and Bedard	[60] /[61]	24			
	USA	Bedard	[55]	19–36			
	USA	Oregon	[62]	16	1.4		14
	Europe	Batten	[46]		5		
	Canada	Dunnett and Wallace	[37]		2		
	USA	EPRI	[29]	6	4.5		
	UK	Carbon Trust	[63]	16[c]	1.5		
	UK	Carbon Trust	[35]			57	
Onshore wind	USA	Bedard	[55]	6			
	USA	Bolinger and Wiser	[64]	20 (1980)			
	USA	Bolinger and Wiser	[64]	6 (2000s)			
	Europe	EWEA	[77]				25
	Germany	Albers	[65]	8–16	1.8–3.6		
	Europe	EWEA	[66]	40			
	Europe	Lemming and Morthorst	[67]		1–7		
	Denmark	Morthorst	[68]	5–15–45			
Offshore wind	Europe	EWEA	[69]	16	3.3		26
	UK	Van Bussel	[70]		4–4.5		
	Netherlands	Rademakers *et al.*	[71]	8–16			25–30
	Europe	EWEA	[73]			26	
	UK	Dale *et al.*	[72]	3[b]			

[a]% Quoted are assumed to be a % of the TIC, although not clearly defined in reports.
[b]Result based on $24/kw/year quoted in the paper.
[c]Costs were quoted in $ and have been converted to € at the conversion rate of €1 to $1.25.

as an input cost in cash flow analysis and also can be used to calculate a percentage relationship to the total cost of electricity (COE) per MWh. However, €/MWh is not the simplest metric as it requires that the total energy output be already calculated. Its disadvantage is that the O/M result is location specific, and will change for the same device used in different locations, which can result in confusion if quoted in reports without qualification.

2. *% of initial cost (IC)*: The next most popular metric is O/M calculated as a percentage of the TIC of a project. The advantage of this method is its simplicity, and that it is uniform in operation in any location, and thus easier to use in cash

flow sheets. It can be used as an input cost in cash flow analysis and the rate can be a variable in sensitivity analysis. The metric has many disadvantages:

i. In a review of the literature where this metric is used, it is often not clear whether % of the initial cost of the device (IC_{WEC}) or the total project initial cost (TIC).
ii. The metric does not reflect costs specific to a location.
iii. The % of IC figure is often arbitrarily chosen for economic analysis and not based on actual evidence.

3. *% of the total OPEX*: This metric defines O/M as a percentage of total OPEX. The advantage of this metric is that it is useful for comparative analysis. The disadvantage is that it cannot be used as an input cost in cash flow analysis.
4. *% of cost of electricity (COE)*: The final method compares the % O/M cost in €/MWh to the total COE. It requires both the O/M and COE based on €/MWh. The metric is useful if COE forms a major component of cost analysis of a report. The disadvantages of this metric are that it requires that COE should be already calculated. The metric consequently cannot be used as an input cost in cash flow analysis.

Scheduled maintenance is estimated on the basis of assumed device reliability (P_{OM1}) and the average duration of a maintenance task (t_{OM1}), while unplanned maintenance could be related to the frequency of occurrence of extreme wave conditions at the design site $\left(P_{OM2} = Q(H_s > 5\ m)\right)$, where Q is the probability of a certain wave height being exceeded and an assumed repair time (P_{OM2}) [59].

A review of O/M costs via the various metrics types already discussed is presented in Table 10.8.

10.2.3.2 Insurance

The cost of insurance is an under-researched area for the whole offshore renewable energy sector. There are two main ways of quoting insurance costs for cash flow analysis: % of IC and €/MWh. The Carbon Trust produced two reports quoting insurance. The first report quotes a list of insurance types and expenses as follows [63]:

- All risk insurance at 2% of IC.
- Cost overrun insurance at 3% of the first year revenue.
- An operational insurance of 0.8% of the IC.
- Business interruption insurance of 2% of energy revenue.

The Carbon Trust report [35] quotes the insurance component of total OPEX at 14%. EWEA quotes insurance as 13% of total OPEX [73]. The EPRI report [29] quotes €27 MWh and 2% of IC for insuring the Pelamis in the Oregon project. The Irish Wind Energy Association (IWEA)[2] use another metric, €/MW, and ‘insurance costs typically work out around €15,000 per MW for the development of the project and the first year of operation with a progressive reduction in cost after year one’.

[2]http://www.iwea.com/index.cfm/page/planning_regulationsandadminis

10.2.4 Decommissioning and salvage costs

The salvage value is a positive credit to the costs once the project has completed. The value assumes a linear depreciation meaning that the salvage value of a component is directly proportional to its remaining life. Salvage value is calculated by using the following equation [74]:

$$S = \frac{RC \times RL}{Lt} \tag{10.3}$$

where S is the salvage value, RC is the replacement cost of the component, RL is the years remaining and Lt is the components' lifetime. The replacement cost is calculated for each replacement time using the discount factor, and thus inflation is factored out of the equation.

If the life of the item is less than the project lifespan, then the remaining life (RL) of the component at the end of the project lifetime is given by

$$RL = Py - CY \tag{10.4}$$

where Py is the project years and CY is the total years up to last replacement of the component, and is calculated by

$$CY = Lt \times \text{trunc}\left(\frac{Py}{Lt}\right) \tag{10.5}$$

If lifetime of the item is greater than the project lifespan,

$$RL = Lt - Py \tag{10.6}$$

Any decommissioning costs are subtracted from the final salvage value.

Salvage value can be substantial especially if the replacement of an item is due close to the termination of a project.

A full schedule of procedures for offshore wind decommission is provided in the paper by Kaiser and Snyder [81].

10.2.5 Revenue methods: tariffs and ROCS

Europe has many separate revenue support methods for wave energy, similar to the variety currently in place for offshore wind. The revenue support systems are intended to stimulate the growth and development of renewables and are thus artificial supports. The intention of these supports is that the sector will mature to a stage where these supports could be gradually withdrawn.

The European commission decreed in 2007 that regional areas in Europe will be required to conform and harmonise their electricity markets, in the European Internal Energy Market (IEM).[3] Seven areas in Europe have been created, e.g. FUI (France, UK and Ireland). There appears to be confusion as to whether this single

[3]http://www.eirgrid.com/europeanelectricityforum/internalenergymarket/

market will have any influence over individual countries' ability to designate separate feed-in tariff (FIT) and support systems. It is the opinion of the author that this will not be the case.

10.2.5.1 Feed-in tariff (FIT)

Feed-in tariff (FIT) refers to the regulatory, minimum guaranteed price per kWh that an electricity utility has to pay to a private, independent producer of renewable power fed into the grid [75]. It is defined as the full price per kWh received by an independent producer of renewable energy including the premium above or additional to the market price, but excluding tax rebates or other production subsidies paid by the government.

Grid sales, or revenue, is defined by

$$\text{Grid Sales} = \text{revenue} \times \text{AEO} \tag{10.7}$$

where AEO is the annual energy output. Revenue is a credit, and is added to other annual cost values for the each year. An example of tariff rates in Europe is presented in Table 10.7.

Limitation of a fixed tariff is that most are not index or inflation linked. Projects starting e.g. in 2015 are at a disadvantage than those that avail of the tariff at the start of the project, 2010.

10.2.5.2 Renewable Obligation Certificates (ROCs)

ROCs are the revenue support mechanism used in the UK. Wave and tidal will receive 5 ROCs per MWh for each project up to a cap of 30 MW with 2 ROCs per MWh above the cap. This enhanced level of support will only be available for capacity installed and operational prior to 1 April 2017 [76]. After April 2017 it is envisaged that the Renewable Obligations scheme will be close to new entrants and be replaced by a FIT support mechanism.

Grid sales or revenues in the UK under the current support system are made up of three elements. The electricity produced is sold on the wholesale market, the ROCs received by the project for each MWh generated are sold. In addition to ROCs the project also receives Levy Exemption Certificates (LEC) for each MWh of electricity it produces and these are sold.

ROCs are based on the daily variation in electricity price, and are thus difficult to provide an exact cost. Estimates for future wholesale electricity, ROC and LEC price, have been calculated by a number of authors, and summary of the findings of those studies is presented in Table 10.9.

At the time of publishing, the ROC system is guaranteed until 2017, and thereafter will be supplanted by an FIT and contract for difference model, the details of which are yet to be announced.[4] The change from one model to another has substantially increased the uncertainty of the sector by the market.

[4]https://www.gov.uk/government/uploads/system/uploads/attachment_data/file/65634/7090-electricity-market-reform-policy-overview-.pdf

Table 10.9 List of wholesale electricity, ROC and LEC prices used in the analysis

Wholesale electricity price	ROC price	LEC price	Source
£84.7/MWh	£41.41/MWh	£2.45/MWh	[78]
£69.6/MWh	£41.37/MWh	£2.16/MWh	[78]
£54.5/MWh	£41.33/MWh	£1.86/MWh	[78]
£69.6/MWh	£40.69/MWh	£4.72/MWh	[79]
£47.4/MWh	£52.46/MWh		[80]

Table 10.10 List of some global revenue support systems [83]

Country	Revenue support mechanism	Price
France	FIT	€0.15/kWh
Italy	FIT	€0.34/kWh
Spain	FIT	€0.07/kWh
Denmark	FIT	€0.08/kWh+extra supplement
Germany	FIT	€0.06/kWh
US	Production tax credits	

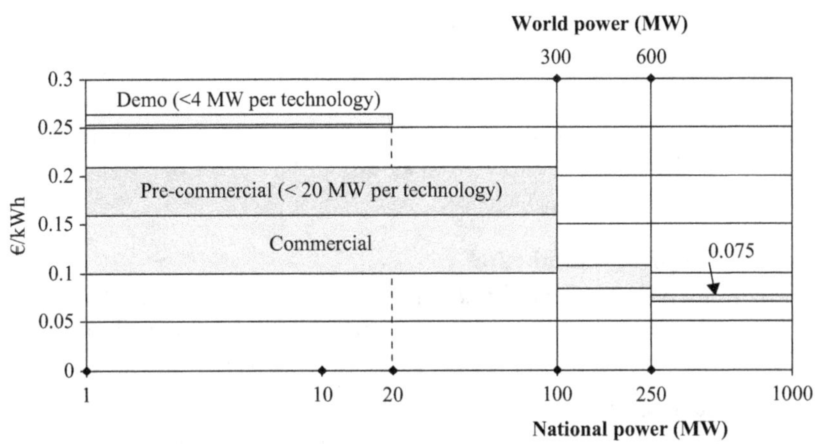

Figure 10.13 Image taken from [82]

10.2.5.3 Other revenue support mechanisms

Table 10.10 lists some other the countries and their respective mechanisms. As can be seen, FIT is most common at present. The scope of this chapter does not permit a detailed discussion on their details and variations. Figure 10.13 details the detailed reducing FIT proposed by Portugal [82].

10.2.6 Economic input factors

10.2.6.1 Discount rate

The discount factor (DF) translates expected financial benefits or costs in any given future year into present value terms. It is important to remember that the discount factor assumes that the project first year starts in the present year and pertains to the life of the project only. A project to be started in the future will also have discount factors for its duration, but will also have a future cash factor added, which is discussed in a later section. The total nominal profit is adjusted for cash depreciation by multiplying the total nominal profit by a discount factor.

DF is calculated using the discount rate, and is calculated by

$$\mathrm{DF} = \frac{1}{(1 + \mathrm{DR})^n} \tag{10.8}$$

The discount rate (DR) is an interest rate commensurate with perceived risk used to convert future payments or receipts (within a project lifetime) to present value. There are two methods of using and defining discount rates:

1. *Discount rates where inflation is factored into the figure*. By defining the discount rate in this way, *inflation is factored out of the economic analysis* during the project lifetime. All costs, therefore, become *real costs*, meaning that they are defined in terms of constant Euros. The assumption is that the rate of inflation is the same for all costs.
2. Discount rates, where inflation is not factored in.

The advantage of method 1 over method 2 is simplicity of use. Method 2 requires that all annual costs and revenues should have an inflation rate factor included, adding to the complexity of the spreadsheet. The disadvantage of the method is that it restricts the user from assessing the impact of varying inflation with a constant discount risk factor.

10.2.6.2 Supply/demand rate

The law of supply and demand defines the effect that the availability of a particular product and the desire (or demand) for that product have on price. There are two aspects of supply/demand that are relevant to offshore wind:

1. Supply/demand of offshore wind products.
2. Supply/demand of materials used in the construction of offshore wind products (see section 10.2.2, which refers to the price of steel).

Both these factors can contribute to price increases for offshore wind projects. Both can occur simultaneously, as happened from 2005 to 2008, which can increase the price of product siginifacntly. Cost-decreasing effects of scaling and learning for offshore wind power can be partly or entirely offset by cost-increasing effects supply/demand surges [57].

This is particularly relevant to the offshore wind sector. In 2000, the cost of installing a wind turbine was ~£1.5 M/MW [84]. Over a 10-year period, the cost per

MW has over doubled to ~£3–4 M/MW [84], due to the huge surge in global demand and inability of supply to cater for it. A similar scenario could happen to wave energy once the technology is established and global demand for devices increases.

Project economic analysis, which assesses project deployment in the future, should ideally factor in supply/demand rates in the Capex.

10.2.6.3 Future purchase costs

The purchase cost of products increases with time due to the consumer price index. Thus, a product costing €1 now will be valued at ~€1.20 in 5-year time for example. Future cost of cash (FCC) is estimated using the interest or borrowing rate, rather than the inflation rate. Therefore, cost of starting projects will increase with time even if there is deflation. The IC of project delayed till the future will have a higher cost than if they were implemented in the present.

The equation to estimate the FCC percent is presented in

$$\mathrm{FCC} = (1 + i)^n \tag{10.9}$$

where i is the borrowing interest rate, and n is the number of years.

The FCC percent is multiplied by the WEC initial cost, to estimate the cost of purchasing the device in the future. Once a project is started at a predetermined time in the future, inflation is thereafter accounted for that project lifetime using the discount rate with or without inflation factored in as explained in the previous section.

10.2.7 Debt/equity

The debt-to-equity ratio is a financial metric used to assess a company's capital structure, or 'capital stack'. Specifically, the ratio measures the relative proportions of the firm's assets that are funded by debt or equity. The debt to equity ratio (also called the risk ratio or leverage ratio) provides a quick tool for financial analysts and prospective investors for determining the amount of financial leverage a company is using, and thus its exposure to increases in interest rate or insolvency.

The advantages of levered investments are that principal and interest obligations are known amounts, which can be forecasted and planned for. Interest on the debt can be deducted on the company's tax return, lowering the actual cost of the loan to the company.

The inclusion of debt increases the cost of the project because it adds interest during construction, which is not there in case of an "all equity funding" project; thus initial outflows goes up. Interest during construction is generally capitalised and capitalisation of interest also implies higher book value of assets and thus higher depreciation and consequently lower taxation. The impact of higher initial cash, outflows on account of interest during construction, outweigh the impact of lower tax outgoings in subsequent years.

It is generally believed that inclusion of debt will have no impact on a project's IRR (discussed later). But in practice inclusion of debt generally changes the project IRR, albeit marginally. This is due to the fact that free cash flow to the firm (FCFF) is not affected with interest payment and debt repayment outflows.

The yearly debt payment D is calculated using

$$D = Cf_d\left(\frac{i_a}{1-\dfrac{i}{(1+i_a)^{n'}}}\right) \tag{10.10}$$

where C is the total initial cost of the project, f_d is the debt ratio, i_a is the effective annual debt interest rate and n' is the debt term in years.

The yearly debt payment can be broken down into payment towards the principal Dp, n and payment of interest Di, n:

$$D = Dp{,}n + Di{,}n$$

Both Dp, n and Di, n vary from year to year.

Debt payments will be subject to tax and depreciation. There is not the scope for discussion of these two aspects in the chapter.

10.2.8 Economic indicators

An economic indicator is a statistic about the business, project or an economy. Economic indicators allow analysis of economic performance and predictions of future performance. Indicators provide data for assessment on commercial viability of projects and other business entities.

10.2.8.1 Cost of electricity

The Cost of Electricity can be defined by two methods:

1. Cost of electricity (COE).
2. Levelised cost of electricity (LCOE).

Cost of electricity (COE) (simple)

The simple form of COE is a basic calculation of CAPEX versus electrical output as per

$$\text{COE} = \frac{\text{CAPEX} + \text{OPEX}}{\text{AEO}} \tag{10.11}$$

This method of COE is not often used, and should not be confused with the predominant method, levelised cost of electricity.

Levelised cost of electricity (LCOE)

The (levelised) cost of electricity (COE) is defined as the average cost per kWh of useful electrical energy produced by the system and is the price at which electricity must be generated from a specific source to break even. It is calculated by dividing the total annualised cost (TAC) of producing electricity by the annual electric output (AEO). COE calculation is given in

$$\mathrm{COE} = \frac{\mathrm{TAC}}{\mathrm{AEO}} \tag{10.12}$$

The total annualised cost (TAC) is the sum of the annualised costs of each system component. It is calculated by multiplying the NPV by the capital recovery factor (CRF). CRF is a ratio used to calculate the present value of an annuity (a series of equal annual cash flows). The equation for the capital recovery factor is

$$\mathrm{CRF} = \frac{1(1+r)^n}{(1+r)^n - 1} \tag{10.13}$$

where r is the discount rate and n is the number of years. The discount rate is the reward an investor demands for accepting a delayed payment.

Another way of defining the LCOE calculation is the present discounted value of energy produced times the levelised cost equals the present discounted value of the fixed and variable costs over the life of the investment.

Table 10.11 provides a review of the LCOE estimates for wave energy over the last decade.

10.2.8.2 Net present value

The net present value is defined as the present value of investments' future net cash flows minus the initial investment [85]. It is derived by summing the discounted cash flows over the project lifetime, defined in

$$\mathrm{NPV} = \sum \mathrm{TNC} \times \mathrm{DF} \tag{10.14}$$

where TNC is the total net cash for that year and DF is the discount factor, which is defined as

$$\mathrm{DF} = \frac{1}{(1+r)^n} \tag{10.15}$$

where r is the discount rate and n is the number of years. The discount rate is the reward an investor demands for accepting a delayed payment.

In general, only positive NPVs are considered viable commercial projects. NPV is a simple mathematical concept that doesn't include any arbitrary variables.

10.2.8.3 Internal rate of return

Internal rate of return indicates the business return according to alternative return that may be gained on the same investment. The internal rate of return is the discount rate that will create a zero net present value. The IRR is based on the NPV formula (10.14), and is solved iteratively for when NPV = 0.

$$\mathrm{NPV}(0) = -\mathrm{IC} + \sum \frac{c(n)}{(1+\mathrm{IRR})^n} \tag{10.16}$$

where c is the cost for year n and NPV(0) is the NPV value equal to zero.

Table 10.11 Review of COE ($ tot €=0.8, € to £=0.85)

Study	Reference	Device	Location	MW	COE original currency	COE €/kWh	Subsidy
EPRI (Previsic)	[29]	Pelamis	California	160	$0.11	0.088	0.06
ESBI	[86]	Pelamis	Ireland	156		0.105	
St. Germain	[87]	Pelamis	Canada	11.25	$0.14–0.18	0.11–0.144	
EPRI (Bedard)	[88]	Pelamis	California	33	$0.08–0.16	0.06–0.128	
Carbon Trust	[35]	Pelamis	UK		£0.05–0.06	0.058–0.07	
Pelamis and Carcas	[53]	Pelamis	UK	.75	£0.20–0.40	0.235–0.47	
Dunnet and Wallace	[37]	Pelamis	Canada	11–20	$0.23–.38	0.18–0.30	
SQW	[89]		review			0.24	
Carbon Trust	(2010)		UK	1 1000	£0.42 £0.12	0.52 0.14	
Oregon Wave	[90]		Oregon	100	$0.47 $0.21	0.37 0.17	
Allan *et al.*	[91]		UK		£0.19	0.22	
ARUP	[38]		UK[a]	180–280	£0.27–0.21	0.31–0.24	
Carbon Trust	[26]		UK	(10% disc%)	£0.20–0.12	0.23–0.14	
Offshore Valuation Group	[92]		UK	4566[a]	£0.195	0.23	
Oregon Wave Energy Trust	[93]		US		$0.25	0.2	
Behrens *et al.*	[94]		Australia		$0.26–0.08	0.21–0.06	

[a]forecast for 2020

IRR threshold of 10–12% is a minimum normally required by financiers for a project to be considered financially viable.[5,6]

10.2.8.4 Payback time

Payback period in capital budgeting refers to the period of time required for the return on an investment to 'repay' the sum of the original investment. The payback

[5]Personal communication Tony Dalton, chief financial accountant for LET Systems, Cork.

[6]IRR requirement for investors is a different scenario (not applicable here). Here, investors judge IRR on their initial investment made and the exit price of the company. The investment is riskier and thus IRR would be expected to be higher, at around 30%.

Table 10.12 Techno-economic software tools for wave energy project commercial feasibility analysis

Country base	Company name	Product name	Reference	Commercial product	Techno		Economic		Results	Results
					Device – power matrix	Resource – hourly annual data	Published Capex costs	Device build cost	Sensitivity	Optimisation
Ireland	HMRC, UCC[a]	NAVITAS	[1]	√	√	√	√	X	√	√
UK	Garrad Hassan	Wavefarmer	[7]	√	√	√			√	
Denmark	Energinet		[8]		√	√				
CA, USA	RE Vison		[9]	√	√	√			√	

[a] http://www.ucc.ie/en/hmrc/projects/navitas/

period is considered a method of analysis with serious limitations and qualifications for its use, because it does not account for the time value of money, risk, financing or other important considerations, such as the opportunity cost. While the time value of money can be rectified by applying a weighted average cost of capital discount, it is generally agreed that this tool for investment decisions should not be used in isolation. Payback period is usually expressed in years, and is the year when the Net cash flow in the cash flow sheet changes from a negative to a positive.

10.2.9 Techno-economic analysis methods and tools

The description of the ingredients of techno-economic analysis is now complete. What remains is to perform an analysis. Analyses are in the majority of cases conducted in-house by companies on Excel, and range in complexity from basic one spreadsheet page calculators to 30–40 page multi-level complexity workbooks, with built-in macros and VB coding. These Excel sheet analysis tools are excellent at providing the results required for the project in questions. However there are many downsides to using excel:

- Huge multiple spreadsheets = slow.
- Expert user only.
- Prone to increasing errors = time to track and correct.
- User has to update continually or out of date.
- Significant time to develop for each project = money.
- Not scalable once built = new spreadsheet every time.
- Static, not dynamic tool.
- Multiple silos of information.
- No standards typically built in – e.g. IEC and DOE standards.

There are a number of software tools now been generated for the market which will assist the user in overcoming these problems (Table 10.12). These tools will have been rigorously tested and validated, and thus should add a degree if confidence in the results issued.

10.3 References

[1] F. Sharkey, M. Conlon and K. Gaughan, 'Investigation of wave farm electrical network configurations', *World Renewable Energy Congress*, Linkoping, 2011.

[2] R. Green and N. Vasilakos, 'The economics of offshore wind', *Energy Policy*, vol. 39, pp. 496–502, 2010.

[3] RenewableUK – 'Challenging the Energy- A Way Forward for the UK Wave & Tidal Industry Towards 2020', October 2010.

[4] BVG Associates, *A Guide to an Offshore Wind Farm*, The Crown Estate, London, UK, 2011.

[5] D. O'Sullivan and T. Lewis, 'Electrical machine options in offshore floating wave energy converter turbo-generators', *World Renewable Energy Congress*, Glasgow, 2008.

[6] D. O'Sullivan and G. Dalton, 'Challenges in the grid connection of wave energy devices', *8th European Wave and Tidal Energy Conference*, Uppsala, Sweden, 2009.

[7] L. Fingersh, M. Hand and A. Laxon, *Wind Turbine Design Cost and Scaling Model*, National Renewable Energy Laboratory, Golden, CO, USA, 2006.

[8] S. Lundberg, *Performance Comparison of Wind Park Configurations*, Technical Report, Chalmers University, 2003.

[9] J. Green, A. Bowen, L. Fingersh and Y. Wan, 'Electrical collection and transmission systems for offshore wind power', *Offshore Technology Conference*, Houston, TX, USA, 2007.

[10] M. Kenny, *Electrical Connection Issues for Wave Energy Arrays*, Masters thesis, University College Cork, 2010.

[11] P. Djapic and G. Strbac, *Cost Benefit Methodology for Optimal Design of Offshore Transmission Systems*, Centre for Sustainable Energy and Distributed Generation (SEDG), London, UK, 2008.

[12] Offshore Design Engineering (ODE), *Study of the Costs of Offshore Wind Generation*, Department of Trade and Industry (DTI), London, UK, 2007.

[13] P. Hopewell, F. Castro-Sayas and D. Bailey, 'Optimising the design of offshore wind farm collection networks', *Universities Power Engineering Conference*, Brighton, UK, 2007.

[14] E. Stoutenburg and M Jacobson, *Optimizing Offshore Transmission Links for Marine Renewable Energy Farms*, OCEANS, Seattle, WA, USA, 2010.

[15] ESB Networks, *Standard Prices for Generator Connections Rev. 2*, ESB Networks, 2012.

[16] N. Boccard, 'Capacity factor of wind power realized values versus estimates', *Energy Policy*, vol. 37, pp. 2679–2688, Nanjing, China, 2009.

[17] J. Beale, 'Transmission cable protection and stabilisation for the wave and tidal energy industries', *10th European Wave and Tidal Energy Conference*, Southampton, UK, 2011.

[18] J. Yang, J. O'Reilly and J. Fletcher, 'Redundancy analysis of offshore wind farm collection and transmission systems', *International Conference on Sustainable Power Generation and Supply*, Nanjing, China, 2009.

[19] F. Sharkey, M. Conlon and K. Gaughan, 'Practical analysis of key electrical interfaces for wave energy converter arrays', *International Conference on Ocean Energy*, Dublin, 2012.

[20] F. Sharkey, E. Bannon, M. Conlon and K. Gaughan, 'Dynamic electrical ratings and the economics of capacity factor for wave energy converter arrays', *European Wave and Tidal Energy Conference*, Southampton, 2011.

[21] International Electrotechnical Committee (IEC), *IEC 60287 Electrical Cables – Calculation of Current Rating*, IEC, 2013.

[22] T. Seppa, 'Increasing transmission capacity by real time monitoring', *IEE Power Engineering Society Meeting*, vol. 2, pp. 1209–1211, 2002.

[23] J. Downes and H. Leung, 'Distributed temperature sensing worldwide power circuit monitoring applications', *International Conference on Power System Technology*, Singapore, 2004.

[24] M. Luton, G. Anders, J. Braun, N. Fujimoto, S. Rizzetto and J. Downes, 'Real time monitoring of power cables by fibre optic technologies tests, applications and outlook', *International Conference on Insulated Power Cables*, Versailles, France, 2003.

[25] M. O'Connor and G. J. Dalton, 'Operational expenditure costs for wave energy projects and impacts on financial returns', *Renewable Energy*, vol. 50, pp. 1119–1131, 2012.

[26] Carbon Trust, *Marine Cost of Energy Methodology*, 2012. Available at: http://www.carbontrust.com/media/54781/mec_cost_estimation_methodology_-_spreadsheet.xls

[27] B. H. Liebman, 'Safeguards, China, and the price of steel', *Review of World Economics*, vol. 142, no. 2, pp. 354–373, 2006.

[28] B. Snyder and M. J. Kaiser, 'Ecological and economic cost-benefit analysis of offshore wind energy', *Renewable Energy*, vol. 34, no. 6, pp. 1567–1578, 2009.

[29] M. Previsic, *System Level Design, Performance, and Costs of California Pelamis Wave Power Plant*, EPRI. Available at: http://oceanenergy.epri.com/attachments/wave/reports/006_San_Francisco_Pelamis_Conceptual_Design_12-11-04.pdf

[30] Carbon Trust and J. Callaghan, *Future Marine Energy – Results of the Marine Energy Challenge: Cost Competitiveness and Growth of Wave and Tidal Stream Energy Carbon Trust – CTC601*, 2006. Available at: http://www.carbontrust.co.uk/Publications/publicationdetail.htm?productid=CTC601&metaNoCache=1

[31] L. Fingersh, M. Hand and A. Laxson, *Wind Turbine Design Cost and Scaling Model*, National Renewable Energy Laboratory, 2006. Available at: http://www.scribd.com/doc/3862797/NREL-Wind-Turbine-Design-Costs-and-Scale-Model.

[32] SEI, *Cost Benefit Analysis of Government Support Options for Offshore Wind Energy*, SEI, 2002. Availabel at: http://www.sei.ie/uploadedfiles/InfoCentre/offshorewindenergy.pdf

[33] R. Barthelmie, S. Frandsen, C. Morgan, A. Henderson, H. S. Christian, C. Garcia, *et al. State of the Art and Trends Regarding Offshore Wind Farm Economics and Financing*, RISO, 2008. Available at: http://www.offshorewindenergy.org/ca-owee/indexpages/downloads/Brussels01_Economics.pdf

[34] T. Luypaert, S. Lee and P. E. Ellefsen, *Energy 101: Offshore Wind*, MIT & LSPR, 2008. Available at: http://www.mitenergyclub.org/assets/2010/3/4/Energy_101_Offshore_Wind_Energy_1.pdf

[35] Carbon Trust, *Cost Estimation Methodology*, ENTEC, 2006. Available at: http://www.carbontrust.co.uk/SiteCollectionDocuments/Various/Emerging%20technologies/Technology%20Directory/Marine/MEC%20cost%20estimation%20methodology%20-%20report.pdf

[36] J. C. Weiss, B. B. Boehlert and J. R. Baxter, *Fiscal Cost-Benefit Analysis to Support the Rulemaking Process for 30 CFR 285*. U.S. Department of the Interior, Minerals Management Service OCS Study MMS 2007-050,

2007. Available at: www.mms.gov/offshore/alternativeenergy/PDFs/Final_Technical_Report_IEc_MMS_2008_0627.pdf

[37] D. Dunnett and J. S. Wallace, 'Electricity Generation from Wave Power in Canada', *Renewable Energy*, vol. 34, no. 1, pp. 179–195, 2009.

[38] ARUP, *Review of the Generation Costs and Deployment Potential of Renewable Electricity Technologies in the UK*, Department of Energy and Climate Change (DECC), 2011. Available at: http://www.decc.gov.uk/assets/decc/11/consultation/ro-banding/3237-cons-ro-banding-arup-report.pdf

[39] G. J. Dalton, R. Alcorn and T. Lewis, 'A 10 year installation program for wave energy in Ireland, a sensitivity analysis on financial returns', *Renewable Energy*, vol. 40, no. 1, pp. 80–89, 2012.

[40] G. Hassan, *UK Offshore Wind: CHARTING the Right Course. Scenarios for Offshore Capital Costs for the Next Five Years*, BWEA, 2011. Available at: http://www.bwea.com/pdf/publications/ChartingtheRightCourse.pdf

[41] R. Walker, L. Johanning and R. Parkinson, 'Weather windows for device deployment at UK test Site: Availability and cost implications', *Weather Windows for Device Deployment at UK test Site: Availability and Cost Implications, European Wave and Tidal Energy Conference, EWTEC2011*, Southampton, 2011.

[42] A. Beyene and D. MacPhee, 'Integrating wind and wave energy conversion', *International Conference on Electrical and Control Engineering, ICECE 2011 – Proceedings*, IEEE Computer Society, Yichang, China, 2011. Available at: http://dx.doi.org/10.1109/ICECENG.2011.6058252

[43] M. O'Connor, T. Lewis and G. Dalton, 'Weather window analysis of Irish west coast wave data with relevance to operations & maintenance of marine renewables', *Renewable Energy*, 52.0: 57–66, 2013.

[44] BVG, *Wave and Tidal Energy in the Pentland Firth Area Crown Estate*, 2011. Available at: http://www.bvgassociates.co.uk/Portals/0/publications/BVG%20TCE%20Pentland%20Firth%20and%20Orkney%20waters%201105.pdf

[45] L. Neij, 'Cost dynamics of wind power', *Energy*, vol. 24, no. 5, pp. 375–389, 1999.

[46] W. M. J. Batten and A. B. Bahaj, *An Assessment of Growth Scenarios and Implications for Ocean Energy Industries in Europe*, Report for CA-OE, Project no. 502701, WP5, Sustainable Energy Research Group, School of Civil Engineering and the Environment, University of Southampton, 2006. Available at: http://eprints.soton.ac.uk/53003/

[47] R. Gross, M. Leach and A. Bauen, 'Progress in renewable energy', *Environment International*, vol. 29, no. 1, pp. 105–122, 2003.

[48] IEA, *Experience Curves for Energy Technology Policy*, 2000. Available at: http://reaccess.epu.ntua.gr/LinkClick.aspx?fileticket=Driek6XFOMs%3D&tabid=582&mid=1096

[49] A. McDonald and L. Schrattenholzer, 'Learning rates for energy technologies', *Energy Policy*, vol. 29, no. 4, pp. 255–261, 2001.

[50] M. Junginger, A. Faaij and W. C. Turkenburg, 'Cost reduction prospects for offshore wind farms', *Wind engineering*, vol. 28, no. 1, pp. 97–118, 2004.

[51] M. Junginger, A. Faaij and W. C. Turkenburg, 'Global experience curves for wind farms', *Energy Policy*, vol. 33, no. 2, pp. 133–150, 2005.

[52] W. Hoffmann, 'PV solar electricity industry: Market growth and perspective'. 14th International Photovoltaic Science and Engineering Conference, *Solar Energy Materials and Solar Cells*, 90. 18–19, 3285–3311, 2006.

[53] M. Carcas, *The Pelamis Wave Energy Converter*, Ocean Power Delivery Ltd., 2007. Available at: http://hydropower.inl.gov/hydrokinetic_wave/pdfs/day1/09_heavesurge_wave_devices.pdf

[54] G. J. Allan, I. Bryden, P. G. McGregor, T. Stallard, J. Kim Swales, K. Turner, *et al.*, 'Concurrent and legacy economic and environmental impacts from establishing a marine energy sector in Scotland', *Energy Policy*, vol. 36, no. 7, pp. 2734–2753, 2008.

[55] R. Bedard, *Ocean Wave Power/Energy Economics*. EPRI, 2009. Available at: http://hinmrec.hnei.hawaii.edu/wp-content/uploads/2010/01/EPRI-Wave-Energy-Economics.ppt#839,5,'Learning

[56] ENTEC, *RenewableUK Wave and Tidal Report*, RenewableUK, 2010. Available at: http://www.bwea.com/pdf/publications/RenewableUK_Wave_tidal_report_%202010.pdf

[57] B. van der Zwaan, R. Rivera-Tinoco, S. Lensink, and P. van den Oosterkamp, 'Cost reductions for offshore wind power: Exploring the balance between scaling, learning and RD', *Renewable Energy*, vol. 41, pp. 389–393, 2012.

[58] E. Hau, *Wind Turbines: Fundamentals, Technologies, Application, Economics*. Springer, Berlin; New York, 2006.

[59] T. Stallard, R. Rothschild and G. A. Aggidis, 'A comparative approach to the economic modelling of a large-scale wave power scheme', *European Journal of Operational Research*, vol. 185, no. 2, pp. 884–898, 2008.

[60] R. Bedard, M. Previsic and G. Hagerman, *North American Ocean Energy Status*, EPRI, 2007. Available at: http://www.oceanrenewable.com/wp-content/uploads/2008/03/7th-ewtec-paper-final-032907.pdf

[61] O. Siddiqui and R. Bedard, *Feasibility Assessment of Offshore Wave and Tidal Current Power Production: A Collaborative Public/Private Partnership*, IEEE Power, Piscataway, NJ, 2005.

[62] Oregan Wave, *Value of Wave Power*, Oregan Wave Energy Utility Trust, 2010a. Available at: http://www.oregonwave.org/our-work-overview/market-development/utility-market-initiative/

[63] Carbon Trust, *Oscillating Water Column Wave Energy Converter Evaluation Report*. Marine Energy Challenge, ARUP, EON, 2005. Available at: http://www.carbontrust.co.uk/SiteCollectionDocuments/Various/Emerging%20technologies/Technology%20Directory/Marine/Other%20topics/OWC%20report.pdf

[64] M. Bolinger and R. Wiser, *Annual Report on U.S. Wind Power Installation, Cost, and Performance Trends*, NREL, 2007. Available at: http://eetd.lbl.gov/ea/ems/reports/ann-rpt-wind-06-ppt.pdf

[65] A. Albers, *O&M Cost Modelling, Technical Losses and Associated Uncertainties*, German Windguard Consulting, 2008. Available at: http://www.ewec2009proceedings.info/allfiles2/554_EWEC2009presentation.pdf

[66] EWEA, *Operation and Maintenance Costs of Wind Generated Power*, 2010b. Available at: http://www.wind-energy-the-facts.org/en/part-3-economics-of-wind-power/chapter-1-cost-of-on-land-wind-power/operation-and-maintenance-costs-of-wind-generated-power.html

[67] J. Lemming and P. E. Morthorst, 'O&M costs and economical costs of wind turbines', *European Wind Energy Conference*, 1999. Available at: http://books.google.com/books?hl=en&lr=&id=AhuLVrUMP7UC&oi=fnd&pg=PA294&dq=/mwh+operation+and+maintenance+wave+wind&ots=QExHLII_4U&sig=d0OwGU1S2MlgmxV1CCMTAerqvTQ#v=snippet&q=maintenance%20&f=false

[68] P. E. Morthorst, *Costs and Prices*, EWEA and RISO, 2003. Availabel at: http://www.ewea.org/fileadmin/ewea_documents/documents/publications/WETF/Facts_Volume_2.pdf

[69] EWEA, *The Cost of Energy Generated by Offshore Wind Power – Wind Energy the Facts*, RISO, 2010a. Available at: http://www.wind-energy-the-facts.org/en/part-3-economics-of-wind-power/chapter-2-offshore-developments/the-cost-of-energy-generated-by-offshore-wind-power.html

[70] G. J. W. van Bussel and M. B. Zaaijer, 'Reliability, availability and maintenance aspects of large-scale offshore wind farms, a concepts study', *Marine Renewable Energies Conference*, Newcastle, UK, 2011. Available at: http://ocw.tudelft.nl/fileadmin/ocw/courses/OffshoreWindFarmEnergy/res00055/MAREC_2001_OM_Paper.pdf

[71] L. Rademakers, H. Braam, M. Zaaijer and G. Van Bussel, *Assessment and Optimisation of Operation and Maintenance of Offshore Wind Turbines*, EWEC, 2003. Available at: http://www.ecn.nl/docs/library/report/2003/rx03044.pdf

[72] L. Dale, D. Milborrow, R. Slark and G. Strbac, 'Total cost estimates for large-scale wind scenarios in UK', *Energy Policy*, vol. 32, no. 17, pp. 1949–1956, 2004.

[73] Intelligent Energy Europe, *Wind Energy – The Facts DEWI*, 2010. Available at: http://www.wind-energy-the-facts.org/en/part-3-economics-of-wind-power/chapter-1-cost-of-on-land-wind-power/operation-and-maintenance-costs-of-wind-generated-power.html

[74] T. Lambert and P. Lilienthal, *HOMER: The Micro-Power Optimisation Model*, Software produced by NREL, 2004. Available at: http://www.pspb.org/e21/media/HOMERModelingInformation.pdf

[75] J. P. M. Sijm, *The Performance of Feed-in Tariffs to Promote Renewable Electricity in European Countries*, ECN, 2002. Available at: http://www.wind-works.org/FeedLaws/Netherlands/ECNFeedLawsc02083.pdf

[76] Departmen of Enterprise Trade and Investment Northern Ireland, *Proposed Changes to the Northern Ireland Renewables Obligation*, Department of Enterprise, Trade and Investment Northern Ireland, 2011. Available at: http://www.detini.gov.uk/niro_2012_consultation.pdf

[77] EWEA, *Wind Power Economics*, EWEA, 2008. Available at: http://www.ewea.org/fileadmin/ewea_documents/documents/publications/factsheets/factsheet_economy2.pdf

[78] FIM Services Ltd, *UK Renewable Energy Pricing Analysis*, 2008. Available at: http://www.fimltd.co.uk/downloads/UK%20Renewable%20Energy%20Pricing%20Paper%20-%20December%202010%20-%20FINAL.pdf

[79] DECC, *Consultation on Proposals for the Levels of Banded Support under the Renewables Obligation*, 2011. Available at: http://www.decc.gov.uk/en/content/cms/consultations/cons_ro_review/cons_ro_review.aspx#

[80] T. Prassler and J. Schaechtele, "Comparison of the financial attractiveness among prospective offshore wind parks in selected European countries." *Energy Policy*, vol. 45, pp. 86–101, 2012.

[81] M. J. Kaiser and B. Snyder, 'Offshore wind decommissioning regulations and workflows in the Outer Continental Shelf United States', *Marine Policy*, vol. 36, no. 1, pp. 113–121, 2011.

[82] A. J. N. A. Sarmento, *Wave Energy: from demonstration to commercialization* 2008. http://www.oestecim.pt/_uploads/AntonioSarmentoAMOeste.pdf

[83] EREC, *National Policy – Overview of EU Member States*, European Renewable Energy Council, 2012. Available at: http://www.erec.org/policy/national-policies.html

[84] O. B. Nielsen, 'The Ormonde Wind Farm: construction process for The UKs first full scale 5 MW offshore wind project', *Offshore Wind Construction and Installation Conference*, London, 2010. Available at: http://www.windenergyupdate.com/offshoreinstallation/index.shtml

[85] H. Khatib, *Economic Evaluation of Projects in the Electricity Supply Industry*, vol. 4, Institution of Electrical Engineers (IEE) power and energy series London, 2003. Available at: http://www.sciencedirect.com/science/article/pii/S0301421505001709

[86] ESBI (2005). *Accessible Wave Energy Resource Atlas*. ESBI Report 4D404A-R2 for Marine Institute (MI) and Sustainable Energy Ireland (SEI). Available at: http://www.marine.ie/NR/rdonlyres/90ECB08B-A746-4247-A277-7F9231BF2ED2/0/waveatlas.pdf

[87] L. A. St. Germain, *A Case Study of Wave Power Integration into the Ucluelet Area Electrical Grid*, Department of Mechanical Engineering, University of Victoria, Canada, 2005. Available at: http://www.iesvic.uvic.ca/pdfs/Dissertation-StGermain.pdf

[88] R. Bedard, *EPRI Ocean Energy Program, Possibilities in California*, EPRI, 2006. Available at: http://oceanenergy.epri.com/attachments/ocean/briefing/June_22_OceanEnergy.pdf

[89] G. Connor, *Economic study for ocean energy development in Ireland*, SQW and SEAI and OEDU, 2010. Available at: http://www.investni.com/ocean_energy_development_ireland_economic_study_sd_july-2010.pdf

[90] Oregan Wave, *Value of Wave Power-Summary of Results*, Pacific Energy Ventures, 2010b. Available at: http://www.oregonwave.org/wp-content/uploads/Task-3.3.1-Value-of-Wave-Power_Summary.pdf

[91] G. Allan, M. Gilmartin, P. McGregor and K. Swales, 'Levelised costs of wave and tidal energy in the UK: Cost competitiveness and the importance

of 'banded' Renewables Obligation Certificates', *Energy Policy*, vol. 39, no. 1, pp. 23–39, 2011.

[92] Offshore Valuation Group, *The Offshore Valuation*, Public Interest Research Centre, 2011. Available at: www.offshorevaluation.org/downloads/offshore_vaulation_full.pdf

[93] Oregon Wave Energy Trust, *Ocean Energy Blue Tag Study – Bridging the Gap between the Cost of Wave Energy and Its Market Value*, Robert K. Harmon & Company LLC, 2011. Available at: http://www.oregonwave.org/wp-content/uploads/R.Harmon_Blue-Tag_Final-Report_July-2011.pdf

[94] S. Behrens, J. Hayward, M. Hemer and P. Osman, 'Assessing the wave energy converter potential for Australian coastal regions', *Renewable Energy*, 2012.

Index

www.ingramcontent.com/pod-product-compliance
Lightning Source LLC
LaVergne TN
LVHW011905200926
848484LV00002B/538